Organic Marine Geochemistry

ACS SYMPOSIUM SERIES **305**

Organic Marine Geochemistry

Mary L. Sohn, EDITOR
Florida Institute of Technology

Developed from a symposium sponsored by
the Division of Geochemistry
at the 189th Meeting
of the American Chemical Society,
Miami Beach, Florida,
April 28–May 3, 1985

American Chemical Society, Washington, DC 1986

Library of Congress Cataloging-in-Publication Data

Organic marine geochemistry.
(ACS symposium series; 305)

"Developed from a symposium sponsored by the
Division of Geochemistry at the 189th Meeting of the
American Chemical Society, Miami Beach, Florida,
April 28–May 3, 1985."

Includes bibliographies and index.

1. Organic geochemistry—Congresses. 2. Marine
sediments—Congresses. 3. Estuarine sediments—
Congresses.

I. Sohn, Mary L., 1952–   . II. American Chemical
Society. Division of Geochemistry. III. American
Chemical Society. Meeting (189th: 1985: Miami Beach,
Fla.) IV. Series.

QA516.5.07235   1986   551.46′083   86–3429
ISBN 0–8412–0965–0

ACS Symposium Series

M. Joan Comstock, *Series Editor*

Advisory Board

FOREWORD

The ACS SYMPOSIUM SERIES was founded in 1974 to provide a medium for publishing symposia quickly in book form. The format of the Series parallels that of the continuing ADVANCES IN CHEMISTRY SERIES except that, in order to save time, the papers are not typeset but are reproduced as they are submitted by the authors in camera-ready form. Papers are reviewed under the supervision of the Editors with the assistance of the Series Advisory Board and are selected to maintain the integrity of the symposia; however, verbatim reproductions of previously published papers are not accepted. Both reviews and reports of research are acceptable, because symposia may embrace both types of presentation.

CONTENTS

PREFACE

BOTH ORGANIC AND INORGANIC AREAS of water-column and sediment geochemistry have recently served as the focus of much research. This book deals with major advances in organic marine and estuarine geochemistry, as well as the effects of organic substances on the speciation and distribution of inorganic and organometallic substances. The organization of this book is such that it should prove useful not only as a collection of related research articles, but also as a reference and a text suitable for graduate and advanced undergraduate courses in organic marine geochemistry.

The authors of the chapters include internationally acclaimed experts in their respective areas of specialization. Basic geochemical topics such as structures are included as well as diagenesis of organic natural products. Also discussed are anthropogenic pollutant substances in the marine environment and the effect of diagenesis and cycling on the distribution and fate of both toxic and nontoxic substances. I hope the combination of review and application chapters has produced a book that will be useful to many people in diverse fields.

Without the enthusiasm and cooperation of the exceptional authors and the encouragement of the former chairmen of the Geochemistry Division (F. Miknis, T. Weismann, and G. Helz), this book never would have materialized. In addition, I would like to thank my husband, Rolf, for assisting in the drafting of tables and figures and for helping with the organization of the book. I am also indebted to the Geochemistry Division for its support in sponsoring the symposium from which this book was developed and to The Petroleum Research Fund, administered by the American Chemical Society, for partial support of the symposium.

MARY L. SOHN
Florida Institute of Technology
Melbourne, FL 32901

May 14, 1985

ix

Organic Marine Geochemistry
An Overview

Mary L. Sohn

Department of Chemistry, Florida Institute of Technology, Melbourne, FL 32901

The use of modern hyphenated methods such as GC-MS have proven invaluable to the task of assigning specific sources to specific organic compounds isolated from marine sediments, suspended particulate matter, interstitial water and seawater. Early chapters of this volume discuss the correlation of various classes of compounds isolated largely from marine and brackish sediments with sources consisting of terrigenous plants, bacteria, phytoplankton, zooplankton and nonbiological components such as anthropogenic input, petroleum seepage, and the weathering of "mineral" deposits.

Biomarkers

A biological marker (or "biomarker") may be defined as "a compound the structure of which can be interpreted in terms of a previous biological origin." Since the isolation and identification of metalloporphyrins in bitumens by Treibs ($\underline{2}$) in the 1930's and their correlation with tetrapyrroles of chlorophylls, implying a biological origin for petroleum, an enormous amount of progress has been accomplished in this area of research.

In order to be a useful biomarker, a compound must retain enough of its original structure to be identified as a modified version of the original biological parent compound. The retention of the carbon backbone of biologically generated alcohols, in alkanes isolated from sediments and petroleum fractions, includes the classic cases of pristane and phytane as diagenetic products of phytol (the alcohol which esterifies the carboxylic acid group in chlorophyll-a) and 5 ,-cholestane from cholesterol. Other examples of the usefulness of alkanes as biomarkers include correlations of various n-alkanes to bacterial populations ($\underline{3}$), and the association of C_{30}-steranes with marine versus nonmarine input (originally as C_{30}-sterols) ($\underline{4}$).

Another factor commonly used as a source indicator with respect to alkanes is the carbon number distribution. Briefly, n-alkane distributions with carbon number maxima in the 17-21 range (C_{17}-C_{21}) originate largely from aquatic algal sources, while maxima characterized by higher carbon numbers are

0097–6156/86/0305–0001$06.00/0

indicative of higher plant input (leaf waxes of terrestrial
plants) (1, 5). In addition, C_{23}-C_{33} n-alkanes with a predominance of an odd number of carbons to an even number of
carbons (OEP) is also indicative of terrestrial higher plant
input (1, 6-9) although loss of odd numbered carbon predominance
can occur with maturity (5). Thus, although petroleum is
believed to be of biological origin, petroleum input is
characterized by a lack of odd to even predominance (O.E.P. =
1.0). The isotopic composition of alkanes and of other classes
of compounds can also be used as a general source input indicator. Isotopic fractionation resulting from the metabolic pathways involved in the synthesis of biologically produced
compounds, when preserved in a diagenetic product, is frequently
used to differentiate between terrestrial and aquatic sources.
Hydrogen and carbon isotopic compositions of biogenic methanes
from shallow aquatic environments is discussed in a later chapter
of this volume (R. A. Burke and W. M. Sackett). The applicability of carbon isotopic data to tracing the source of deep-sea
Mesozoic sediments is discussed by R.M. Joyce and E. S. Van
Vleet.

Many other classes of lipids besides alkanes have proven to
be valuable biomarkers, including fatty acids, sterols, amino
acids, polysaccharides, glycerols, wax esters, porphyrins and
many isoprenoids. The greater the extent of alteration of a
compound by diagenetic processes, the harder it is to correlate
the product to the biological precursor. However, the successive
changes undergone by the precursor molecule can sometimes be
deduced from the structure of the product. In these cases the
"biomarker" is also a geochemical marker and may be referred to
as a biogeochemical indicator. For example, stenol to stanol
ratios can be used as an indicator of sediment oxicity. High
stenol/stanol ratios are indicative of oxidizing rather than
reducing environments (1, 20). Perylene provides an interesting
example of a PAH of disputed origin, which may prove to be a
reliable indicator of syn-and post-depositional anoxia when
present in greater than trace amounts (11). Indicators of the
biogeochemistry of chlorophyll and the temperature dependence of
chlorophyll diagenesis is reviewed in the chapter by J.W. Louda
and E. W. Baker. Other chapters in this volume concerned with
source input markers and geochemical indicators include those by
S.C. Brassell and G. Eglinton, J.W. de Leeuw, J. Whelan et al.,
J.J. Boon et al., R.M. Joyce and E. S. Van Vleet.

Humic Substances
Compared to the well defined structures and interrelationships of
the discrete molecular biomarkers and geochemical indicators
discussed above, relatively little is known about marine humic
substances. Whether the major source of input to dissolved and
sedimentary marine humic substances is of terrigenous or marine
origin is still occasionally debated in the literature, although
most interpretations point towards marine planktonic sources
(12-14), based most convincingly on isotopic ratios, GC-MS, ^{1}H
and ^{13}C NMR and infrared spectra (15-17). Stuermer et al (12)
propose a combination of ^{13}C and H/C ratios as a

source input indicator of terrestrial versus marine derived humic
substances.

One of the most useful tools for determining input sources,
functional group content and diagenetic relationships between
humic fractions has proven to be ^{13}C NMR. The early applications
of ^{13}C NMR to humic substances were done by solution NMR and were
thus limited to base soluble humic and fulyic acids (16, 18, 19).
The subsequent application of solid-state ^{13}C NMR to humic
samples has allowed for direct comparisons between soluble and
insoluble fractions as well as between humic substances from
diverse environments. ^{13}C NMR analysis of solid samples is made
possible by the use of cross polarization and magic angle
spinning (CP/MAS). Cross polarization is a sensitivity enhancing
technique in which the application of radio frequency pulses at
the resonance frequencies of ^{1}H and ^{13}C establishes contact
between the two populations allowing for cross polanization (i.e.
transfer of magnetization) from a large ^{1}H population to a minute
^{13}C population. Magic angle spinning (spinning the sample at
several kHz at an angle of 54.7° relative to the externally
applied magnetic field) minimizes line broadening due to chemical
shift anisotropy (20-21) The combination of these techniques
with high power decoupling and fourier transform analysis has
yielded high resolution spectra which can provide reliable
estimates of various carbon types, (21-24).

The important question of the reliability of quantification
of solid state ^{13}C NMR results can be answered in part by
relaxation studies and the measurement of relaxation constants
(25-27) and is discussed in greater detail by M.A. Wilson and
A.H. Gillam in a later chapter of this volume. Quantification of
the aromatic content (f_a) of marine humic and fulvic acids by
CP/MAS ^{13}C NMR has demonstrated that these substances have low
aromaticities relative to terrestrial counterparts (16,28). The
low aromatic (high aliphatic) content of marine humic substances
is a result of their in situ production. Higher plants
associated with terrestrial environments contribute lignin to
soil humic substances. The lack of lignin in marine plants
results in a low aromaticity of marine humic substances although
coastal sediments may contain high levels of vascular plant
residues (29). Lignin is an aromatic based natural polymer
characterized by methoxyl group substitutions on the benzene
ring structure. These lignin-associated methoxyl groups
typically produce a distinctive ^{13}C NMR peak which is normally
interpreted as a source marker for terrestrial input. However,
as discussed in the chapter by M.A. Wilson and A.H. Gillam,
contributions from other carbon types (possibly amino acid
carbon) may contribute to this signal. The relatively new
technique of dipolar dephasing is capable of distinguishing
between terrestrial lignin methoxyl signals and other protonated
carbons which may resonate in the methoxyl region of the
spectrum. Dipolar dephasing takes advantage of different
relaxation rates of various types of carbon and utilizes this
phenomenon to distinquish between protonated and nonprotonated
carbon (26). The utilization of dipolar dephasing in studying
the diagenetic changes undergone by organic matter in marine
sediments is discussed in a chapter by W.T. Cooper et al..

Recent advances in the structural interrelationships among humic
substances of marine and estuarine sedimentary origin is
discussed in the chapter by P.G. Hatcher and W.H. Orem.

Organic Pollutants in the Marine Environment

The biogeochemistry of organic pollutants in marine systems is of
enormous economic and environmental impact. The environmental
behavior of polychlorinated biphenyls (PCB's) has been studied
rather extensively because of their detrimental effects on human
health and on living marine resources (30-32). As discussed in
the chapters by J.W. Farrington et al., due to recent advances in
gas capillary chromotographic methods, it is now possible to
study the biogeochemistry of individual PCB's rather than that of
combined industrial mixtures of PCB's (33-36). In order to
realistically assess the risks to animal health, it is important
to be able to work with individual PCB levels rather than with
unresolved mixtures because individual PCB's can vary greatly in
terms of toxicity (37).

The contamination of coastal marine environments by
anthropogenic hydrocarbon input is also a matter of urgent
concern with respect to human health and preservation of natural
resources. Hydrocarbon contamination by oil spills can often be
traced to sources by fingerprinting techniques. An excellent
review of current developments in this field is presented by E.S.
Van Vleet (38) including discussions on computer interfaced total
fluorescence, transmission infrared spectroscopy, GC and GC-MS,
trace metal analysis etc.. The correlation of hydrocarbons with
petroleum sources is frequently hindered by differential
weathering. For a detailed discussion on the weathering of
petroleum in the marine environment, the reader is referred to a
current literature review (39). In this volume, the
fingerprinting of hydrocarbon contamination from specific land
use activities, supported by GC-MS analysis of polynuclear
aromatic compounds (PNA) are discussed by R.H. Pierce et al. and
by E.B. Overton et al., while fingerprinting of hydrocarbon
contamination from carbonied coal products (creosote and coal
tar) is described by T.L. Wade et al..

Volatile Organic Substances

The decomposition of complex organic substances by microbial
marine populations results in the in situ production of biogenic
methane, which is found in trace amounts in all fresh and salt
waters. The relative importance of acetate assimilation and CO_2
reduction as primary methanogenic pathways is discussed in the
chapter by R.A. Burke and W.M. Sackett. In the past, these
pathways were studied by the use of ^{14}C labelled substrates.
However, the authors demonstrate the advantages of using stable
hydrogen and carbon isotopic compositions of biogenic methanes to
evaluate the relative contributions from these two pathways.
Seasonal distributions of methane are discussed in the chapter by
J.D. Cline et al..

Volatile halogenated methanes are also produced in situ in
the marine environment and have been found to be associated with
marine macroalge (40). The release of these halogenated methanes

into coastal seawater has also been demonstrated (41-43) and is
of environmental concern. A seasonal study of the release of
polybromomethanes is described in the chapter by P.M. Gschwend
and J.K. MacFarlane.

Organic - Inorganic Interactions

The major influence of organic marine substances on the
partitioning and speciation of inorganic and organometallic
substances is discussed in the closing chapters of this volume.
Complexation, adsorption and/or reduction of ions and compounds
by organic matter in marine and brackish systems greatly affect
the cycling and bioavailability of many potential nutrients and
toxic substances.
    The complexation of metal ions by humic substances is a well
documented phenomenon (44-51). Due to the ubiquitous nature of
these naturally occurring ligands, humic substances play an
important role in determining the partitioning of metal ions and
other substances in both terrestrial and aquatic environments
(52-56). In order to predict or model the partitioning of a
particular metal ion in a given environment, it is essential to
be able to evaluate the magnitude of the interaction between the
pertinent components of the environment (the substrates) and the
metal ion. When dealing with the various solid phases which may
be present, adsorption of the ion onto each phase (clay, sand,
silt, hydrous metal oxides, humus) can be studied via the
evaluation of adsorption isotherms (56). In the solution phase,
complexation is an important process which competes with ion-pair
formation and subsequent precipitation. The strength of the
complexation reaction can be measured by evaluation of a
conditional formation constant. (57-59, 46).
    Adsorption of dissolved substances onto suspended matter is
a primary process in the removal of dissolved substance from the
water column and subsequent concentration in sediments. The
major role played by adsorbed organic coatings on particulate
matter is well known (60, 61). The processes of complexation and
adsorption in marine systems and their effects on the speciation
of various ions and compounds are discussed in the closing
chapters.

Literature Cited

1. Eglinton, G. In "Advances in Organic Geochemistry 1968";
   Schenck, P.A.; Havenaar, I., Eds.; Pergamon Press:
   Oxford, 1968 pp. 1-24.
2. Treibs, A. Angew. Chimie 1936, 49, 682-686.
3. Volkman, J.K.; Farrington, J.W.; Gagosian, R.B.;
   Wakeham, S.G. In "Advances in Organic Geochemistry
   1981"; Bjoroy, M., Ed.; John Wiley & Sons Limited: 1982,
   pp. 228-240.
4. Moldowan, J.M. Geochim. Cosmochim. Acta 1984, 48,
   2767-2768.
5. Baker, E.W.; Louda, J.W. In "Advances in Organic
   Geochemistry 1981; Bjoroy, M., Ed.; John Wiley & Sons
   Limited: 1982 pp 295-319.

6. Eglinton, G.; Hamilton, R.J. Science 1967, 156, 1322-1335.

7. Leenheer, M.J.; Flessland, K.D.; Meyers, P.A. Org. Geochem. 1984, 7, 141-150.

8. Brassell, S.C.; Comet, P.A.; Eglinton, G.; Isaacson, P.J.; McEvoy, J.; Maxwell, J.R.; Thomson, I.D.; Tibbetts, P.J.C.; Volkman, J.K. In "Advances in Organic Geochemistry 1981; Bjoroy, M., Ed.; John Wiley & Sons Limited: pp. 375-391.

9. Simoneit, B.R.T. In "Chemical Oceanography" Riley, J.P.; Chester, P., Eds.; Academic Press: New York, 1978; Vol. 7, pp. 233-311.

10. Gagosian, R. B.; Smith, S.O.; Lee, C.; Farrington, J.W.; Frew, N.M. In "Advances in Organic Geochemistry 1979" Douglas, A.G.; Maxwell, J.R., Eds.; Pergamon Press:, 1980; pp.407-419.

11. Louda, J.W.; Baker, E.W. Geochim. Cosmochim Acta 1984, 48, 1043-1058.

12. Stuermer, D.H.; Peters, K.E.; Kaplan, I.R. Geochim. Cosmochim. Acta 1978, 42, 989-997.

13. Nissenbaum, A.; Kaplan, I.R. Limol. Oceanogr. 1972, 17, 570-582.

14. Stuermen, D.H.; Harvey, G.R. Nature, 1974, 250, 480-481.

15. Stuermer, D.H.; Payne, J.R. Geochim. Cosmochim Acta 1976, 40, 1109-1114.

16. Hatcher, P.G.; Rowan, R.; Mattingly, M.A. Org. Geochem 1980, 2, 77-85.

17. Gillam, A.H.; Wilson, M.A. Org. Geochem. 1985, 8, 15-25.

18. Vila, F.J.G.; Lentz, H.; Ludemann, H.D. Biochem. Biophys. Res. Commun. 1976, 672, 1063-1069.

19. Wilson, M.A.; Goh, K.M., J. Soil Sci 1977, 28, 645-652.

20. Andrew, E.R. In "Progress in NMR Spectroscopy", Emsley, J.W.; Feeney, J.; Sutcliffe, L.H., Eds.; Pergamon, 1972; Vol. 8, pp. 1-39.

21. Bartuska, V.J.; Maciel, G.E.; Schaefer, J.; Stejskal, E.O. Fuel 1977, 56, 354-358.

22. Hatcher, P.G.; VanderHart, D.L.; Earl, W.L. Org. Geochem 1980, 2, 87-92.

23. Miknis, F.P.; Netzel, D.A.; Smith, J.W.; Mast, M.A.; Maciel, G.E. Geochim. Cosmochim. Acta 1982, 46, 977-984.

24. Resing, H.A.; Garroway, A.N.; Hazlett, R.N. Fuel, 1978, 57, 450-454.

25. Yoshida, T.; Maekawa, Y.; Fujito, T. Anal. Chem. 1983, 55, 388-390.

26. Wilson, M.A.; Pugmire, R.J.; Grant, D.M. Org. Geochem. 1983, 5, 121-129.

27. Wilson, M.A.; Vassallo, A.M. Org. Geochem. 1985, 8, 299-312.

28. Hatcher, P.G.; Breger, I.A.; Mattingly, M.A. Nature 1980, 285, 560 562.

29. Ertel, J.R.; Hedges, J.I. Geochim. Cosmochim. Acta 1984, 48, 2065-2074.

30. Miller, S. Environ. Sci. Technol 1982, 16, 98A-99A.

31. Bopp, R.F.; Simpson, H.J.; Olsen, C.R.; Kostyk, N. Environ. Sci. Technol. 1981, 15, 210-216.

32. Pierce, R.H.; Olney, C.E.; Felbeck, G.T. Geochim. Cosmochim. Acta 1974, 38, 1061-1073.
33. Ballschmiter, K.; Zell, M. Frezenius Z. Anal. Chem. 1980, 302, 20-31.
34. Duinker, J.C.; Hillebrand, M.T.J. Environ. Sci. Technol. 1983, 17, 449-456
35. Mullin, M.D.; Pochini, C.M.; McCrindle, S.; Romkes, M.; Safe, S.H.; Safe, L.M. Environ. Sci. Technol. 1984, 18, 468-476.
36. Duinker, J.C.; Hillebrand, M.T.J.; Boon, J.P. Netherlands J. Sea. Res. 1983, 17, 19-38.
37. Kimbrough, R.D. "Halogenated Biphenyls, Terphenyls, Naphthalenes, Dibenzodixins and Related Products. Elsevier/North Holland Biomedical Press: New York, 1980, 406 pp.
38. Van Vleet, E.S. Mar. Techn. Soc. J. 1984, 18, 11-23.
39. Payne, J.R.; McNabb, G.D., Jr. Mar. Techn. Soc. J. 1984, 18, 11-23.
40. Hewson, W.D.; Hager, L.P. J. Phycol 1980, 16, 340.
41. Gschwend, P.M.; MacFarlane, J.K.; Newman, K.A. Science 1985, 227,1033.
42. Dryssen, D.; Fogelquist, E. Oceanol. Acta 1981, 4, 313.
43. Lovelock, J.E.; Maggs, R.J.; Wade, R.J. Nature 1973, 256, 193.
44. Adhikari, M.; Hazra, G.C. J. Indian Chem. Soc. 1972, 49, 947-951.
45. Ardakani, M.; Stevenson, F.J. Soil Sci. Soc. Am. Proc. 1972, 36, 884-890.
46. Gamble, D.S.; Schnitzer, M.; Hoffman, I. Can. Journ. Chem. 1970, 48, 3197-3204.
47. Gamble, D.S. Anal. Chem. 1980, 52, 1901-1908.
48. Mantoura, R.F.C.; Riley, J.P. Analyt. Chim. Acta 1975, 78, 193-200.
49. Mantoura, R.F.C.; Dickson, A.; Riley, J.P. Est. Coast. Mar. Sci. 1978, 6, 387-408.
50. Mills, G.L.; Hanosn, A.K. Jr.; Quinn, J.G.; Lammela, W.R.; Chasteen, N.D. Mar. Chem. 1982, 11, 355-377.
51. Piotrowicz, S.R.; Harvey, G.R.; Boran, D.A.; Weisel, C.P.; Springer-Young, M. Mar. Chem. 1984, 14, 333-346.
52. Nissenbaum, A.; Swaine, D.J. Geochim.Cosmochim. Acta. 1976, 40, 809.
53. Rashid, M.A.; Leonard, J.D. Chem. Geol. 1973, 11, 89-97.
54. Theis, T.L.; Singer, P.C. In "Trace Metals and Metal-Organic Interactions in Natural Waters"; Singer, P.C., Ed.; Ann Arbor Science Publishers, 1973.
55. Wallace, G.T. Mar. Chem. 1982, 11, 379-394.
56. Davies-Colley, R.J.; Nelson, P.O.; Williamson, K.J. Environ. Sci. Technol. 1984, 18, 491-499.
57. Gamble, D.S.; Schnitzer, M.; Kerndorff, H. Geochim. Cosmochim. Acta 1983, 47, 1311-1323.
58. Clark, J.S.; Turner, R.C. Soil Sci. 1969, 107, 8-11.
59. Elgala, A., El-Damaty, A.; Abel-Latif, I. Z. Pflanz.Rodenk. 1976, 3, 293-300.

60. Davis, J.A.; Leckie, J.O. Environ. Sci. Technol. 1978,
    12, 1309-1315.
61. Balistrieri, L.S.; Brewer, P.G.; Murray, J.W. Deep-Sea
    Res. 1981, 28A, 101-121.

RECEIVED December 10, 1985

MOLECULAR MARKERS

2

Molecular Geochemical Indicators in Sediments

S. C. Brassell and G. Eglinton

Organic Geochemistry Unit, School of Chemistry, University of Bristol, Bristol BS8 1TS, United Kingdom

Sediments from contemporary aquatic environments contain a diversity of compounds that provide an assessment of the sources of their organic matter. These components include lipids with structural features that are indicative of their biological origins. Thus, specific diterpenoids and triterpenoids are markers for sediment contributions from terrigenous vegetation. Similarly, among the numerous sterols recognised in sediments many are diagnostic of their algal origins, notably 4α-methylsterols derived from dinoflagellates. Several lipid types characterise contributions from bacteria; for example, acyclic isoprenoid alkanes arising from methanogens. Illustrative examples of such diagnostic lipid distributions show the possibilities for differentiating between sediments receiving allochthonous terrigenous organic matter and those dominated by autochthonous algal and bacterial contributions. Within the hydrocarbon distributions, by contrast, several features can denote non-biological sources of organic components, such as the weathering of ancient sediments, oil seepage and oil spillage. A further development in environmental assessment using molecular indicators stems from the recent recognition that the unsaturation of specific lipids contributed to sediments by coccolithophorid algae provide a measure of water temperatures.

This paper concentrates on three aspects of the application of molecular organic geochemistry to the interpretation of the biological origins of sedimentary organic matter and the use of such information in the evaluation of depositional environments. First, the basis for the assignment of specific compounds as diagnostic markers for the contributions of particular biota to sediments is considered and illustrative examples of such molecular indicators are given. Second, the sedimentary distributions of such components are discussed and the possibilities for their use in the qualitative assessment of contributions from different biological sources to sediments is addressed. Third, one group of lipids which occur widely in marine sediments, namely long-chain (C_{37} to C_{39}) alkenones, are considered in terms of their potential to provide an assessment of oceanic water temperatures in the shallow subsurface from the sedimentary molecular record.

The standard methodology used in investigations of the organic constituents of sediments typically involves their extraction, then fractionation according to polarity or compound class and finally their evaluation and identification by gas chromatography (gc) and computerised gas chromatography-mass spectrometry (gc-ms). Details of such procedures are not given here, but can be found in the references cited.

Molecular Indicators in Sediments

The development of molecular markers as indicators of biological contributions to sedimentary organic matter relies on the information from the lipid composition of appropriate organisms. Such information is often scant. Therefore, existing data is used to propose working hypotheses that can be modified or amended when additional information becomes available. The underlying rationale for this approach is the tacit assumption that the biochemical pathways of lipid biosynthesis in different organisms are not necessarily uniform at the present time, nor have they been over geological history. Rather, it seems that the lipid compositions of biota have been tailored throughout evolution to meet their environmental needs. Hence, the discrepancies and similarities between the lipids of different organisms can be used to assess their generic relationships leading to chemotaxonomy. Such chemotaxonomic description of organisms using lipid components relies on the information obtained from the analysis of both laboratory cultures and natural populations of individual species. The assignment of a given compound recognised in sediments to a specific biological source is based on such classifications aided by information from environmental analyses, such as the investigation of sediments thought to receive dominant inputs from a particular species or class of organism.

Indeed, the verification of the biological source and significance of individual lipid components of geochemical interest is often best served by studies designed to evaluate their origins in a chosen environment.

For some years it has been held that dinoflagellates are the biological source of the 4α-methylsterols that occur, frequently as abundant components, in marine sediments (1-3). Such assignments were initially based on the widespread literature concerning 4α-methylsterols in cultured dinoflagellates (4) and the abundance of these compounds in sediments known to receive major contributions from this class of algae. The distributions of 4α-methylsterols, and also 4α-methylstanones, observed in the bottom sediments of Priest Pot, a small lake in the English Lake District, were almost identical to those recognised in a bloom of the dinoflagellate Peridinium lomnickii collected from the overlying waters (5). This study demonstrated that P. lomnickii is the source of the 4-methylsteroids in Priest Pot sediments and, in more general terms, provided convincing evidence in support of the link between sedimentary 4α-methylsteroids and their presumed biological source, dinoflagellate algae.

Overall, there are clear indications that particular organic components may be specific to single or multiple biological sources or, alternatively, may be non-diagnostic of their biological origins. In simple terms an individual component, or the distribution of a specific compound class, may be representative of contributions to sedimentary lipids from either algal, terrigenous higher plant or bacterial sources. Not all components, however, can be placed into one of these three classes; for example, some occur in all three, some may only occur in animals, whereas the origins of others are unknown. In addition to considerations of their ultimate biological origins individual compounds in sediments can be either unaltered biosynthetic products or, alternatively, derivatives formed by diagenetic / catagenetic processes, which retain structural elements that attest to their original biological source. Most biological marker compounds occurring in sediments can be considered as markers of one of the following five groups:

     (i)   non-diagnostic components
     (ii)   algae
     (iii)   terrigenous higher plants
     (iv)   bacteria
     (v)   'unknown'

Examples of individual components held to be representative of these five categories are shown in Figures 1-5. The majority are known lipid constituents of organisms, whereas others are either diagenetic products or of an as yet undefined origin. The compounds chosen provide an illustrative, rather than a comprehensive, indication of the range of structural types in each category. Most of them have been discussed elsewhere in a fuller review paper (6), together with other structurally similar examples. Also, the individual compounds and compound distributions biosynthesised by, and held to be diagnostic for, algal, terrigenous higher plant and bacterial inputs to recent sediments are published elsewhere (3). Within categories (ii), (iii) and (iv) individual marker compounds can be uniquely representative of a particular species, genus or class of organism. Alternatively, they can be non-specific components which occur widely in different, but perhaps distantly related, families. Clearly, similarities in the lipid compositions of related species or genera may reflect their ancestral links, whereas differences may stem from the divergence of their biosynthetic pathways.

Non-diagnostic biological markers. The components considered to be non-diagnostic indicators (Figure 1) are likely to be those that play a fundamental role in biosynthetic processes. For example, essential constituents of cell structures (e.g. membranes) or similar physiological units might be expected to be common to many different types of organism and therefore occur ubiquitously. Hexadecanoic acid, for example, is a typical constituent of the membranes of numerous land and aquatic plants, animals and bacteria. It may be present in the form of wax esters or triacylglycerols but is generally of little value in differentiating between contributions to sediments from its various sources. Similarly squalene is the biosynthetic precursor of many triterpenoids and, thus, occurs widely in organisms. Cholesterol, like hexadecanoic acid, is a constituent of the cell membranes of many different families of organisms, with the notable exception of bacteria. It acts as a 'rigidifier' in cell membranes. It is a prominant sterol of many algae and land plants, and is often the only sterol component of copepods. Tocopherols are held to play an important role in photosynthetic processes. Hence, they are abundant in many photosynthetic organisms, including higher plants, algae and cyanobacteria (7), making them non-diagnostic markers of the sources of sedimentary organic matter.

Algal indicators. Individual marker compounds indicative of sediment contributions from algae (Figure 2) may occur in both marine and lacustrine environments, although their source species will differ. Indeed, botryococcene is the only component shown which has yet to be recognised in both environmental regimes. Carotenoids are perhaps the most source specific lipid constituents of organisms. For example, diatoxanthin occurs only in diatoms and can therefore be regarded as a highly specific marker for their contributions to sediments. Its absence in a sediment, however, cannot be taken as evidence for the lack of diatom contributions to organic matter, due to its lability. Indeed, this lability makes carotenoids of limited use as markers of the sources of organic matter in most sedimentary environments, with the notably exception of relatively shallow water systems, such as post-glacial lakes (8). Recent work (2,5) has provided convincing

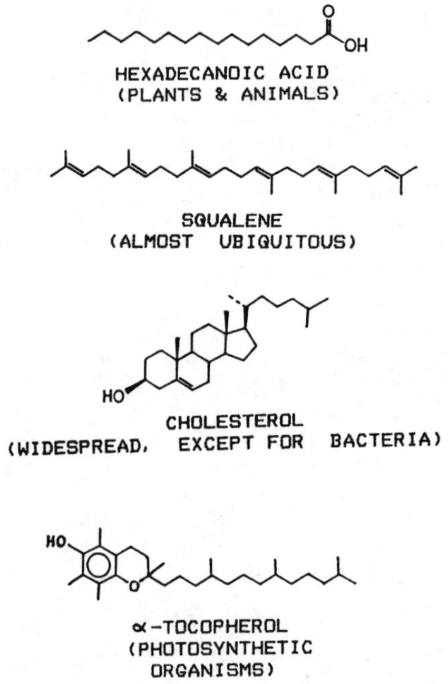

Figure 1. Compound structures, names and biological origins of examples of non-diagnostic indicators of the sources of sedimentary organic matter.

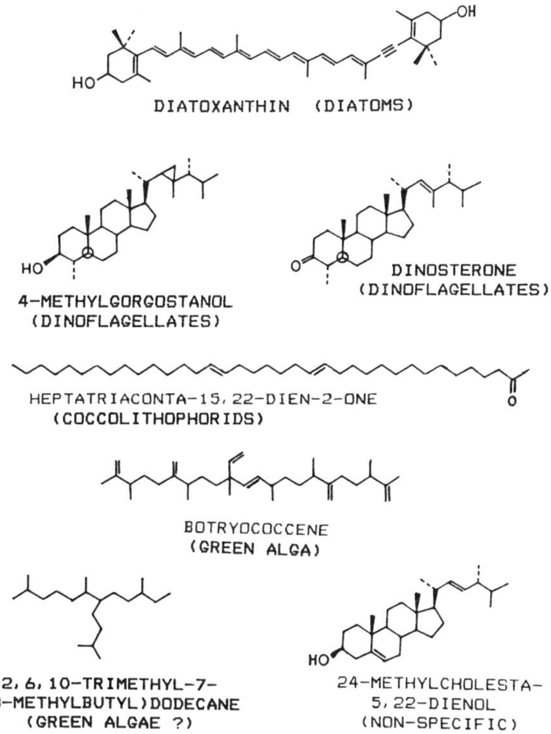

Figure 2. Compound structures, names and biological origins of examples of indicators of algal contributions to sedimentary organic matter.

evidence that the 4α-methylsteroids found in both marine
and lacustrine sediments are derived from dinoflagellate
algae.   Hence,   both   4α-methylgorgostanol
(22,23-methylene-4α,23,24-trimethyl-5α(H)-cholestan-3β-ol)
and   dinosterone
(4α,23,24R-trimethyl-5α(H)-cholest-22E-en-3-one) can be
regarded as examples of steroidal compounds that are
diagnostic of dinoflagellate contributions to sediments
(e.g. 6). In addition, dinoflagellates are the only
organisms in which these components have been recognised
(4).   Long-chain   alkenones,   such   as
heptatriaconta-15,22-dien-2-one, occur widely in both
recent and ancient oceanic sediments (9) and are generally
regarded as markers of inputs from prymnesiophyte algae,
most   notably   coccolithophorids   (9).   Their   recent
recognition in lacustrine sediments (10) is consistent
with an origin from such organisms. Botryococcene is the
single marker compound included among these examples which
has yet to be identified in sedimentary organic matter.
However, its saturated analogue, botryococcane, and
various homologues, have been identified in various
petroleums and oil shales (11,12) where their structural
specificity provides an unambiguous indication of
contributions from the green alga Botryococcus braunii.
The   branched   alkane
2,6,10-trimethyl-7-(3-methylbutyl)dodecane has also been
recognised in a green alga, Enteromorpha prolifera (13),
and occurs in both recent and ancient sediments and in
petroleums   (14).   24-Methylcholesta-5,22-dienol
(24R-methylcholesta-5,22-dien-3β-ol) is an abundant sterol
constituent   of   various   species   of   diatoms,
coccolithophorids and other algae (15). Although its
presence in several different algae precludes its use a
specific marker for contributions from a single class of
organism, this sterol is a general indicator of inputs
from planktonic algae to sediments. The structures of
many other 4-desmethylsterol constituents, differing most
notably in their side chains, provide similar evidence of
their derivation from an algal source.

Terrigenous   indicators.   Components   derived   from
terrigenous sources (Figure 3) tend to be more prominant
constituents of lacustrine, coastal or continental shelf
sediments. Aeolian transportation of land-derived dusts,
however, also results in significant inputs of such
material in the open ocean (16,17). Prominant among the
components of aeolian dusts are the n-alkanes in the C_{25}
to C_{33} range derived from higher plant waxes (18). n-C_{31}
is a member of this homologous family of compounds, with
which it will commonly co-occur. Hence, the distribution
pattern of these n-alkanes is more significant than the
presence of any individual member. Their persistence in
the sediment record and in aeolian materials apparently
stems from their resistance to microbial degradation.
Many pentacyclic triterpenoids, namely components with

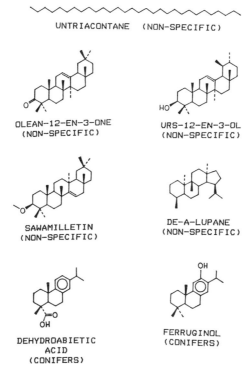

UNTRIACONTANE   (NON-SPECIFIC)

OLEAN-12-EN-3-ONE
(NON-SPECIFIC)

URS-12-EN-3-OL
(NON-SPECIFIC)

SAWAMILLETIN
(NON-SPECIFIC)

DE-A-LUPANE
(NON-SPECIFIC)

DEHYDROABIETIC
ACID
(CONIFERS)

FERRUGINOL
(CONIFERS)

Figure 3. Compound structures, names and biological origins of examples of indicators of terrigenous higher plant contributions to sedimentary organic matter.

oleanane, ursane, taraxerane, lupane and friedelane
skeletons, are characteristic of contributions to
sediments from terrigenous higher plants. They generally
occur as functionalised components in organisms, notably
as alcohols, ketones, acids and esters, and are not
specific to individual classes of biota. The most
frequently encountered members in sediments are those with
α- and β-amyrin skeletons (ursane and oleanane,
respectively), notably the Δ^{12} alcohols and ketones (e.g.
olean-12-en-3-one and urs-12-en-3β-ol). Sawamilletin
(3β-methoxytaraxer-14-ene) appears to be one of the more
refractory terrigenous triterpenoids, since it has been
observed in certain sediments devoid of other
triterpenoids of similar origins ([19]). One feature of
general significance is that the presence of a
functionality at C-3 in such triterpenoids appears to make
them succeptible to photochemical or microbial
degradation, giving rise to ring-A opened carboxylic acids
and, subsequently, to ring-A degraded triterpenoids (e.g.
de-A lupane; [20]). Both they and their intact precursors,
tend to be especially abundant in deltaic sediments from
tropical regions ([20,21]). Diterpenoids, such as
dehydroabietic acid and ferruginol ([22,23]) are marker
components present in many species of terrigenous higher
plants, but especially prevelant in conifers. Hence, they
tend to be most abundant in sediments from temperature
climes, especially those receiving inputs from extensively
forested areas such as the Astoria Fan ([22]).

Bacterial indicators. Bacteria influence the composition
of sedimentary organic matter in two ways: they degrade
components derived from other organisms and they

contribute their own, often characteristic, biosynthetic
products (Figure 4). It is the latter effect that is
considered here. iso-Branched carboxylic acids have been
long regarded as markers for contributions from bacteria
to sediments ([24]), although they are of little use in
distinguishing between inputs from different bacterial
types. Perhaps the most source specific lipid components
found in bacteria are the glycerol ether-linked acyclic
isoprenoids that are characteristic of the cell membranes
of archaebacteria (e.g. [25]). Such components have been
recognised among the polar lipids of sediments and
petroleums ([26]). The neutral lipids of archaebacteria also
contain a number of unique acyclic isoprenoid structures,
such as 2,6,10,15,19-pentamethyleicosane ([27]). The
occurrence of this component and related isoprenoid
alkanes in marine sediments is thought to reflect
contributions from methanogens, since inputs from
halophiles and thermoacidophiles in such environments is
highly unlikely ([28]). A number of C_{30} pentacyclic
triterpenes and triterpanes with 5-membered E-rings,
including fern-9(11)-ene, neohop-13(18)-ene and
hop-17(21)-ene, have been identified in an anaerobic
photosynthetic bacterium ([29]). Their occurrence together

with other triterpenes in marine sediments has been taken
to reflect bacterial contributions, although they are also
found in ferns (28). In addition to the C_{30} triterpenoids,
a whole family of extended (>C_{31}) hopanoids exist which
are diagnostic bacterial compounds, since they play a role
in the membrane structures of these primative organisms
(30). Bacterial carotenoids, like those characteristic of
algae, are highly specific. For example, spirilloxanthin
is restricted to, and therefore a marker for, purple
photosynthetic bacteria. It is perhaps noteworthy that
all bacterial carotenoids, unlike those of higher
organisms, possess identical functionalities and ring
systems at either end of the polyisoprenoid chain. This
inherent structural symmetry presumably reflects the
simplicity of the biosynthetic pathways producing these
compounds.

'Unknown' indicators. The precise biological origins of a
significant number of components recognised in sedimentary
organic matter is not known (Figure 5). All such 'unknown'
compounds possess structures that retain obvious links to
biosynthetic components, yet none has been identified in
living organisms. In some cases it may be unclear whether
the compound is a direct biosynthetic product or is
generated by the diagenetic transformation of an
unidentified precursor. Several 'unknown' indicators
occur widely in geological materials; for example, the
triterpane gammacerane has been recognised in numerous
sediments and petroleums, yet its origin, whether
diagenetic or biosynthetic, remains enigmatic. The only
biological triterpenoid of related structure that might be
a precursor is tetrahymenol (i.e. gammaceran-3β-ol), a
component of the protozoan Tetrahymena (31). This
triterpenol, however, has yet to be identified in any
recent or ancient sediment; hence, this particular
possible precursor/product relationship lacks appropriate
supporting evidence in terms of the geological occurrence
of the two compounds. A similar dichotomy exists with
28,30-bisnorhopane (32,33); either it is a diagenetic
product of some other hopanoid or it originates from an as
yet unknown biological source. Overall, a bacterial
origin for 28,30-bisnorhopane seems most probable, given
its hopanoid structure and its dominance of the
hydrocarbon distributions in certain sediments (33). On
the basis of their widespread occurrence in sediments and
petroleums it has been proposed (34) that the tricyclic
terpenoids related to tricyclohexaprane are derived from a
probable bacterial membrane lipid, tricyclohexaprenol.
This C_{30} carbon skeleton, however, has yet to be
identified in any organism and the link between tricyclic
terpenoids and bacteria remains speculative, based solely
on the geological evidence. The A-nor sterones found in
many marine sediments (35) have not been reported in
living organisms, although A-nor steranes do occur in
sediments (36). Sponges contain such altered steroidal

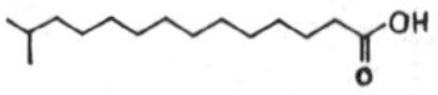

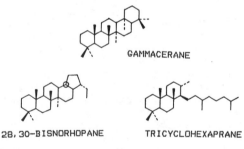

ISO-PENTADECANOIC ACID
(NON-SPECIFIC)

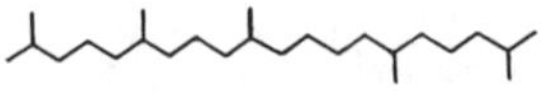

2, 6, 10, 15, 19-
PENTAMETHYLEICOSANE
(METHANOGENS)

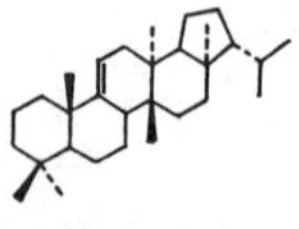

FERN-9(11)-ENE
(ANAEROBES ?)

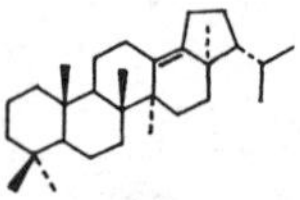

NEOHOP-13(18)-ENE
(NON-SPECIFIC)

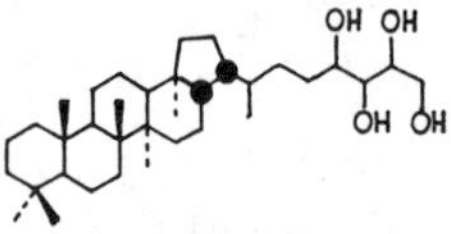

BACTERIOHOPANETETRAOL
(AEROBIC BACTERIA)

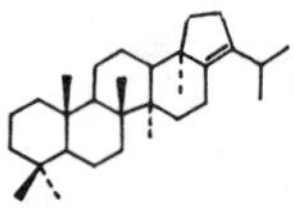

HOP-17(21)-ENE
(ANAEROBES ?)

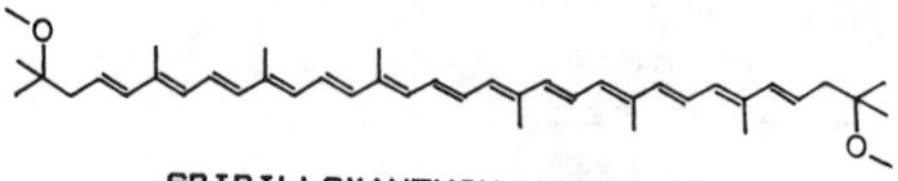

SPIRILLOXANTHIN (PURPLE
PHOTOSYNTHETIC BACTERIA)

Figure 4. Compound structures, names and biological origins of examples of indicators of bacterial contributions to sedimentary organic matter.

GAMMACERANE

28, 30-BISNORHOPANE

TRICYCLOHEXAPRANE

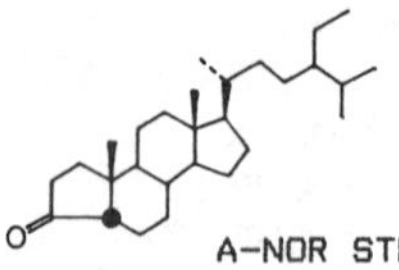

A-NOR STERONE

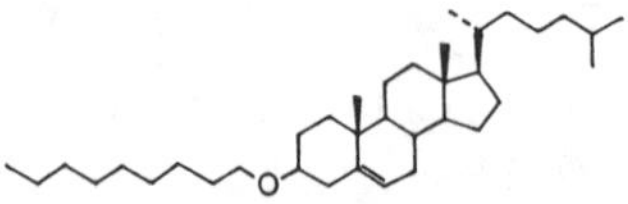

STEROL ETHER
(3-NONYLOXYCHOLEST-5-ENE)

Figure 5. Compound structures and names of examples of sedimentary components derived from unidentified sources.

skeletons (37), but an origin from sponges cannot be verified solely on the evidence of the sedimentary occurrence of these compounds. Similarly, the source of sterol ethers is unclear. These compounds occur in diatomaceous oozes from many locations and ages (38,19), but the dominance of C_{27} components among their steroidal moieties makes an origin from diatoms unlikely, since such a distribution would be atypical for these algae. It is also evident, however, that sterol ethers are not diagenetic products of steroids because the carbon number distributions of their steroidal moieties do not match those of other steroids in the same sediments. Also, there are few potential sources of C_9 to C_{12} alkyl moieties required to form these ethers among sedimentary lipids.

<u>Compound Class Distributions in Biota</u>. General considerations of the biological occurrence of individual components make it evident that particular skeletal types are restricted to specific classes of organism. For example, 4-methylsteroids appear to be resticted to dinoflagellates, whereas 4-desmethylsteroids are biosynthesized by a wide range of algae, including dinoflagellates, and higher plants. Similarly, among the pentacyclic triterpenoids hopanoids and fernenes are diagnostic of bacteria, whereas oleananoid and other related triterpenoids are specific markers for terrigenous higher plants. By contrast, the biological origin of gammacerane is unknown. These differences in the biological occurrence of pentacyclic triterpenoids is a clear manifestation of the divergence in the biosynthetic processes between such families of organisms.

<u>Anthropogenic Hydrocarbon Indicators</u>. All of the compounds cited above represent individual markers for different biological origins of sedimentary organic matter. Such compounds can become pollutants in sedimentary environments when they arise from, for example, high algal productivity induced by lake eutrophication caused by sewage and urban wastewater influxes. More generally anthropogenic influences on sedimentary organic matter stem from industrial effluents, such as chlorinated aromatic hydrocarbons (especially PCBs), and hydrocarbons derived from fossil fuels. In recent sedimentary environments the presence of an anachronous distribution of hydrocarbons, for example, thermally mature steranes (39), can indicate inputs of compounds from petroleum combustion and spillage. It is also possible, however, for such compounds to originate from the natural weathering of ancient sediments (39). A major diagnostic feature indicating petroleum contamination of sediments is the presence of an unresolved complex mixture of aliphatic hydrocarbons produced by the extensive biodegradation or weathering of crude oil (39). The use of such characteristics contrasts sharply with the principle of

utilizing individual marker compounds in the assessment of
sediment inputs.

Relative Contributions from Different Biota to Sediments.
In the above discussions the occurrence of specific marker
compounds or their distributions is shown to provide an
indication of the biological or anthropogenic origins of
sedimentary organic matter. This approach does not,
however, permit the direct assessment of the relative
proportions of contributions from different biological
sources for various reasons. First, the marker compounds
reflect only a small, and perhaps non-representative,
portion of the total organic matter. Second, individual
components undoubtedly differ significantly in their
resistance to microbial degradation, both in the water
column and in the sediment. Hence, their comparative
abundances are influenced by the extent and nature of such
alteration. Third, the concentrations of the diagnostic
compounds for different classes of organisms can vary by
orders of magnitude. In summary, the use of marker
compounds for the absolute quantitative assessments of the
various biological origins of sedimentary organic matter
is problematic; however, comparative data can be
obtained.

Distributions of Marker Compounds in Marine Sediments

In the investigation of organic compounds in sediments the
experimental procedures invariably include a separation
scheme to divide and simplify total sediment extracts into
suitable fractions of different polarity. Typically, this
experimental procedure will yield a number of fractions
containing principally hydrocarbon, ketone, carboxylic
acid, alcohol or polar components. The reconstituted ion
chromatograms (RIC) from gc-ms analysis of three such
fractions are discussed herein to illustrate the observed
distributions of marker compounds in marine sediments and
discuss their inferred biological origins.

Hydrocarbon fraction. The aliphatic hydrocarbon
distribution (Figure 6) of a sediment from the Cariaco
Trench (DSDP 15-147C-3-3) off Venezuela contains two
structural families of diagnostic compounds in addition to
the steroids and triterpenoids in the polycyclic region.
First, the n-alkane distribution is dominated by
odd-numbered higher homologues (i.e. C_{29}, C_{31}) which are
indicative of sediment contributions from terrigenous
higher plants. At this site such components may originate
from aeolian transported Saharan dust (16). Second, and
more significantly, the distribution of hydrocarbons is
dominated by various acyclic isoprenoids. Two of these
components 2,6,10,15,19-pentamethyleicosane and lycopane
(Peaks E and F in Figure 6) may be attributed to
archaebacterial inputs, most probably from methanogens
(28), although the evidence for lycopane is

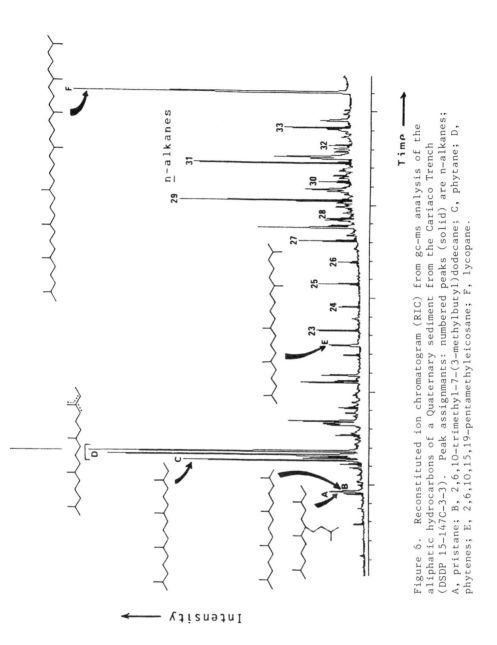

Figure 6. Reconstituted ion chromatogram (RIC) from gc-ms analysis of the aliphatic hydrocarbons of a Quaternary sediment from the Cariaco Trench (DSDP 15-147C-3-3). Peak assignmants: numbered peaks (solid) are n-alkanes; A, pristane; B, 2,6,10-trimethyl-7-(3-methylbutyl)dodecane; C, phytane; D, phytenes; E, 2,6,10,15,19-pentamethyleicosane; F, lycopane.

circumstantial. Both phytane and the phytenes (Peaks C and D, respectively, in Figure 6) may also be of archaebacterial origin (27), or they may be products of phytol diagenesis (e.g. 41). Pristane (Peak A in Figure 6) may be contributed by zooplankton or it may be formed diagenetically from phytol (e.g. 41 or tocopherols (7). In all cases it is an autochthonous component deriving from photosynthetic organisms or their predators. 2,6,10-Trimethyl-7-(3-methylbutyl)dodecane (Peak B in Figure 6) probably reflects contributions from green algae (13).

Ketone / Carboxylic Acid fraction. In the investigation of the lipid composition of a Pliocene (2Ma) diatomaceous ooze from the Walvis Ridge (DSDP 75-532-21-2) the total extract was methylated prior to compound class separation, thereby giving a fraction containing both ketones and carboxylic acid methyl esters (Figure 7). The dominant components of this fraction were recognised by gc and gc-ms as C_{37} to C_{39} alkadienones and alkatrienones presumed to derive from coccolithophorid algae (9). Other significant components of this fraction included a series of C_{22} to C_{30} n-alkanoic acids, which are indicative of inputs from terrigenous higher plants (18). 17β(H),21β(H)-Bishomohopanoic acid (Peak A in Figure 7) may be taken as a marker for bacterial contributions to the sedimentary organic matter since it is the expected major oxidation product of bacteriohopanetetraol. This fraction also contains tocopherols (α-, γ- and δ-; Figure 7) which are non-diagnostic components that may originate from any photosynthetic organism. Typically, therefore, they will be of mixed origins from terrigenous higher plants, algae and bacteria.

Alcohol fraction. Sterols are a major class of compound in immature and unconsolidated sediments. In most open marine environments they are presumed to principally derive from marine algal sources (3,15). Contributions from terrigenous higher plants may also be evident where the sediments are, in part land-derived, such as in coastal locations (15,42). The major component in the sterol composition of a sediment from the Cariaco Trench (DSDP 15-147B-4-2; analysed as TMS ethers; Figure 8; Table I) is dinosterol (4α,23,24-trimethyl-5α(H)-cholest-22E-en-3β-ol; Peak #18), a well-established marker for dinoflagellate algae (1-3,5). Other 4α-methylsterols (Peaks 15 and 19-21 in Figure 8) similarly provide evidence of dinoflagellate inputs to the sediment. None of the other sterols can be linked directly to a specific source, but all may be attributed to a variety of planktonic algal sources, such as diatoms, dinoflagellates and coccolithophorids. One noteworthy feature of this sterol distribution is the high proportion of 5α(H)-stanols relative to Δ^5-sterols. It may arise from preferential degradation of Δ^5-stenols (43),

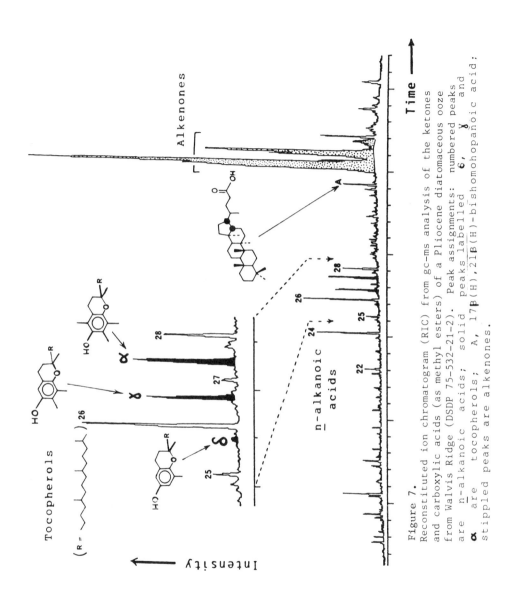

Figure 7.
Reconstituted ion chromatogram (RIC) from gc-ms analysis of the ketones and carboxylic acids (as methyl esters) of a Pliocene diatomaceous ooze from Walvis Ridge (DSDP 75-53-21-2). Peak assignments: numbered peaks are n-alkanoic acids; solid peaks labelled **6**, **ɣ** and **ɑ** are tocopherols; **A**, 17β(H),21β(H)-bishomohopanoic acid: stippled peaks are alkenones.

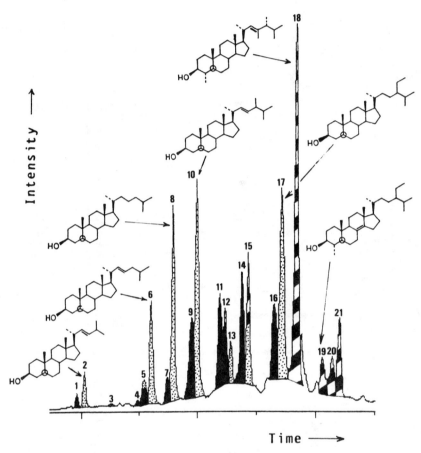

Figure 8. Reconstituted ion chromatogram (RIC) from gc-ms analysis of the sterols (as TMS ethers) of a Quaternary sediment from the Cariaco Trench (DSDP 15-147B-4-2). Peak assignments are given in Table I. Solid, stippled and hatched peaks are Δ^5-stenols, 5α(H)-stanols and 4α-methylstanols, respectively.

Table I. Peak Assignments of Sterols in Figure 8.

Peak[a]	Assignment[b]
1	24-norcholesta-5,22E-dien-3β-ol
2	24-nor-5α(H)-cholest-22E-en-3β-ol
3	24-nor-5α(H)-cholestan-3β-ol
4	27-nor-24-methylcholesta-5,22E-dien-3β-ol
5	27-nor-24-methyl-5α(H)-cholest-22E-en-3β-ol + cholesta-5,22E-dien-3β-ol
6	5α(H)-cholest-22E-en-3β-ol
7	Cholest-5-en-3β-ol
8	5α(H)-cholestan-3β-ol
9	24-methylcholest-5,22E-dien-3β-ol
10	24-methyl-5α(H)-cholest-22E-en-3β-ol
11	24-methylenecholest-5-en-3β-ol
12	24-methylene-5α(H)-cholestan-3β-ol + 24-methylcholest-5-en-3β-ol
13	24-methyl-5α(H)-cholestan-3β-ol
14	23,24-dimethylcholesta-5,22E-dien-3β-ol
15	23,24-dimethyl-5α(H)-cholest-22E-en-3β-ol + 24-ethylcholesta-5,22E-dien-3β-ol + 4α,24-dimethylcholesta-5,22E-dien-3β-ol
16	24-ethylcholest-5-en-3β-ol
17	24-ethyl-5α(H)-cholestan-3β-ol
18	4α,23,24-trimethyl-5α(H)-cholest-22E-en-3β-ol
19	24-ethyl-4α-methyl-5α(H)-cholest-8(14)-en-3β-ol
20	4α,23S,24R-trimethyl-5α(H)-cholestan-3β-ol
21	4α,23R,24R-trimethyl-5α(H)-cholestan-3β-ol

a   Number of peak in Figure 8, where analysed as
    TMS ethers.
b   Sterol assignments are based on individual gc
    retention behaviour and mass spectra by
    comparison with reference or literature
    standards (e.g. 2) or by spectral inter-
    pretation.

or, alternatively, it may reflect the high proportion of sterols derived from dinoflagellate sources, since these organisms can be major sources of 5α(H)-stanols (5).

Long-chain Alkenones - Climatic Indicators?

Following the recognition of series of C_{37} to C_{39} alkenones in the coccolithophorid Emiliania huxleyi and in sediments from the Japan Trench and elsewhere (44,45), the same series of components, but with a markedly greater dominance of the alkadienones, was identified in sediments from the Middle America Trench (46). From these data it seemed plausible that the difference in the unsaturateion of these components might be a reflection of the water temperatures of their environment (47). The ability to vary their lipid unsaturation with growth temperature is a well known characteristic of algae and bacteria which is designed to maintain their membrane fluidity. Thus, it seemed plausible that the changes in the unsaturation of long-chain alkenones might also be temperature dependent. Subsequent studies of a wide variety of Quaternary sediments from locations of different latitude and water temperature have confirmed this general observation (48). Further evaluation of this molecular tool for climatic assessment has been directed towards its comparison with an established measure of palaeoclimatic fluctuations, namely $\delta^{18}O$ values for foraminifera (48). In a study of sediments from the Kane Gap region in the eastern equatorial atlantic the unsaturation of C_{37} alkenones (b in Figure 9) shows a variability which broadly corresponds to the glacial / interglacial cycles assessed from $\delta^{18}O$ of planktonic foraminifera (Figure 9). In particular the lower values for the C_{37} alkenone unsaturation index (U^K_{37}) correspond to the glacial maxima, most markedly in the upper section of the core. The n-alkane concentrations in the core (c in Figure 9) vary independently of the alkenone unsaturation index, suggesting that they are influenced by factors other than water temperatures. Indeed, since their distributions are typical of those characteristic of higher plants (18), it is probable that they reflect the input of terrigenous organic matter, perhaps as an aeolian input (16). Both of the molecular trends (b and c in Figure 9) show little relationship to the carbonate content of the sediments, which also shows downhole fluctuations (a in Figure 9). These data, discussed in detail elsewhere (48, 49), demonstrate that molecular geochemical information can play an important role in the assessment of the climatic record in sediments.

Conclusions

The structural specificity of organic compounds in different organisms enables their occurrence and distributions to be used in the assessment of the

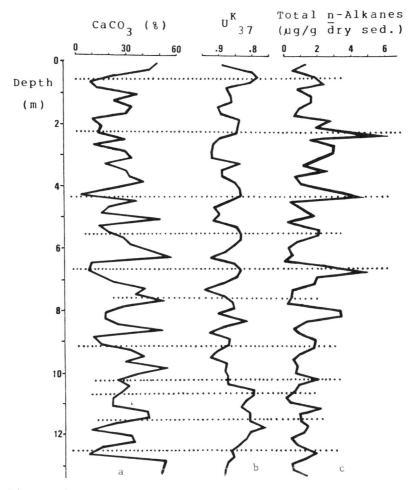

Figure 9. Downhole plots in a 13m gravity core from the Kane Gap region (Ml6415-2) of: a, carbonate contents; b, alkenone unsaturation (U^K_{37} = $[C_{37:2}]/[C_{37:2}]+[C_{37:3}]$); c, total n-alkane concentrations. The dotted horizons denote the levels corresponding to glacial maxima, as indicated by preliminary $S^{18}O$ data on foraminiferal carbonate (cf. 48,49).

biological origins of sedimentary organic matter. In
particular, such information can distinguish contributions
from algae, terrigenous higher plants and bacteria. It
also appears that the unsaturation of alkenones provides a
molecular measure of palaeoclimate insofar as it records
water palaeotemperatures. Overall, organic molecules in
sediments are both sensitive and unique indicators of
depositional palaeoenvironments, often recording
information that is not accessible by other techniques.

Acknowledgments

We thank Dr. Ian Marlowe for his invaluable efforts and
contributions to the study of the Kane Gap core. We are
grateful to Drs. M. Sarnthein and U. Pflaumann for the
preliminary information on the glacial maxima inferred
from the $\delta^{18}O$ values of foraminiferal carbonate. We
acknowledge financial support from the N.E.R.C. (GR3/2951
and GR3/3758) for gc-ms facilities and from the Royal
Society for a V.G. Minichrom gc data system.

Literature Cited

1. Boon, J. J.; Rijpstra, W. I. C.; de Lange, F.; de
    Leeuw, J. W.; Yoshioka, M.; Shimizu, Y. Nature,
    Lond. 1979, 277, 125.
2. de Leeuw, J. W.; Rijpstra, W. I. C.; Schenck, P. A.;
    Volkman, J. K. Geochim. Cosmochim. Acta 1983, 47,
    455.
3. Brassell, S. C.; Eglinton, G. In "Coastal Upwelling:
    Its Sediment Record"; Suess, E.; Thiede, J., Eds.;
    Plenum: New York, 1983; Part A, p. 545.
4. Withers, N. In "Marine Natural Products"; Scheuer,
    P. J., Ed.; Academic: New York, 1983; Vol. 5, p.87.
5. Robinson, N.; Eglinton, G.; Brassell, S. C.;
    Cranwell, P. A. Nature, Lond. 1984, 308, 439.
6. Brassell, S. C.; Eglinton, G.; Maxwell, J. R.
    Biochem. Soc. Trans. 1983, 11, 575.
7. Goossens, H.; de Leeuw, J. W.; Schenck, P. A.;
    Brassell, S. C. Nature, Lond. 1984, 312, 440.
8. züllig, H. Schweiz Z. Hydrol. 1982, 44, 1.
9. Marlowe, I. T.; Brassell, S. C.; Eglinton, G; Green,
    J. C. In "Advances in Organic Geochemistry 1983";
    Schenck, P. A.; de Leeuw, J. W.; Lijmbach, G. W. M.,
    Eds.; Org. Geochem. 1984, 6, 135.
10. Cranwell, P. A. Geochim. Cosmochim. Acta 1985, 49,
    1545.
11. Moldowan, J. M.; Seifert, W. K. J. Chem. Soc., Chem.
    Commun. 1980, 912.
12. Brassell, S. C.; Eglinton, G.; Fu Jiamo In "Advances
    in Organic Geochemistry 12"; Leythaeuser, D.;
    Rullkotter, J., Eds.; Pergamon, Oxford, in press.
13. Rowland, S. J.; Yon, D. A.; Lewis, C. A.; Maxwell,
    J. R. Org. Geochem. 1985, 8, 207.
14. Yon, D. A.; Ryback, G.; Maxwell, J. R. Tetrahedron
    Lett. 1982, 23, 2143.

15. Gagosian, R. B.; Nigrelli, G. E.; Volkman, J. K. In "Coastal Upwelling: Its Sediment Record"; Suess, E.; Thiede, J., Eds.; Plenum, New York, 1983; Part A, p. 241.

16. Simoneit, B. R. T.; Eglinton, G. In "Advances in Organic Geochemistry 1975"; Campos, R; Goni, J., Eds.; Enadimsa, Madrid, 1977; p. 415.

17. Schneider, J. K.; Gagosian, R.B.; Cochran, J. K.; Trull, T. W. Nature, Lond., 1983, 304, 429.

18. Eglinton, G.; Hamilton, R. J.; Science, 1967, 156, 1322.

19. Brassell, S. C., unpublished data.

20. Corbet, B.; Albrecht, P.; Ourisson, G. J. Am. Chem. Soc. 1980, 102, 1171.

21. Hoffmann, C. F.; Mackenzie, A. S.; Lewis, C. A.; Maxwell, J. R.; Oudin, J. L.; Durand, B.; Vandenbroucke, M. Chem. Geol. 1984, 42, 1.

22. Simoneit, B. R. T. Geochim. Cosmochim. Acta 1977, 41, 463.

23. Brassell, S. C. ; Eglinton, G.; Howell, V. J. In "Marine Petroleum Source Rocks"; Brooks, J.; Fleet, A.J., Eds.; Blackwells, Oxford, in press.

24. Cranwell, P. A. Chem. Geol. 1973, 11, 307.

25. Holzer, G. Colorado Sch. Mines Quat. 1983, 9.

26. Chappe, B.; Albrecht, P.; Michaelis, W. Science 1982, 217, 65.

27. Holzer, G.; Oró, J.; Tornabene, T. G. J. Chromatog. 1979, 186, 795.

28. Brassell, S. C.; Wardroper, A. M. K.; Thomson, I. D.; Maxwell, J. R.; Eglinton, G. Nature, Lond. 1981, 290, 693.

29. Howard, D. L.; Simoneit, B. R. T.; Chapman, D. J. Arch. Microbiol. 137, 200.

30. Ourisson, G.; Albrecht, P.; Rohmer, M. Pure Appl. Chem. 1979, 51, 709.

31. Tsuda, Y.; Morimoto, A.; Sano, T.; Inubishi, Y.; Mallory, F. B.; Gordon, J. T. Tetrahedron Lett. 1965, 1427.

32. Seifert, W. K.; Moldowan, J. M.; Smith, G. W.; Whitehead, E. V. Nature, Lond. 1978, 271, 456.

33. Katz, B. J.; Elrod, L.W. Geochim. Cosmochim. Acta 1983, 47, 389.

34. Ourisson, G.; Albrecht, P.; Rohmer, M. Trends Biochem. Sci. 1982, 7, 236.

35. McEvoy, J.; Maxwell, J. R.; Brassell, S. C. unpublished data.

36. van Grass, G.; de Lange, J.; de Leeuw, J. W.; Schenck, P.A. Nature, Lond. 1982, 296, 59.

37. Minale, L.; Sodano, G. J. Chem. Soc., Perkin I 1974, 2380.

38. Boon, J. J.; de Leeuw, J.W. Mar. Chem. 1979, 7, 117.

39. Brassell, S. C.; Eglinton, G. In "Analytical Techniques in Environmental Chemistry"; Albaiges, J., Ed.; Pergamon, Oxford, 1980; p. 1.

40. Rowland, S. J.; Maxwell, J. R. Geochim. Cosmochim. Acta 1984, 48, 617.

41. Didyk, B. M.; Simoneit, B. R. T.; Brassell, S. C.; Eglinton, G. Nature, Lond. 1978, 272, 216.

42. Huang, W. Y.; Meinschein, W. G. Geochim. Cosmochim. Acta 1976, 40, 323.

43. Nishimura, M. Geochim. Cosmochim. Acta 1978, 42, 349.

44. Volkman, J. K.; Eglinton, G.; Corner, E.D.S.; Sargent, J. In "Advances in Organic Geochemistry 1979"; Douglas, A. G., Maxwell, J. R., Eds.; Pergamon, Oxford, 1980; p. 219.

45. de Leeuw, J. W.; van der Meer, F. W.; Rijpstra, W. I. C.; Schenck, P. A. In "Advances in Organic Geochemistry 1979"; Douglas, A. G.; Maxwell, J. R., Eds.; Pergamon, Oxford, 1980; p. 211.

46. Brassell, S. C.; Eglinton, G.; Maxwell, J. R. In "Initial Reports of the Deep Sea Drilling Project"; U.S. Government Printing Office, Washington, D.C., 1981; Vol. LXVI, p. 557.

47. Brassell, S. C.; Eglinton, G. In "Heterotrophic Activity in the Sea"; Hobbie, J.; Williams, P. J. leB., Eds.; Plenum, New York, 1984; p. 481.

48. Brassell, S. C.; Eglinton, G.; Marlowe, I. T.; Sarnthein, S.; Pflaumann, U. Nature, Lond., in press.

49. Brassell, S. C.; Brereton, R. G.; Eglinton, G.; Grimalt, J.; Liebezeit, G.; Pflaumann, U.; Sarnthein, M. In "Advances in Organic Geochemistry 12"; Leythaeuser, D; Rullkötter, J., Eds.; Pergamon, Oxford, in press.

RECEIVED October 31, 1985

Sedimentary Lipids and Polysaccharides

As Indicators for Sources of Input, Microbial Activity, and Short-Term Diagenesis

J. W. de Leeuw

Department of Chemistry and Chemical Engineering, Delft University of Technology, De Vries van Heystplantsoen 2, 2628 RZ Delft, The Netherlands

Organic compounds present in recent marine sediments reflect the biochemical composition of the overlying water column as well as the in situ benthic activity. Suites of sedimentary compounds are also considered as starting compounds in chemical diagenetic pathways. Samples from several locations have been analyzed (e.g. the Namibian Shelf (S.W.-Africa), the Black Sea, Solar Lake (N. Sinai) and the Gavish Sabkha (S. Sinai)). Selected groups of sedimentary compounds such as sterols, 4-methylsterols, hydroxy fatty acids, carotenoids, long chain unsaturated methyl- and ethylketones, sterolethers, long chain 1,15-diols and 15-keto-monools, tocopherols, thiophenes and polysaccharides are discussed. These compounds have been selected in such a way that attention can be paid to the origin of sedimentary compounds, how they reflect the composition of the living communities in the overlying water column and the sediment itself, to symbiotic relationships, to microbial activity and to early diagenetic processes.

One of the ultimate goals in organic geochemistry is the detailed reconstruction of the environment of deposition using organic matter characteristics. Organic molecules isolated from sediment samples and subsequently identified carry information about their biological origin or about the state of diagenesis of the sediment or both. Therefore scientists working in the molecular organic geochemistry field have used and still use terms like biomarkers, chemical fossils, biological markers, geochemical fossils, guide molecules, biotracers, etc. to indicate the information content of individual organic compounds or suites of sedimentary organic compounds. Due to recent developments in the organic geochemistry of recent sediments in particular, these terms can be rather confusing and it is no longer clear what kind of information content of the organic molecules is being referred to. Therefore it is necessary to reevaluate the "marker concept" in organic geochemistry.

0097–6156/86/0305–0033$08.25/0

Figure 1 schematically shows an approach by which sedimentary organic
molecules are considered as information carriers, how the various
kinds of information contents can be discriminated and how the
assembled information is evaluated. The different ways by which the
information content is expressed are shortly discussed hereafter.

Structures. Structural features such as the skeleton, the stereo-
chemical configuration and the nature of functional groups of
individual compounds can be highly informative. If the structures
encountered are identical or reflect an origin from naturally
occurring compounds, the degree of information is determined by the
uniqueness of the occurrence in the natural environment which in its
turn is determined by the uniqueness of the biosynthetic machinery of
a certain group of organisms. Well-known examples are the extended
hopanoids (1), dinosterol (2) and the very long unsaturated ketones
(3). Sometimes the structures of the sedimentary organic molecules as
such do not occur in the biosphere but are (based on specific
structural features) easily traced back to natural precursors which
are not entirely unique for certain groups of organisms. In this case
the structural information can be used to unravel diagenetic pathways
and to determine the degree of diagenesis of the sediment under study.
Examples hereof are steroid hydrocarbons such as sterenes, dia-
sterenes, spirosterenes, steranes, diasteranes and aromatized
steroids which all reflect specific biochemical and chemical transformations of the originally present sterols (4). If the structural
features still reflect structures of unique natural compounds, then
they contain both information about their specific natural origin and
diagenetical pathways. Examples hereof are the extended hopanes, 4-
methyldiasterenes, steranes and certain diterpenes (5).

Distribution patterns. A distribution pattern of a certain class of
compounds can be highly informative in several aspects, while the
individual components do not have any information content. The most
common example to illustrate this type of information content are n-
alkane patterns observed in sediment extracts and oils. If the
envelope of n-alkanes maximizes in the C_{27}-C_{31} range and there is a
strong odd even predominance, an origin from higher plant waxes is
assumed. Based on exactly the same distribution pattern one also can
conclude that the sediment has not undergone severe diagenesis. If
the n-alkane patterns are smooth without any odd over even
predominance only some general conclusions can be made about the
degree of maturation. An even over odd predominance of n-alkanes can
occur in sediments or oils which originate from hypersaline
depositional environments (6).

Mode of occurrence. It is only in recent years that in some organic
geochemical investigations attention has been paid to the fact that
organic molecules in sediments (especially in recent sediments) occur
in different modes (e.g. 7,8). Since molecules in organisms are
present as such or as parts of larger structural entities it is again
the uniqueness of the biosynthetic machinery in specific groups of
organisms which determines the various modes by which the molecules
are present in the cells. To benefit from this potential source of
information present in living systems sequential or specific
procedures to isolate suites of organic molecules from sediment

samples are required. Even when the molecules as such or the distribution patterns of certain classes of compounds do not yield much information, their mode of occurrence might be highly informative. Straight chain fatty acids, for example, do not carry much information. If, however, it is analyzed which fatty acids occur free, which are esterified, or which are amide bound, we can obtain a lot of information from these compounds in terms of their biological origin and/or of early stage diagenetical pathways. This kind of information has been used to trace back the origin of alcohols and sterols in several sediments (7).

Total profiles. By analysis of total extracts, total hydrolyzates or total pyrolysis products, etc. without any preseparation, one can obtain so called total profiles. When gas chromatography or gas chromatography-mass spectrometry are applied for these kinds of analyses, we can study total profiles of lipid compounds which are amenable by GC. However, monitoring fractions of sedimentary organic matter can also be performed by IR, NMR, UV-VIS and other spectroscopic techniques. In this paper we will limit ourselves to GC and GC-MS profiles of lipid fractions obtained from sediments. The profiles might be considered as bird's eye views of the sedimentary organic matter. One can compare relative concentrations of individual compounds or classes of compounds directly, and by using the above mentioned information contents characterize the origin and/or diagenesis of the sediment in terms of relative contributions from different groups of organisms such as higher plants vs. dino-flagellates vs. coccolithophores vs. bacteria. Furthermore, if one observes in these total profiles similar or even identical distribution patterns of different classes of compounds, a correlation between these compound classes must exist, due to either the occurrence of similar patterns in certain groups of organisms or due to distinct diagenetic pathways by which one class of compounds is transformed to another compound class without affecting the distribution patterns. In this way individual molecules or distribution patterns which do not contain much information as such are becoming information carriers of some importance. No examples of this line of information are reported in the literature.

In this paper a number of examples will be given to illustrate the above mentioned concept. For that purpose first a critical evaluation of a selected part of our work in Delft from the last five or six years is presented. Subsequently the results of some ongoing research will be discussed to further illustrate the application of organic molecule information expressed in modes of occurrence and as total profiles of organic compounds from sediments. Finally the finding of new organic sulphur compounds and their possible significances are reported. In this report, mainly data from recent and subrecent sediments are discussed. Although some attention will be given to carbohydrates, the majority of the organic molecules considered here are lipids. It should be noted that this paper is not at all meant to give a complete review of organic information carriers in sediments; the investigations discussed are selected from our work at Delft to illustrate the above mentioned concept.

Results and Discussion

Table I summarizes some of the sediments of which samples have been
analyzed or are being analyzed in our group. Results obtained from
extracts of these sediment samples are discussed in this paper.

Table I. Sediments Investigated

Sample	Age	Location
1. Solar Lake	0 - 2500 yr. BP	North Sinai
2. Gavish Sabkha	very recent	South Sinai
3. Sarsina	6×10^6 yr.	North Apennines
4. Sapropel S_7	225×10^3 yr.	Eastern Mediterranean
5. Black Sea Unit 1	0 - 3000 yr. BP	Black Sea
6. Black Sea Unit 2	3000 - 7000 yr. BP	Black Sea
7. Livello Bonarelli	Cretaceous	Central Apennines
8. Sapropels	Very recent	Mediterranean
9. DSDP 362	Pleistocene/Miocene	Walvis Ridge
10. Namibian Shelf	1000 - 2500 yr. BP	off S.W.-Africa
11. Rozel Point Oil	Miocene	Utah
12. Wadden Sea	Very recent	The Netherlands

Figure 2 shows their geographical position, except for the Rozel
Point oil which comes from the Uinta basin in Utah, USA, and the mud
samples from the Dutch Wadden Sea. The occurrence and possible
significance of several suites of organic compounds isolated from a
number of recent sediments have been reported by us over the last
years. Firstly, results of this work will be critically evaluated and
secondly, a few new compounds series and analytical approaches are
discussed.

Midchain ketones and sterol ethers. Figure 3 shows the upper part of
a Total Ion Current (TIC) trace obtained from a TLC fraction of an
extract of the Namibian Shelf diatomaceous ooze. Long midchain
ketones and sterolethers together with a series of wax esters were
shown to be present in this fraction (9). The sterolethers are
composed of common Δ^5-sterol moieties and C_8 and C_9 alkylmoieties(I).

R_1 = H, CH$_3$, C$_2$H$_5$; R_2 = H, CH$_3$; M = 6 OR 7

$CH_3-(CH_2)_m-CH_2-O$ I

Although these compounds are encountered in several other sediments
(10), their unique structural information cannot be validated since
these compounds are not (yet) discovered by the natural product
chemists in living systems. The reported occurrence of cholesteryl

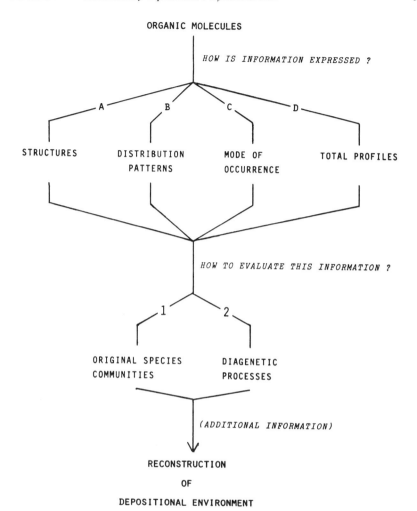

Figure 1. Sedimentary organic compounds as information carriers.

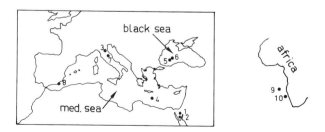

Figure 2. Geographical location of sediments investigated.
Numbers refer to sediments listed in Table I

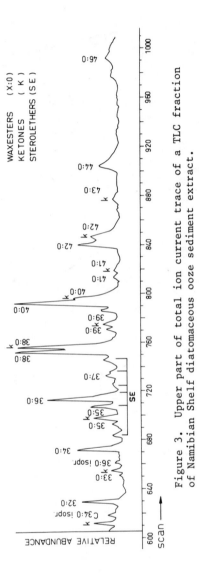

Figure 3.   Upper part of total ion current trace of a TLC fraction
of Namibian Shelf diatomaceous ooze sediment extract.

hexadecyl ether in bovine cardiac muscle (11) indicates that sterol-ethers as such occur in nature. Long-chain ketones as such are known to occur in plant waxes (12). In plants they are biosynthesized via the corresponding hydrocarbons and the distribution patterns of the hydrocarbons and mid-chain ketones in plants are therefore very similar and are characterized by a strong odd over even predominance. The mid-chain ketones present in this sediment extract do not at all reflect the hydrocarbon pattern and do have a strong even over odd predominance (II).

$$CH_3-(CH_3)_m-\overset{\overset{O}{\|}}{C}-(CH_2)_n-CH_3$$

$$|m-n| \leqslant 2 \; ; \; C_{31}-C_{43} \; ; \; C_{38,40,42}$$

II

Thus a plant cuticle origin is highly unlikely. Their distribution pattern to some extent mimics the wax ester pattern, which might indicate a common biological origin of these compound classes. A diagenetical relationship seems less likely. Since in the Namibian Shelf sediment sample investigated virtually no organic compounds are encountered which are derived from terrestrial organisms, both the sterolethers and the long mid chain ketones probably originate from a group of marine organisms. At this moment nothing is known or can be said about the geological fate of these compounds.

Alkan-1,15-diols and alkan-15-one-1-ols. The occurrence of long chain 1,15-diols (III) and alkan-15-one-1-ols (IV) have been reported in the sterol and stanol fractions of Black Sea Unit I and Unit II sediment extracts for the first time (13).

$$CH_3-(CH_2)_m-\overset{\overset{OH}{|}}{\underset{H}{C}}-(CH_2)_{13}-CH_2OH \qquad CH_3-(CH_2)_m-\overset{\overset{O}{\|}}{C}-(CH_2)_{13}-CH_2OH$$

III $m = 14, 15, 16 \; ; \; C_{30} \; (n = 14)$ IV

The upper part of two GC-traces are shown in Figure 4. Although these and similar compounds have been reported to occur in other sediments ever since (14), as yet nothing can be said firmly about their biological origin. The identical distribution patterns of the diols and keto-ols point to a tight biochemical relationship in the source organisms whatever they are, although in this case a diagenetical relationship via a partial reduction or oxidation cannot fully be excluded.

Although it is very likely that due to their specific structures and typical distribution patterns the mid-chain ketones, the sterolethers and the long chain 1,15-diols and keto-ols have a high information content and occur as such in living species, we are completely unable to use this information at this moment since we do not know what organisms they reflect and what the geological fate of these compounds is.

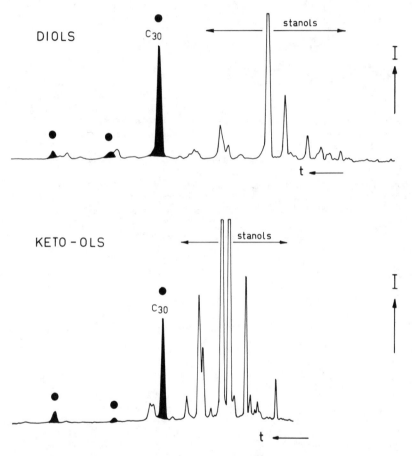

Figure 4. Upper part of GC-traces obtained for two sterol fractions isolated from Black Sea sediment extracts.

Alkyl- and alken-1-enyl-diacylglycerides. Alkyl- and alken-1-enyl-diacylglycerides (V, VI) were encountered in both the tissues of the lugworm Arenicola marina and in the intertidal flat sediment of the Wadden Sea (The Netherlands) where these lugworms live (15).

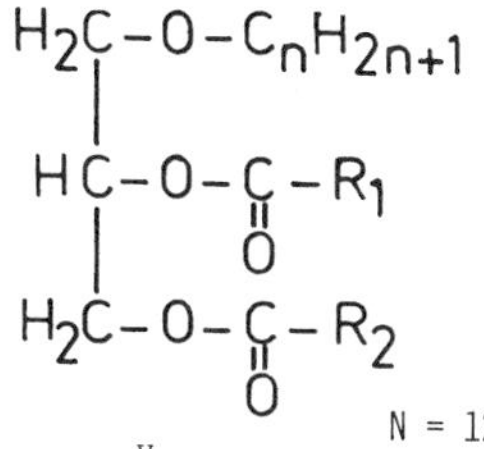

$$H_2C-O-C_nH_{2n+1}$$
$$HC-O-\underset{O}{\overset{||}{C}}-R_1$$
$$H_2C-O-\underset{O}{\overset{||}{C}}-R_2$$
V

$$H_2C-O-C=C-C_mH_{2m+1}$$
$$HC-O-\underset{O}{\overset{||}{C}}-R_1$$
$$H_2C-O-\underset{O}{\overset{||}{C}}-R_2$$
VI

N = 12-20; M = 10-18

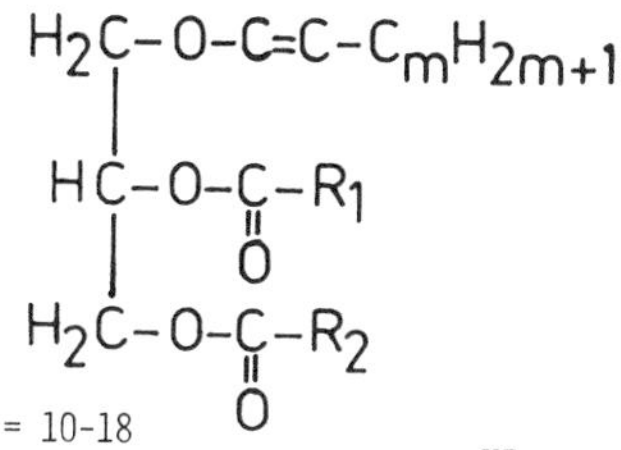

Figure 5 shows the GC-trace of a TLC fraction corresponding with the alkyldiacylglycerides separated from a total extract from the intertidal mud sediment. After TLC-separation the appropriate fraction was hydrolysed and subsequently silylated before GC and GC-MS analysis (15). The relative high abundance of iso-, anteiso and mid chain-methyl branched structures in the ether moieties supports a bacterial origin. Whether bacteria biosynthesize these compounds themselves or whether bacterially produced fatty acids are incorporated by other organisms higher up in the food chain, cannot be concluded at this moment. The geological fate of these compounds is as yet completely unknown, although it might be speculated that part of the alkyl moieties released from more ancient sediments after BBr$_3$ treatment originate from these alkylglycerides (16).

Very long chain di- and tri-unsaturated methyl and ethyl ketones. The very long polyunsaturated methyl and ethylketones (VII) were first encountered in high concentrations in extracts of Walvis Ridge sediments (17).

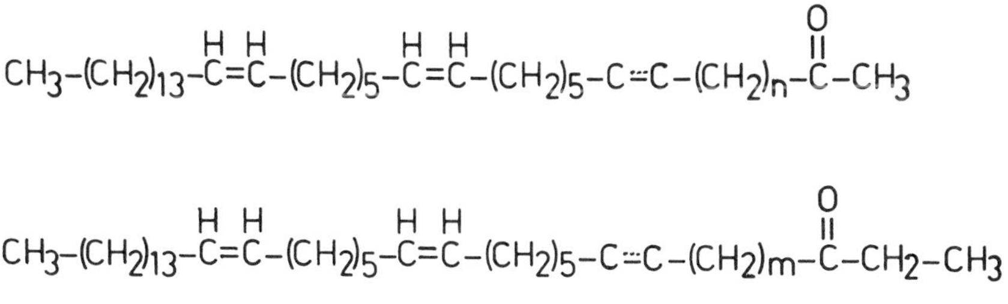

$$CH_3-(CH_2)_{13}-\overset{H}{C}=\overset{H}{C}-(CH_2)_5-\overset{H}{C}=\overset{H}{C}-(CH_2)_5-C=C-(CH_2)_n-\overset{O}{\overset{||}{C}}-CH_3$$

$$CH_3-(CH_2)_{13}-\overset{H}{C}=\overset{H}{C}-(CH_2)_5-\overset{H}{C}=\overset{H}{C}-(CH_2)_5-C=C-(CH_2)_m-\overset{O}{\overset{||}{C}}-CH_2-CH_3$$

N = 6, 7; M = 5, 6

VII

Later on, their complete structural identification followed (3). Ever since, they have been encountered in many sediment extracts, sometimes together with smaller amounts of the corresponding hydro-carbons and fatty acids (18). By mere luck, these compounds were also

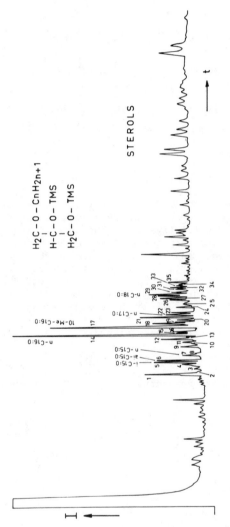

Figure 5. GC-trace of the alkylglycerol fraction of the Wadden
Sea anaerobic tidel flat sediment.

encountered in the widespread coccolithophore Emiliania huxleyi (19).
Therefore, these very long chain ketones in sediments were thought to
reveal the presence of coccolithophores in the original environment
of deposition. When sediments are deposited underneath the calcium
compensation depth (CCD) these molecules are then the only entities
in the sediment informing us about the original presence of
coccolithophores. However, a problem arises about the origin of these
compounds since Emiliania huxleyi species first appeared in the
fossil record about 250,000 years ago and many sediments which
contain these ketones are much older. Recent findings (20) show that
these ketones also occur in species of the Isochrysidales, indicating
that they are not strictly limited to the true coccolithophorids.
Although we at least have some useful ideas now where these
sedimentary compounds come from, nothing at all is known about their
geological fate.

Methoxy sugars, deoxy sugars and heptoses. Recently we have
investigated the occurrence of polysaccharides in a number of recent
sediments (21,22). By using the procedure schematically indicated in
the upper part of Figure 6, GC and GC-MS traces are obtained which
reveal peaks corresponding to alditolacetates obtained from polysaccharides. The GC trace obtained from the deepest sample from a
Solar Lake core (Figure 6) exhibits a number of major peaks derived
from very common sugars which as far as their structures are
concerned are not very informative. The smaller peaks, however,
reveal the presence of various methoxy- and deoxy sugars (VIII and
IX), which are thought to originate mainly from bacterially produced
polysaccharides.

VIII: 3-O-ME-ARA (24)   IX: 6-DEOXY-GLU (26)   X: 3-O-ME-HEPTOSE (60)

The heptoses (X) in the upper part of the trace (after "IS") reveal
the presence of heptoses which probably indicate an origin from
lipopolysaccharides (LPS) occurring in gram-negative bacteria. This
possible origin from LPS is supported further by other experiments
(see hereafter). Almost nothing is known about the geological fate of
these specific sugars or about the polysaccharides in general. This
is remarkable since polysaccharides seem to make up a major part of
the organic matter in recent sediments (23).
The examples given above for the alkylglycerides, the long chain
unsaturated ketones and the specific sugar components show that in
these cases relatively firm conclusions can be made about their biological origin, but as yet nothing is known about their geological fate.

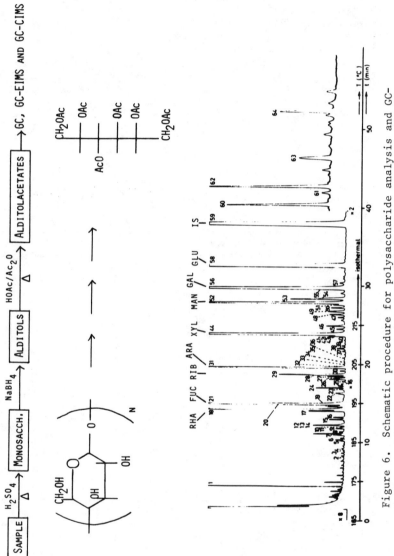

Figure 6. Schematic procedure for polysaccharide analysis and GC-trace of alditolacetates obtained from Solar Lake sediment.

4-Methylsteroids. Although 4-methylstanols and 4-methylsteranes are known to occur in recent and ancient sediments, for some time their origin was not very well understood (24,25). However, after the structural elucidation of the so called "Black Sea Sterol" as 4α-methyl-$5\alpha(H)$-Δ^{22}-23,24-dimethyl-cholesten-3β-ol (dinosterol, XI) in extracts from the Black Sea sapropel layer (Unit II) an origin from dinoflagellates became obvious (2).

R_1 = H, CH_3 ; R_2 = H, CH_3, C_2H_5

XII   XI   XIII

Ever since, dinosterol and many other 4-methylsterols (XII), including 4-methylsteroidketones, have been encountered in many sediments (8,26,10). The large variety of 4-methylsteroids encountered in a Black Sea Unit I sediment and their modes of occurrence very strongly support an origin from dinoflagellates due to the steadily increasing knowledge of steroids occurring in many species of dinoflagellates (8,27). There are good reasons to believe that the ratio of dinosterol to gorgosterols (XIII), calculated from sediment extracts can inform us about the relative amount of free and symbiotically living dinoflagellates in the depositional environment (28). Because diagenetical pathways of steroids have been studied thoroughly (4), it can be concluded that 4-methyldiasterenes and -steranes present in ancient sediments and oils represent the geological fate of these dinoflagellate sterols. This is, unfortunately, one of the few examples where we believe that we understand both the origin and the fate of a group of sedimentary compounds.

In the above discussed examples we have dealt with sedimentary organic compounds which probably occur as such in living organisms. Hereafter several examples will be discussed in which information is also obtained about diagenetic pathways.

Prist-1-ene. Although prist-1-ene (XIV) does not occur in extracts of sediments, it is a major, sometimes the most abundant pyrolysis product of organic matter in immature sediments (29).

XIV

Very recently it was demonstrated that tocopherol moieties in kerogen
are likely precursors of prist-1-ene (Figure 7) (30). This idea was
supported by the fact that tocopherols are widely distributed in
photosynthetic tissues and that they also occur as such in several
recent sediments (31). It is tempting to conclude that during
"natural pyrolysis" the generated pristene will be transformed to the
well known component, pristane, in ancient sediments and oils. This
example nicely illustrates that we have to be very careful when we
conclude that acyclic isoprenoid hydrocarbons such as pristane
originate from the chlorophyll side chain, phytol, based solely on
structural similarities.

Loliolides and dihydroactinidiolide. Although loliolides as such
occur in living systems the recently discovered sedimentary
loliolides (XV, XVI, XVII) isolated from Namibian Shelf diatomaceous
ooze are thought to be compounds which have been generated by very
early stage diagenetical pathways from carotenoids in the oxic zone
of the water column (32).

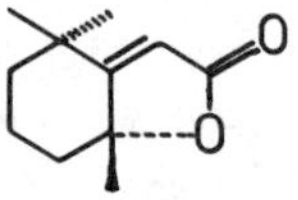

DIH-ACT.   ISO-LOL.   LOL.

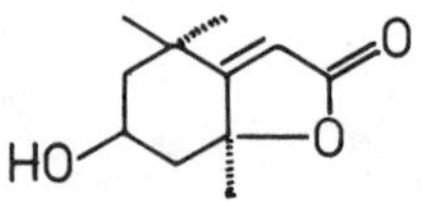

      XV   XVI   XVII

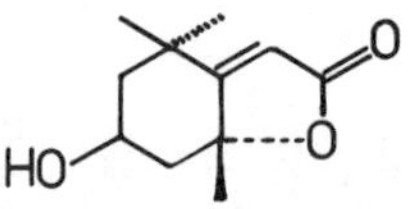

The conversion suggested in Figure 8 might proceed via the 5,8
furanoxides reported earlier (33). Although the structural features
inform us about a possible origin, nothing can be said about
subsequent transformation of these molecules upon increasing
maturation.

A-nor-steranes and De-A-steranes. De-A-steroid ketones (XVIII),
hydrocarbons (XIX) and A-norsteranes (XX) were shown to be present in
extracts from Cretaceous black shales in the North Central
Apennines (34,35). The De-A-steroidal compounds are thought to be
indicative of a diagenetical pathway as shown in Figure 9. A similar
pathway has been suggested for 3-oxy-triterpenoids (36). Although
A-nor-steranes are not known to occur in living systems, some sponges
can biosynthesize 3β-hydroxymethyl-A-norsteranes from dietary sterols
(37). Therefore it is tempting to suggest that the A-nor-steranes
originate from certain sponges and that their existance in sediments
is the result of a specific diagenetic route (Figure 10).

XVIII   XIX

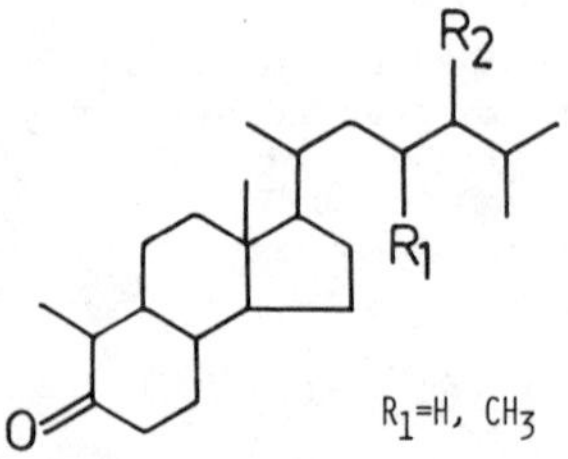

R_1=H, CH_3

R_2=H, CH_3, C_2H_5

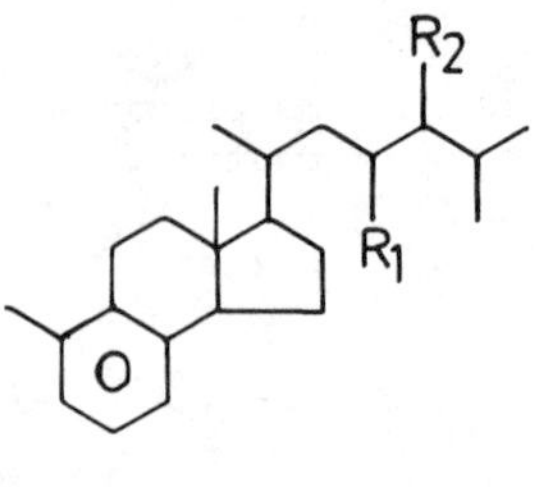

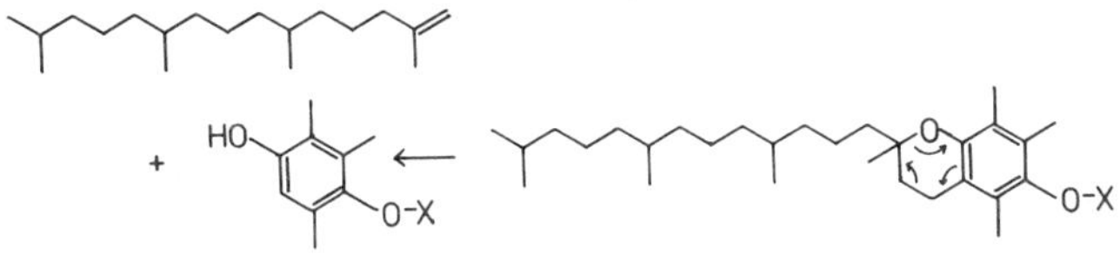

Figure 7.   Tocopherols as precursors for pristene and pristane.

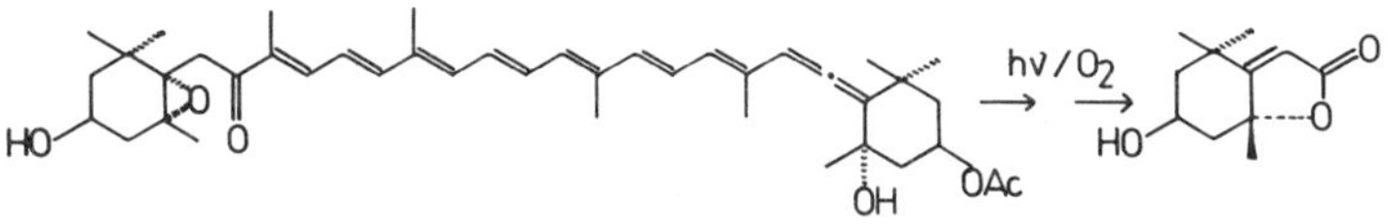

Figure 8.   Fucoxanthin as a possible precursor of loliolide.

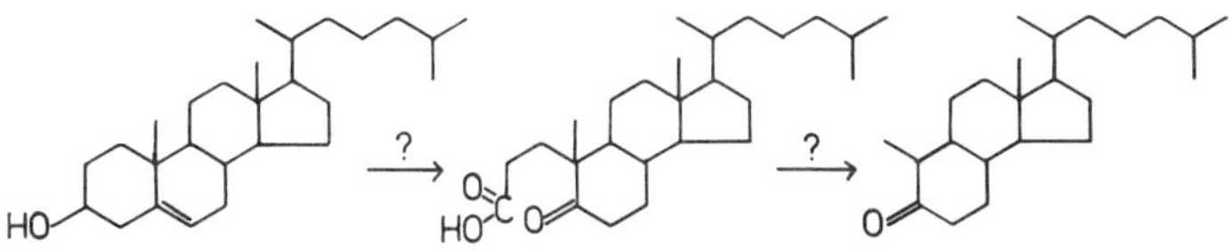

Figure 9.   Possible diagenetic pathway for De-A-steroids.

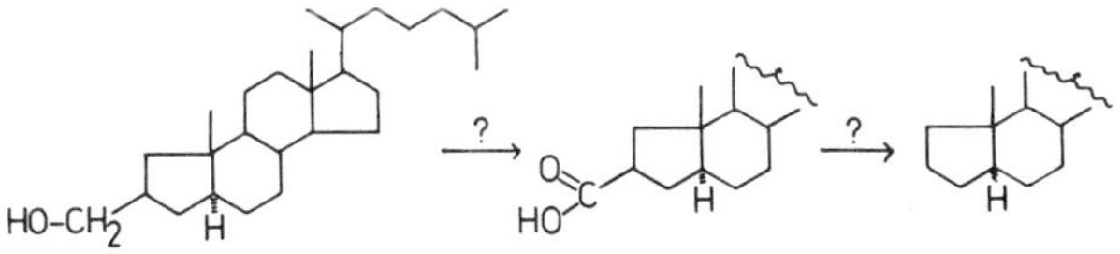

Figure 10. Possible diagenetic pathway for A-nor steroidal hydrocarbons.

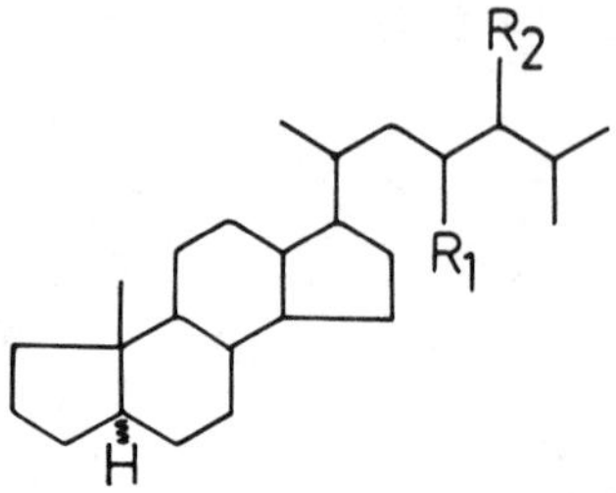

$R_1 = H, CH_3$

$R_2 = H, CH_3, C_2H_5$

XX

The above discussed loliolides, A-nor-steranes and De-A-steranes, are thought to carry information about particular diagenetic pathways since it is assumed that these compounds do not reflect compounds which, as such, occur in living biota. Once a certain pathway is proven, the precursor molecules are known and in this way additional information about the biological origins can be obtained.

As mentioned in the introduction, the information content of sedimentary organic compounds can also be expressed by their mode of occurrence. A general analytical procedure was set up to discriminate lipids which can be extracted as such, which are released after subsequent base treatment, and those which are released after subsequent acid treatment (Figure 11). Complex lipids present in the first extract can be further studied after saponification. The final residue can be investigated by other chemical degradation reactions or by flash pyrolysis. Extracts obtained are not further separated but are derivatized and analyzed by GC and GC-MS (38,39). Since bacterial contributions to sediments are thought to be much more important than originally expected in the organic geochemistry field (40), exactly the same procedure as sketched above was applied to numerous bacteria and to several recent sediments (41). Results are shown here for the photosynthetic sulphur bacterium Rhodobacter sulfidophilus and for a Mediterranean sapropel (S_7, 250,000 yr.). Figure 12 shows the GC-traces of the directly extractable lipids (-Free-), the lipids released after base treatment of the first residue (-OH⁻-) and the lipids after acid treatment of the second residue (-H⁺-). Major compounds in both the free and OH⁻-fractions are the mono unsaturated and saturated C_{16}- and C_{18}-fatty acids and phytol. In this bacterium there obviously is not very much difference between the free and esterified lipids. The third GC-trace (-H⁺-), however, shows a completely different pattern. The major components are β-hydroxy fatty acids representing a very particular distribution pattern ($C_{14:0}$, $C_{20:1}$ and $C_{22:0}$ are major components). The fatty acid composition is also different when compared with the first and second extracts. Besides the C_{16} and C_{18} fatty acids, α,β-unsaturated $C_{14:1}$ and $C_{20:2}$ fatty acids are encountered. An isomeric mixture of dihydroxy fatty acids is also present.

This phenomenon was encountered in all gram-negative and cyano-bacteria investigated. In most cases, however, the β-hydroxy fatty acid patterns are restricted to C_{10}, C_{12}, C_{14} and C_{16}-components (41). The selective release of the β-hydroxy fatty acids upon acid treatment indicates that they are linked via amide bonds to complex substances. From numerous investigations it is well known that amide

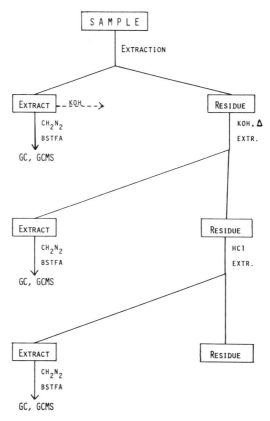

Figure 11. Analytical procedure to discriminate lipids by their mode of occurrence.

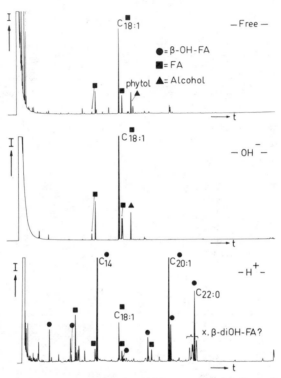

Figure 12. GC-traces of three different lipid fractions of
Rhodobacter sulfidophilus.

bound β-hydroxy fatty acids occur in bacterial substances like Lipopolysaccharides (LPS) and ornithin lipids (42,43). The schematical drawing of a partial LPS-structure and its occurrence in the bacterial cell (Figure 13) shows the presence of amide linked β-hydroxy fatty acids and esterified fatty acids in this substance. In a separate study of Acinetobacter calcoaceticus it was indeed demonstrated that β-hydroxy fatty acids from LPS almost exclusively occur in the $-H^+-$ fraction when the mentioned isolation procedure is used (38). Using this approach we have studied several recent sediments. One study is discussed here. The fractions "-Free-", "-OH$^-$-" and "-H$^+$-" were obtained from a 225,000 yr old Mediterranean sapropel and the GC-traces generated contain all kinds of information (Figure 14). First of all, β-hydroxy fatty acids in the $C_{12}-C_{20}$ range, including straight chain-, iso- and anteiso components, are almost exclusively present in the $-H^+$-fraction. Based upon their mode of occurrence (amide-bound), their distribution pattern ($C_{12}-C_{20}$) and structural features (iso- and anteiso-branching) it is evident that these compounds originate from LPS and/or ornithine lipid substances and thus represent a bacterial contribution to the sedimentary organic matter. The almost exclusive presence of phytol in the $-OH^-$-fraction indicates that phytol is present in an esterified mode, probably as the phytol moiety in chlorophylls. The C_{30}-diol and keto-ol as well as the very long unsaturated methyl- and ethylketones ("VLUK") occur as free lipids only, which is in agreement with their mode of occurrence noticed before in sediments and organisms. The fatty acids show different distribution patterns in the three fractions indicating that in this case, much information is hidden in the mode of occurrence.

Due to the GC and GC-MS analyses of whole extracts without any further separation, several classes of compounds with their own distribution patterns are visualized simultaneously in one gas chromatogram or total ion current trace. Comparison of these "total profiles" obtained from organic compounds of different sediments, from organisms, from recognizable geological macrostructures, etc., is another way by which information can be expressed. Figure 15 illustrates this approach. Extracts were obtained from a Mediterranean sapropel (S_1, 8000 yr BP) and from fossil oak cuticles (Miocene) after base treatment of both samples. Without separation, these extracts were derivatized and analyzed by GC and GC-MS. Comparison of the total profiles shows that there is a striking similarity in the n-alkane, the fatty acid and the alcohol distribution patterns and in the alkane/fatty acid/alcohol ratios within the $C_{22}-C_{32}$ chain length region. Such a similarity, which is easily observed using this "total profiles" approach, indicates that a considerable part of the organic matter present in the sapropel is derived from plant cuticles which have been fluvially transported to this marine sediment. If this reasoning is correct, the mechanism of sapropel formation, as proposed by Rossignol-Strick (44), is firmly supported by these observations.

Finally, some ongoing investigations are discussed dealing with the search for new information carriers among the organic compounds in sediments. One way to search for new informative compounds is to study organic compounds isolated from sediments which are well documented by other scientific disciplines. As an example, some results are shown from a Miocene sediment sample from the Northern

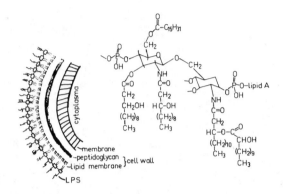

Figure 13. Schematical drawing of partial LPS-structure.

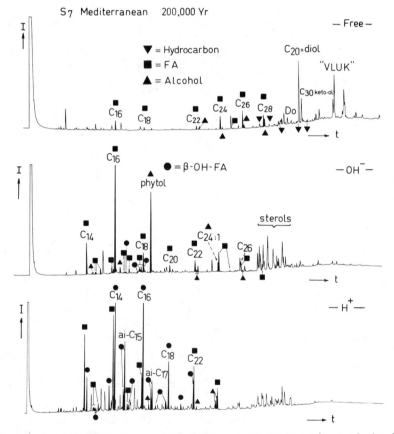

Figure 14. GC-traces of three different lipid fractions of the S₇ Mediterranean sapropel extracts.

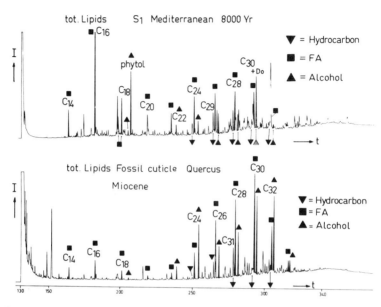

Figure 15. Total profiles of lipids from an S_1 Mediterranean sapropel and from a fossil cuticle.

Apennines which consists mainly of gypsum and marl layers and which
has been deposited under hypersaline conditions (6). The gas
chromatogram of the hydrocarbon fraction (urea non adduct) of the
marl is shown in Figure 16. Figure 17 shows the hydrocarbon fraction
of the gypsum layer. A number of phenomena are observed. The phytane-
pristane ratio is very high (≥ 10) in both samples and a small even
over odd predominance of the n-alkanes is noted in the gypsum sample.
Extended $\Delta^{17(21)}$-hopenes are present in the marl. The most abundant
compounds in the gypsum sample are 14α(H),17α(H)- and 14β(H),17β(H)-
pregnanes and homopregnanes. Further, a more or less typical pattern
of α,β-norhopane, α,β-hopane and gammacerane is present in this
sample. Several of these phenomena have been noticed before (45) in
sediments which might have been deposited under hypersaline
conditions. Some features were also observed in a recent Sabkha
environment (39). Table II summarizes the phenomena mentioned.

Table II. Characteristics of Hypersaline Depositional
Sedimentary Environments

Phenomena:
- phytane >> pristane (phythenes; diphytanylethers)
- isoprenoid and other thiophenes and thiolanes?
- pregnanes and homopregnanes

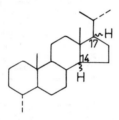

- even-odd predominance n-alkanes
- norhopane/hopane/gammacerane distribution
- $C_{30}-C_{35}$ 17(21) hopenes $\triangleq$ $C_{30}-C_{35}$ hopanes

Origin:
halophylic (archae)bacteria; photosynthetic sulphur bacteria;
cyanobacteria; sulphate reducing bacteria; dunaliella.
(sabkha's; lagoons; "Solar Lake", evaporitic basins).

Further work is necessary to better understand and apply the obvious
information present. For example, one wonders whether the high
phytane/pristane ratio is the result of relatively high amounts of
phytenes and/or diphytanylethers present in (halophilic) archaebacteria and is not at all an indication for anoxicity (46,47).

The isoprenoid thiophene compound shown in Figure 16 has been
observed in other sediments also (48). Inspection of the so called
aromatic hydrocarbon fraction of the marl sample with GC-FID, GC-FPD
and GC-MS revealed the abundant presence of this compound. Moreover,
this fraction is composed, for the most part, of organic sulphur
compounds, as can be judged from the GC-trace shown in Figure 18.
Other isoprenoid thiophenes, short chain and long chain benzo-

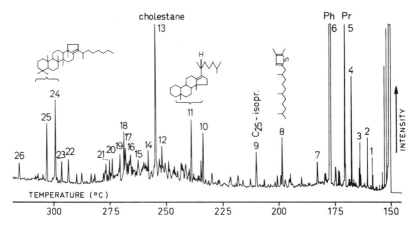

Figure 16. GC-trace of the hydrocarbon fraction (non adduct) of the marl extract of Sarsina sediment.

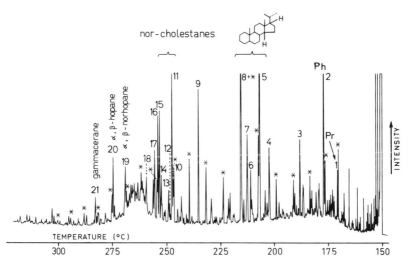

Figure 17. GC-trace of the hydrocarbon fraction of gypsum extract of Sarsina sediment.

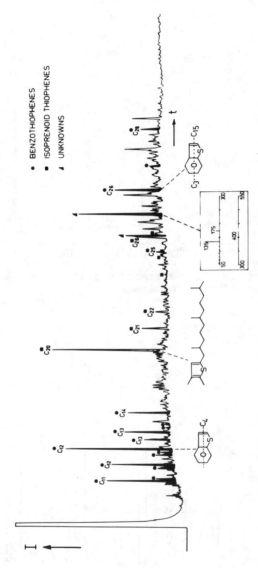

Figure 18. GC-trace of the aromatic hydrocarbon fraction of the marl extract of Sarsina sediment.

thiophenes are also major compounds in this fraction. At the moment
we can only speculate about the origin of these new organic sulphur
compounds. It was shown by comparison with a synthesized standard,
that one of these isoprenoid thiophenes occurred as a single stereo-
isomer in a Cariaco Trench sediment sample (48). Therefore, we assume
that these compounds are formed during the very early stages of
diagenesis. Even a direct origin from organisms (sulphur reducing or
oxidizing bacteria) cannot be ruled out at this stage.

   A further separation by column chromatography yielded various
fractions of the Sarsina aromatic hydrocarbon fraction. Homologous
series of n-alkyl thiophenes with peculiar distribution patterns were
observed by mass - chromatography generated after analysis of a
particular fraction by GC-MS (see Figure 19). The C_1 substituted n-
alkylthiophenes (m/z = 111) exemplified an identical pattern as the
n-alkane pattern; a clear odd/even predominance and C_{29} as the major
component. The C_2-substituted n-alkyl thiophenes (m/z = 125) showed
similar patterns but in this case there is an even/odd predominance
and C_{30} is now the major component. The same shift is observed when
the C_3- and C_4-substituted n-alkyl thiophenes (m/z = 139 and 153
respectively) traces are observed. Although this kind of distribution
pattern information is very clear in itself, the explanation has to
await further studies. The characteristic patterns indicate, however,
that an occurrence of compounds possessing long chain n-alkyl
thiophene moieties in living biota is not unlikely.

   Another striking distribution pattern was observed for isomeric
mixtures of n-alkyl thiolanes observed in a fraction of Rozel Point
oil. Figure 20 shows the GC-FPD trace of this fraction. Every major
peak consists of numerous isomers (different values for n an m but
per peak n + m = constant) of these thiolanes. It is remarkable to
see that the overall distribution pattern is very similar to fatty
acid patterns in young sediments. This might be another indication
that these thiolanes are either formed during early stages in the
diagenesis or that they are present as such or in a functionalized
form in organisms.

The above mentioned organic sulphur compounds represent new
information but more work is necessary to trace their origin. To
summarize, it can be stated that at this moment our knowledge about
the information content of sedimentary organic compounds is
increasing steadily. However, much more work is necessary to better
understand the different ways by which the information contents are
expressed before all information can be used efficiently to
reconstruct paleo-environments. To better understand the information
carried by sedimentary organic molecules we have to tune our
analytical extraction and isolation procedures in such a way that
modes of occurrences of lipids and total lipid profiles are optimally
expressed without losing structural information of individual
compounds or compound class distribution patterns.

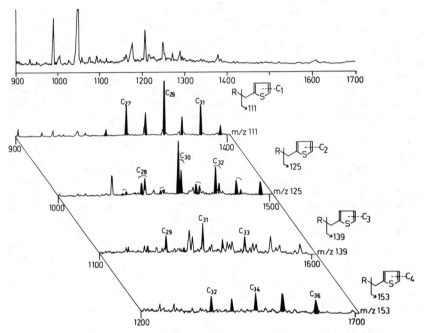

Figure 19. Mass chromatograms showing the distribution patterns of several groups of n-alkylthiophenes present in a Sarsina marl aromatic hydrocarbon fraction.

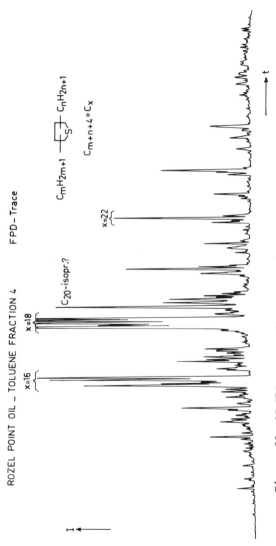

Figure 20. GC-FPD-trace of an aromatic hydrocarbon fraction revealing the distribution patterns of n-alkylthiolanes in Rozel Point oil.

Literature Cited

1. Ourisson, G.; Albrecht, P.; Rohmer, M. Pure Appl. Chem. 1979, 51, 709-729.
2. Boon, J.J.; Rijpstra, W.I.C.; de Lange, F.; de Leeuw, J.W.; Yoshioka, M.; Shimizu, Y. Nature 1979, 227, No. 5692, 125-127.
3. de Leeuw, J.W.; van der Meer, F.W.; Rijpstra, W.I.C.; Schenck, P.A. In "Advances in Organic Geochemistry 1979"; Douglas, A.G.; Maxwell, J.R., Eds.; Pergamon Press, Oxford, 1980; p. 211.
4. Mackenzie, A.S.; Brassell, S.C.; Eglinton, G.; Maxwell, J.R. Science 1982, 217, 491-504.
5. Simoneit, B.R.T. Geochim. Cosmochim. Acta 1977, 41, 463.
6. ten Haven, H.L.; de Leeuw, J.W.; Schenck, P.A. Geochim. Cosmochim. Acta 1985, in press.
7. Cranwell, P.A. Org. Geochem. 1984, 7, 25-37.
8. de Leeuw, J.W.; Rijpstra, W.I.C.; Schenck, P.A.; Volkman, J.K. Geochim. Cosmochim. Acta 1983, 47, 455-465.
9. Boon J.J.; de Leeuw, J.W. Mar. Chem. 1979, 7, 117-132.
10. Brassell, S.C. Ph.D. Thesis, Bristol University, Bristol, 1980.
11. Funasaki, J.; Gilbertson, J.R. J. Lipid Res. 1968, 9, 766-768.
12. Kolattukudy, P.E.; Croteau, R.; Buckner, J.S. In "Chemistry and Biochemistry of Natural Waxes"; Kolattukudy, P.E., Ed.; Elsevier; Amsterdam, Oxford, New York, 1976; p. 294.
13. de Leeuw, J.W.; Rijpstra, W.I.C.; Schenck, P.A. Geochim. Cosmochim. Acta 1981, 45, 2281-2285.
14. Smith, D.J. Ph.D. Thesis, Bristol University, Bristol, 1984.
15. Liefkens, W.; Boon, J.J.;de Leeuw, J.W. Neth. J. of Sea Res. 1979, 13(3/4), 479-486.
16. Chappe,B.; Michaelis, W.; Albrecht, P. In "Advances in Organic Geochemistry 1979"; Douglas, A.G.; Maxwell, J.R., Eds.; Pergamon Press, Oxford, 1980, p. 265.
17. Boon, J.J.; van der Meer, F.W.; Schuyl, P.J.W.; de Leeuw, J.W.; Schenck, P.A. Init. Rep. of the Deep Sea Drill. Proj. XL. 1978, 627-637.
18. Marlowe, I.T.; Brassell, S.C.; Eglinton, G.; Green, J.C. In "Advances in Organic Geochemistry 1983"; Schenck, P.A.; de Leeuw, J.W.; Lijmbach, G.J.W., Eds.; Pergamon Press, Oxford, 1984, p. 135.
19. Volkman, J.K.; Eglinton, G.; Corner, E.D.S.; Sargent, J.R. In "Advances in Organic Geochemistry 1979"; Douglas, A.G.; Maxwell, J.R., Eds.; Pergamon Press, Oxford, 1980, p. 219.
20. Marlowe, I.T.; Green, J.C.; Neal, A.C.; Brassell, S.C.; Eglinton, G.; Course, P.A. Br. Phycol. J. 1984, 19, 203-216.
21. Klok, J.; Cox, H.C.; Baas, M.; Schuyl, P.J.W.; de Leeuw, J.W.; Schenck, P.A. Org. Geochem. 1984, 7, No. 1, 73-84.
22. Klok, J.; Cox, H.C.; Baas, M.; de Leeuw, J.W.; Schenck, P.A. Org. Geochem. 1984, 7, No. 2, 101-109.
23. Klok, J.; Baas, M.; Cox, H.C.; de Leeuw, J.W.; Rijpstra, W.I.C.; Schenck, P.A. In "Advances in Organic Geochemistry 1983"; Schenck, P.A.; de Leeuw, J.W.; Lijmbach, G.J.W.; Eds.; Pergamon Press, Oxford, 1984, p. 265.
24. Mattern, G.; Albrecht, P.; Ourisson, G. Chem. Comm. 1970, 1570-1571.
25. Rubinstein, I.; Strausz, O.P. Geochim. Cosmochim. Acta 1979, 43, 1387-1392.

26. Gagosian, R.B.; Smith, S. Nature 1979, 277, 287-289.
27. Kokke, W.C.M.C.; Fenical, W.; Djerassi, C. Phytochem. 1981, 20, 127-134.
28. Djerassi, C. Pure Appl. Chem. 1981, 53, 873-890.
29. Larter, S.R.; Solli, H.; Douglas, A.G.; de Lange, F.; de Leeuw, J.W. Nature 1979, 279, No. 5712, 405-408.
30. Goossens, H; de Leeuw, J.W.; Schenck, P.A.; Brassell, S.C. Nature 1984, 312, No. 5993, 440-442.
31. Brassell, S.C.; Eglinton, G.; Maxwell, J.R. Biochem. Soc. Trans. 1983, 11, 575-586.
32. Klok, J.; Baas, M.; Cox, H.C.; de Leeuw, J.W.; Schenck, P.A. Tetr. Hedr. Lett. 1984, 25, No. 48, 5577-5580.
33. Repeta, D.J.; Gagosian R.B.; In "Advances in Organic Geochemistry 1981"; Bjorøy, M., Ed.; John Wiley, Chichester, 1983; p. 380.
34. van Graas, G.; de Lange, F.; de Leeuw, J.W.; Schenck, P.A. Nature 1982, 296, No. 5852, 59-61.
35. van Graas, G.; de Lange, F.; de Leeuw, J.W.; Schenck, P.A. Nature 1982, 299, No. 5882, 437-439.
36. Schmitter, J.M.; Arpino, P.J.; Guiochon, G. Geochim. Cosmochim. Acta 1981, 45, 1951-1955.
37. Bohlin, L.; Sjöstrand, U.; Djerassi, C. JCS Perkin I 1981, 1023-1028.
38. Goossens, H.; de Leeuw, J.W.; Rijpstra, W.I.C.; Meijburg, G.J.; Schenck, P.A. Geochim. Cosmochim. Acta, submitted.
39. de Leeuw, J.W.; Sinninghe Damsté, J.S.; Klok, J.; Schenck, P.A.; Boon, J.J. In "Hypersaline Ecosystems - The Gavish Sabkha"; Krumbein, W.E., Ed.; Springer, Heidelberg; 1985, p. 350.
40. Ourisson, G.; Albrecht, P.; Rohmer, M. Scient. Amer. 1984, 251, 34-41.
41. Goossens, H.; Rijpstra, W.I.C.; Düren, R.R.; de Leeuw, J.W.; Schenck, P.A. Org. Geochem. submitted.
42. Lechevalier, M.P. Crit. Rev. Microbiol. 1977, 7, 109-210.
43. Galanos, C.; Lüderitz, O.; Rietschel, E.T.; Westphal, O. In "Biochemistry of Lipids II"; Goodwin, T.W., Ed.; University Park Press, Baltimore, 1977; Vol. 14, p.239.
44. Rossignol-Strick, M.; Nesteroff, W.; Olive, P.; Vergnaud-Grazzini, C. Nature 1982, 295, 105-110.
45. Sinninghe Damsté, J.S.; ten Haven, H.L.; de Leeuw, J.W.; Schenck, P.A. Org. Geochem., submitted.
46. Didyk, B.M.; Simoneit, B.R.T.; Brassell, S.C.; Eglinton, G. Nature 1978, 272, 216-222.
47. Holzer, G.; Oro, J.; Tornabene, T.G. J. of Chromat. 1979, 186, 795-809.
48. Brassell, S.C.; Lewis, C.A.; de Leeuw, J.W.; de Lange, F.; Sinninghe Damsté, J.S. Nature, submitted.

RECEIVED September 23, 1985

4

Phenolic and Lignin Pyrolysis Products of Plants, Seston, and Sediment in a Georgia Estuary

Jean K. Whelan[1], Martha E. Tarafa[1], and Evelyn B. Sherr[2]

[1] Chemistry Department, Woods Hole Oceanographic Institution, Woods Hole, MA 02543
[2] University of Georgia Marine Institute, Sapelo Island, GA 31327

Phenolic and lignin pyrolysis products measured in
plants, seston, and sediment in a Georgia estuary
together with isotopic data suggest that vascular
plant material is present in estuarine organic matter
pools. The pyrolysis products together with isotopic
patterns of this plant material do not resemble those
obtained from Spartina. The pyrolysis–GC–MS data
also gave indications of changes in distribution of
methoxy phenolic compounds in going from plant materials to soils and seston, which could be due to
either degradation processes or to a change in source
in seston and sediments. These phenolic and methoxy-
phenolic pyrolysis products have generally not been
detected to date in either surface or sub–surface
deep ocean sediments from several areas of the world.
The combination of isotopic data, together with lignin and higher plant pyrolysis products, appears to
provide a useful method of determining plant sources
of organic matter in estuaries.

Determining the plant origins of organic matter in coastal waters
and sediments is of interest to geochemists and to ecologists in
answering specific questions regarding the sources and fates of
carbon in these systems. For Georgia salt marsh estuaries, the
initial view was that most of the organic carbon present in the
suspended material (seston) in the water column, as well as in the
marsh soils and estuarine sediment, and which supported the estuarine food web, was derived from production of the marsh cordgrass,
Spartina alterniflora (1-2). This concept has had to be modified
based on stable carbon isotope analysis (3-5). Although the distinctive isotopic signature of Spartina ($\delta^{13}C$ = -12 to -13
$^o/_{oo}$) is present in the salt marsh soils and marsh animals,
there is little isotopic evidence for significant amounts of
Spartina-derived organic matter in the seston and sediment of the
open estuary. Instead, there appears to be a large background of

0097–6156/86/0305–0062$06.00/0

organic carbon with an isotopic composition ($\delta^{13}C$ of -18 to $-24^o/_{oo}$) depleted in ^{13}C compared to that of Spartina. Stable carbon isotope analysis alone, however, cannot be used to determine whether this material originates primarily from in situ phytoplankton production ($\delta^{13}C$ of -20 to $-26^o/_{oo}$) or represents a mixture of algal carbon and Spartina detritus with terrestrial organic matter ($\delta^{13}C$ of around $-26^o/_{oo}$) coming into the estuary via riverflow (5).

It is apparent that other types of geochemical analyses must be combined with stable isotope analysis to clarify the sources of organic matter in Georgia estuaries and similar coastal systems. Here we present preliminary evidence that characterization of higher plant and lignin pyrolysis and CuO oxidation products in the pools of organic carbon in estuaries can provide additional information concerning the relative importance of the potential vascular plant contributors to these pools. Lignin is produced only by vascular plants, and in surface sediments, appears to be more resistant to degradation in natural environments than some other biopolymers (6). When the complex polymer of lignin is chemically oxidized, a series of methoxy phenol derivatives are produced which are quantitatively related to the amount and type of the contribution vascular plant tissues (7). Analysis of lignin-derived phenols via copper oxide oxidation and subsequent gas chromatography has been used to trace terrigenous material in the estuaries and nearshore shelf of the Pacific and Gulf of Mexico coasts (6, 8-10).

Lignin derivatives have also been analyzed qualitatively in isolated, synthetic, and degraded lignins (11), as well as in peats (12) and in sediments and particulates via pyrolysis-gas chromatography—mass spectrometry (PGCMS) both in Chesapeake Bay (13) and in the Rhine delta (14). A high correlation has been found between pyrolysis products and the lignin units from which they arise (15-16). The technique is attractive because of the small amount of sample (50-100 mg dry weight) and the minimal sample preparation required. This report describes an application of PGCMS to the analysis of lignin and phenolic pyrolysis products from vascular plants, sediments, and seston in the Georgia estuary.

Methods

Ground and sieved (1 mm to 52 mm particle size) dry samples were put into a 2 mm i.d. x 10 mm quartz tube, and held in place with quartz wool. The tube was mounted in a desorption probe and screwed into the cool programmable interface of a Chemical Data Systems (CDS) 820 Geological Reaction System (17). The sample was heated from 250°C to 550°C at 30°C per minute. Volatile C7-C28 compounds formed during the pyrolysis were collected on the CDS 820 System Tenax traps and then desorbed onto 1/8 inch o.d. x 3 inch Tenax traps, previously heated overnight in a helium stream at 280°C, which were then either stored in screw cap vials (if the mass spectrum was to be run within 24 hours) or sealed in glass ampules in a nitrogen atmosphere and frozen (if they had to be stored for longer time periods).

Mass spectral analyses were carried out by desorbing the Tenax traps into a GC mass spectrometer (Carlo Erba model 4160 GC equipped with a cryogenic oven attachment connected to a Finnigan 4500 quadrupole mass spectrometer). The traps were desorbed in a helium stream via a heated desorption module which attached to the GC injection port. The GC oven was cooled to -20°C while the desorption heater was heated to 280°C and then maintained at that temperature for 15 min. The GC oven was then ramped to 40°C and the J&W DB-5 (5% methyl phenyl silicone bonded phase) fused silica capillary column (0.32 mm i.d. x 30 mm long) was programmed from 40°C to 280°C at 4°C per minute. Mass spectra were acquired using electron impact ionization at 70 eV and were collected via scanning from 40 to 350 amu at 1 second intervals. The GCMS system is interfaced to a Finnigan Incos 2300 Data System which contains the NBS/EPA/NIH reference library of 31,000 spectra. Compounds were tentatively identified using the forward search algorithm included with the Incos system. Analyses of some of the same samples are being carried out by pyrolysis in the laboratory of Dr. Jap Boon, FOM Institute AMOLF, Kruislaan, The Netherlands.

Results and Discussion

A total of 5 plant, 2 seston, and 5 sediment samples from the Georgia estuary for which the stable carbon isotope composition had previously been determined were analyzed (see Table I).

Table I. Stable carbon isotope compositions of the plant, seston, and sediment samples from the Georgia estuary analyzed for lignin pyrolysis products

Sample	$\delta^{13}C$, $^o/_{oo}$ (P.D.B.)*
Plants:	
Cordgrass, Spartina alterniflora (Spar)	-12.3 + 0.3
Black rush, Juncus roemerlanus (BR)	-25.8 + 0.1
Sedge, Scirpus americanus (Sedg)	-26.0
Wild rice, Zizaneopsis miliaceae (WR)	-28.2 + 0.8
Mixed cypress & hardwood leaves (MCH)	-24.0
Seston:	
Lower Altamaha River seston (R5)	-27.5
Doboy Sound seston (DSS)	-21.3
Sediments:	
Spartina marsh creekbank (S-11)	-19.2 + 0.4
Spartina marsh creekbed (S-12)	-20.7 + 0.7
Tidal creekbed off open estuary (S-13)	-19.7
Mudflat at mouth of Doboy Sound (S-15)	-21.3

* From (4-5)

Pyrolysis products tentatively identified as phenols and methoxy phenols (see discussion below) which appear to be diagnostic of particular inputs, are shown in Figure 1. Some of these, particularly the methoxyphenols, are almost certainly derived from lignin (11-16). The approximate relative percentages of each, as shown in Figures 2-4, were calculated as follows: A specific region of the GCMS containing the peaks of interest was examined in each sample via mass scans, diagnostic of particular compounds (determined via detailed identification of all peaks in the GC-mass spectrum), as shown in Table II. The height of the GC peak for a

Table II. Mass spectral peaks diagnostic of compounds
in Figure 1 (listed in decreasing order of
intensity for each compound)

Compound	Mass Spectral Peaks m/e
A	151, 166, 123
B	151, 182, 108
C	137, 180
D	164, 149
E	164, 149
F	137, 105, 66
G	152, 151
H	164
I	151, 166, 123
J	164, 149
K	194, 179
L	194, 77, 179
M	181, 196, 95
N	58, 108, 178
O	167, 182
P	107, 128, 152
Q1 & Q2	137, 152, 122
R1 & R2	150, 135, 107
S	91, 150, 135
T	154, 139, 111, 93
U	137, 152
V	168, 153

particular mass chromatogram was then divided by the attentuation of the reconstructed total ion chromatogram (RIC) in the same scan range and multiplied by 100. This procedure gives a rough estimate of the relative amount of each pyrolysis product in the sample as normalized to the strongest RIC peak (which generally remained constant throughout the sample set) in the region of the GCMS being examined.

The proposed compound structures shown in Figure 1 are based partly on mass-spectal computerized library searches as compared to the 31,000 compound NIH/NBS library. It was also assumed that

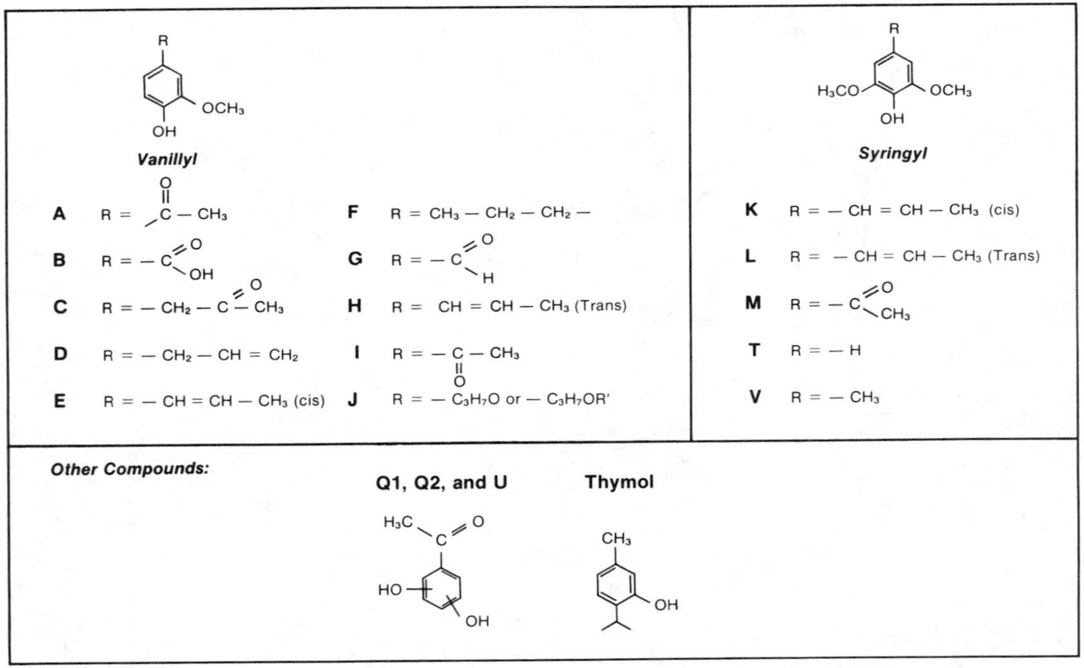

Figure 1. Proposed tentative structure of methoxy phenols and other pyrolysis products which may be derived from lignin or other higher plant material. (Proposed structures based on computerized GCMS Mass spectral identifications only – see text.)

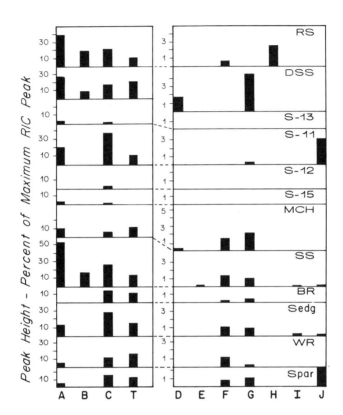

Figure 2. Levels of vanillyl derivatives in pyrolysis products of Georgia estuary samples. Sample code: River seston (>52 μm) (RS); Doboy Sound seston (>52 μm) (DSS); sediment from open creekbed (S-13); sediment from tidal creek (marsh creekbank) (S-11); sediment from tidal creek (marsh creek bed) (S-12); sediment from open estuary (Doboy mudflat) (S-15); mixed cypress and hard-wood (MCH); swamp soil (SS); black rush (BR); sedge (sedg); wild rice (WR); and Spartina (Spar). See Figure 1 for compound identification.

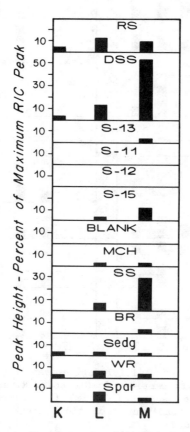

Figure 3. Levels of syringyl derivatives among pyrolysis products from a Georgia estuary. Same sample and compound codes as for Figures 1 and 2.

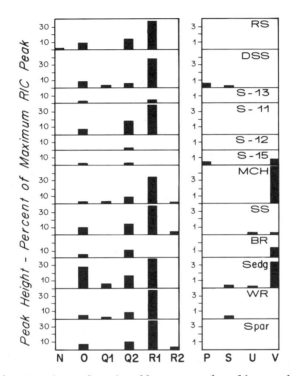

Figure 4. Levels   of   miscellaneous   phenolic   and   methoxy
pyrolysis products.   Same sample and compound codes as in
Figures 1 and 2.

the carbon skeletons of methoxy-phenolic pyrolysis products are
not drastically different from those found in the parent lignin
(15-16, 18-19) and that structures of lignin pyrolysis products
reported here are probably the same as those previously reported
(11, 14). In addition, two of the samples analyzed in this work
are also being analyzed by pyrolysis GCMS in the laboratory of Dr.
Jap Boon in the Netherlands which will allow comparison of our mass
spectra with those of authentic standards. Because the bonded
phase capillary GC columns being used in both laboratories have
the same liquid phase (J. Boon, private communication), it was
asumed in the present work that the relative order of GC elution
of isomeric pyrolysis products was the same as in Ref. (11) and in
other work of Saiz-Jimenez and de Leeuw (unpublished manuscript).

The vanillyl and syringyl pyrolysis products shown in Figure
1 are probably correct, although these structures will have to be
checked by comparison to authentic standards in future work. The
mass spectal computer gave very high confidence ratings (greater
than 900 out of a possible 1000 on Fit, Rfit, and purity parameters
on the Incos data system) to the structures shown in Figure 1 for
compounds A-C, F, G, I, and M. The mass spectra for all of these
compounds are fairly unique, as shown in Table II. The computer
identified compounds D, E and H (which possessed almost identical
mass spectra) as propenyl vanillyl compounds. The proposed structures shown in Figure 1 are based on the fact that the parent
lignin contains only straight (rather than branched) C3 chains
bonded to the phenyl groups (18-19) and on the expected GC elution
order based on Ref. (11). A fourth compound, J in Figure 1, was
also identified by the computer as propenyl benzene. It is proposed that this compound contains a hydroxyl or alkoxyl group (or
some other easily eliminated group) bonded to a straight C3 side
chain which is eliminated during electron impact fragmentation to
produce a mass spectrum very similar to that of the corresponding
olefin (20).

Similar arguments apply to the propenyl syringyl compounds K
and L. Compounds T and V were identified by the mass spectal computer as 3, 4-dimethoxyphenol and trimethoxyphenol, respectively.
However, library spectra do not always contain the correct reference isomers (11). In addition, because pyrolysis product structures T and V in Figure 1 have been identified as abundant syringyl
lignin pyrolysis products (11), and because isomers often give very
similar mass spectra (20), we propose that compounds T and V are
actually the unsubstituted and methyl syringyl derivatives,
respectively.

A number of other pyrolysis products believed to be either
phenols or methoxyphenols derived from higher plants were also
identified in the sample set. Identification of these, including
N through P and S in Figure 4, are less certain than those discussed above so that structures are not proposed. Compounds Q1,
Q2, and U, which produce very similar mass spectra and, therefore,
are proposed to be isomeric (20), were identified with a high
degree of confidence by the mass spec computer as the diphenolic
structure shown in Figure 1. Compounds R1 and R2, which also
appear to be isomeric, gave mass spectra very similar (but not

identical) to thymol shown in Figure 1. Therefore, we propose
that these two compounds, which occurred widely in the sample set,
are terpene derivatives related to thymol.

Computerized GCMS data of other samples obtained from more
marine sediments have also been examined for the possible presence
of lignin-derived pyrolysis products (Whelan, unpublished data).
Three marine gorgonian (coral) samples showed only traces of compound which might be G and of a dialkyl phenol. Marine copepods,
two sediment samples from a sub-bottom depth of 788 m near the
Canary Islands, a sediment sample from a sub-bottom depth of 564 m
in the Japan Trench, and two surface sediment samples from the Peru
upwelling region all showed no traces of the pyrolysis products
shown in Figures 1-4 (determined by careful searches of mass
chromatograms of the masses shown in Table 1) with the possible
exception of compound U. The fact that these marine samples
generally do not produce these phenolic and methoxy phenolic
compounds supports the source of these pyrolysis products as being
non-marine, most probably from higher plants and/or lignin.

As shown in Figure 1, three general groups of phenolic pyrolysis products were found in the samples analysed: vanillyl phenols,
syringyl phenols, and miscellaneous phenolic and methoxy compounds.
The relative proportions of the individual products in the various
samples in each of the three groups are presented respectively in
Figures 2-4.

The vanillyl type of phenolic structure is produced in the
lignin of all vascular plants (18-19). Levels of vanillyl phenolic
pyrolysis products were high in the plants, swamp soil, and in
seston in the Altamaha River and in Doboy Sounds (Figure 2). In
three of the estuarine sediment samples, however, these compounds
were either undetected or found in only trace amounts (Figure 2).
The acetyl derivative, A, was concentrated in swamp soil, in river
and estuarine seston, and in the Spartina creekbank sediment
(S-11). We postulate that the seston and sediment have a significant contribution from swamp soil, or more likely all of these
organic pools have a similar plant-derived component which produces
a higher proportion of the acetyl structure upon pyrolysis. The
latter hypothesis is supported by the presence of the vanillyl
compound J in significant amounts only in Spartina and in the
Spartina marsh creekbank sediment (S-11). The low or non-existent
levels of this compound in the other samples suggests that Spartina
is the major contributor of higher plant carbon to marsh creekbank
sediment. However, if this hypothesis is correct, it is difficult
to see why the syringl products (K-M in, Figure 3) should be
present in Spartina but missing in Sediment S-11.

Five syringyl derivatives (compounds K-M, T, and V, Figure 1)
were detected among the pyrolysis products (Figures 2-4). Syringyl
phenols are also characteristic of lignins, but occur in quantity
only in angiosperm plants (7; 18-19). Small amounts of three
different acetyl and propenyl syringyl pyrolysis products were
produced by wild rice and sedge (Figure 3), while only two of these
structures were obtained from Spartina and the mixed cypress and
hardwood tree leaves. A much larger proportion of the acetyl
derivative, compound M, was found in swamp soil and Doboy Sound

seston, suggesting a common organic source for these two pools of organic matter. Somewhat elevated levels of these three syringyl compounds occurred in river seston. The lower proportion of compound M to L as compared to the Doboy Sound seston may indicate a less degraded or a different plant source contributing to seston in the river. The open estuary sediment samples, S-13 and S-15, had syringyl structures which probably came from plant debris. In contrast, the creekbank sediment, which had abundant vanillyl structures, showed only the unsubstituted syringyl compound, T. The absence of other syringyl structures in sediment S-11 is somewhat surprising if Spartina is the main source of plant material to this sediment, since two additional syringyl structures (L and M) were found among the Spartina pyrolysis products.

Figure 4 shows levels of other phenoxy and methoxy pyrolysis products (N through S and U, some of which are shown in Figure 1) which might be derived from lignin and/or other higher plant material. The specificity of these pyrolysis products as markers of higher plant material will have to be investigated in future work. Q1 and Q2 appear to be acetyl diphenolic structures, while both R1 and R2 were identified by the computer as related to thymol (see Figure 1). Both types of structures may be typical of higher plants, but are not necessarily derived from lignin. Thymol (R1 and R2) is a monoterpene which occurs widely in higher plants (21).

For the Georgia estuary samples, the plants, swamp soil, seston, and sediment S-11 from the tidal creekbank all showed similar distributions of compounds Q2 and R1. Compound O also occurred as a pyrolysis product in these samples, with the amount being more enriched in sedge than in the other plants. These compounds are either absent or present in lower levels in the spectra of pyrolysis products from the other sediment samples: S-13, S-12 and S-15. The results are consistent with the vanillyl product distribution (Figure 2) and suggest that these pyrolysis products also arise from vascular plants and that there is a significant vascular plant contribution to the seston and creekbank sediment. The smaller amounts of compounds O, Q2, and R1 in the open estuary sediments indicate either a lesser influence or a more diagenetically altered higher plant carbon.

Pyrolysis products N, R2, P, S, U, and V, shown in Figure 4, were more specific than those discussed above. Compound V was produced only by sedge, black rush, mixed cypress and hardwood leaves, and in smaller amounts, by swamp soil. The presence of this compound in sediment S-15, but not in other seston or sediment samples, suggests the contribution of specific higher plants to this sediment sample. The occurrence of compound P, in sediment S-15 and Doboy Sound seston, but not in the higher plants analyzed or in swamp soil may indicate a fairly specific marine precursor for this particular pyrolysis product.

In addition to the analyses of lignin products by PGCMS as discussed here, one sample of estuarine sediment, comparable to sample S-15, was analyzed for lignin oxidation products by the CuO oxidation method (22). The distribution of various types of phenolic compounds in the sediment is compared to that of Spartina

alterniflora and other types of plants previously analyzed with the
CuO method (7) in Table III.

Table III. Comparison of three lignin parameters determined
for an estuarine sediment sample with those of Spartina
alterniflora and other vascular plants.
S/V = wt. ratio of syringyl to vanillyl phenols;
C/V = wt. ratio of cinnamyl to vanillyl phenols;
Co/Fe = wt. ratio of the cinnamylphenols p-coumaric
acid and ferulic acid.

Sample	S/V	C/V	Co/Fe
Estuarine sediment	1.24	0.46	2.89
*Spartina alterniflora	1.1	1.0	0.84
Other nonwoody angiosperm tissue	1.5	0.71	2.5
Angiosperm woods	2.5	0	0
Gymnosperm nonwoody tissue	0.02	0.49	5.0
Gymnosperm woods	0	0	0

*plant data from Ref. 7.

The lignin signature in this sediment sample most closely resembles
that of nonwoody angiosperm tissue, but it does not fit well with
the specific distribution of lignin phenols obtained for Spartina.
This result is consistent with those obtained by PGCMS, even though
the specific relation between oxidative and pyrolytic lignin degradation products has not yet been determined. Thus, sediment S-15
does not show the same pyrolysis products as Spartina in Figures
2-4. The strongest phenolic pyrolysis signal in S-15 is most like
those of swamp soil and Doboy Sound seston as shown in Figure 3
and consists of a relatively strong syringyl component in comparison to other sediment samples.
    Several sediment samples from the Mississippi Fan in the Gulf
of Mexico did show traces of compound U. These sediments should
have been heavily influenced by terrigenous organic matter. However, it is not currently known to what depths lignins survive in
deep sea sediments (6). It has recently been shown that vanillyl
and p-hydroxy lignin are partially preserved in some buried woods
for periods of up to 2500 years (23). We plan to reanalyze DSDP
Mississippi Fan samples, along with more hemipelagic types of
marine sediments, in future work to look in detail for the compounds in Figure 1.

Summary

Determining the relative contributions of marine and terrestrial
plant organic carbon to the pools of organic matter in coastal
systems is often difficult. In Georgia salt marsh estuaries, the
major sources of organic matter are in situ phytoplankton produc-

tion, detritus from the marsh grass Spartina alterniflora, and
terrestrial plant material introduced by river flow. Stable carbon
isotope studies indicate that, although Spartina carbon is present
in salt marsh sediment and to some extent in sediment in the open
estuary, there is a large background of ^{13}C-depleted organic
matter in sediment and in suspended particulate material which
could come either from phytoplankton or from terrestrial material.
In order to clarify the source of this material, samples of estuarine vascular plants, seston, and sediment were analysed for content and composition of phenolic and methoxy phenolic pyrolysis
products by pyrolysis and subsequent gas chromatography and mass
spectrometry. Some or all of these products (particularly the
methoxy phenols) are proposed to be diagnostic lignin pyrolysis
products. These products, which are not produced by more marine
samples, appear to be diagnostic of terrigenous plants. Based on
these analyses, Spartina has a unique signature which also appeared
in a sample of surface mud from the bank of a tidal creek draining
a Spartina marsh. Sediment from the tidal creek bed and from two
open estuarine sites showed very little of these pyrolysis products, and the composition did not match that of Spartina. Suspended material in the estuary had a stronger signal which resembled
that of tree leaves and soil in coastal freshwater swamps. Seston
and some sediment in the open estuary also contained phenolic
(probably non-lignin) pyrolysis products which were not present in
the plants and which were probably of marine origin. One estuarine
sediment sample was independently analysed for lignin CuO oxidation
products via gas chromatography. The composition of the lignin
present in this sample resembled that of some non-woody angiosperm
tissues from the swamp, but did not closely match the lignin
products found for Spartina. It appears that organic carbon
derived from vascular plants is present in estuarine seston and
sediment, but Spartina may not be the primary source for this
material.

Acknowledgments

This work was supported by grants from the Sapelo Island Research
Foundation to E. Sherr and by National Science Foundation Grant
OCE83-00485 to J. K. Whelan and J. M. Hunt. We are grateful to
Dr. John Hedges for lignin oxidation product analysis of one of
our sediment samples, and to Drs. Jim Alberts, Chuck Hopkinson,
John Ertel and John Hedges for their comments on the manuscript.
Thanks also go to Dr. Nelson Frew for mass spectra and to Christine
Burton and Richard Sawdo for technical assistance. Contribution
No. 6055 of the Woods Hole Oceanographic Institution and No. 546
of the University of Georgia Marine Institute.

Literature Cited

1. Teal, J. M., Ecology 1962, 43, 614-624.
2. Odum, E. P.; De La Cruz, A. A. In "Estuaries"; Lauff, G. H.,
    Ed.; AMER. ASSOC. ADV. SCIENCE: Washington, D.C., 1967; pp.
    383-388.

3. Haines, E. B. Oikos, 1977, 19, 254–260.
4. Haines, E. B.; Montague, C. L. Ecology 1979, 60, 48–56.
5. Sherr, E. B. Geochim. Cosmochim. Acta 1982, 46, 1227–1232.
6. Hedges, J. I.; Mann, D.C. Geochim. Cosmochim. Acta 1979, 43, 1809–1818.
7. Hedges, J. I.; Mann, D.C. Geochim. Cosmochim. Acta 1979, 43, 1803–1807.
8. Hedges, J. I.; Parker, P.L. Geochim. Cosmochim. Acta 1976, 40, 1019–1029.
9. Hedges, J. I.; Turin, H. J.; Ertel, J. R. Limnol. Oceanogr. 1984, 29, 35–45.
10. Ertel, J. R.; Hedges, H. I. Geochim. Cosmochim. Acta 1984, 48, 2065–2074.
11. Saiz-Jimenez, C.; de Leeuw, J. W. Org. Geochem. 1984, 6, 417–422.
12. Bracewell, J. M.; Robertson, G. W.; Williams, B. L. J. Anal. Appl. Pyr. 1980, 2, 53–62.
13. Sigleo, A. C.; Hoering, T.C.; Helz, G. R. Geochim. Cosmochim. Acta 1982, 46, 1619–1626.
14. Van de Meent, D.; DeLeeuw, J.W.; Schenek, P.A. J. Anal. Appl. Pyr. 1980, 2, 249–263.
15. Obst, J. R. J. Wood Chem. Technol. 1983, 3, pp. 377–397.
16. Martin, F.; Saiz-Jimenez, C.; Gonsalez-Vila, F. J. Holzforschung 1979, 33, 210–212.
17. Whelan, J. K.; Fitzgerald, M. G.; Tarafa, M. Environ. Sci. Tech. 1983, 17, 292–298.
18 Sarkanen, K. V. In "The Chemistry of Wood"; Browning, B. L., Ed.; R. E. Krieger Publishing Co.: Florida, 1963 (reprinted 1981); pp. 249–311.
19. Sarkanen, K. V.; Ludwig, C. H. "Lignins"; Wiley-Interscience: New York, 1981.
20. Biemann, K. "Mass Spectrometry"; McGraw Hill: N.Y. 1962; pp. 87–95.
21. Buchanan, M.A. In "The Chemistry of Wood", Browning, B. L., ed.; R. E. Krieger Publishing Co.: Florida, 1963; pp. 324–325.
22. Hedges, J. I.; Ertel, J. R. Anal. Chem. 1982, 54, 174–178.
23. Hedges, J. I.; Cowie, G. L.; Ertel, J. R.; Barbour, R. J.; Hatcher, P.G. Geochim. Cosmchim. Acta 1985, 49, 701–711.

RECEIVED October 31, 1985

5

Characterization of Particulate Organic Matter
From Sediments in the Estuary of the Rhine and from Offshore Dump Sites of Dredging Spoils

Jaap J. Boon[1], B. Brandt–de Boer[1], Wim Genuit[1], Jan Dallinga[1], and E. Turkstra[2]

[1]FOM–Institute for Atomic and Molecular Physics, Kruislaan 407, 1098 SJ Amsterdam, The Netherlands
[2]D. B. W. RIZA, P.O. Box 17, Lelystad, The Netherlands

Screening of estuarine and marine sediment samples by automated pyrolysis mass spectrometry combined with factor-discriminant analysis leads to a useful classification related to the geographical position and the sources of the organic matter. The mass spectral data give preliminary information about the organic matter composition. Analysis of the characteristic mass peaks m/z=86 and 100 by PMSMS and PGCMS points to bacterial polyalkanoates in the mud fraction of the river sediments. These and other pyrolysis products show that sewage sludge with degraded lignocellulose material and bacterial biomass is an important source of particulate organic matter in river sediment of this heavily populated area in the Netherlands.

Extensive sedimentation of particulate matter from fluvial and marine origin takes place in the estuary of the Rhine, particularly in the harbors along the New Waterway, downstream from Rotterdam. Yearly about 20 million m^3 of this mud is dredged and dumped into the North Sea (1). As these sediments have trapped organic and inorganic pollutants such as Cr, Cu, Hg, Cd, Pb, Zn and Ni, a considerable pollutant load is imposed on the North Sea ecosystem either directly by the outflowing river water or indirectly by dredging and dumping (2).

Detailed knowledge on the molecular level about the fate of the riverine particulate organic matter and its associated pollutant cocktail is still limited. The complex processes which occur in the mixing area between fluvial and marine water in the estuary, result in sediments with characteristics of aquatic and marine origin, mixed up or maybe even reacted with each other in some ways. When these sediments are dredged and dumped offshore, the sedimentary material is exposed to open sea conditions with increased chances of resuspension and resedimentation and consequently a new diagenetic regime for the organic matter.

In order to study aspects of the fate of these sediments, we have studied sedimentary particulate matter from sites in the estuary of the Rhine downstream from Rotterdam, from several dredging spoil dumpsites offshore (Loswal Noord), and from sediments of the Flemish

0097–6156/86/0305–0076$06.00/0
© 1986 American Chemical Society

Banks in the Southern North Sea. The organic composition of the mud fraction from these sediments was investigated by Curie-point pyrolysis mass spectrometric techniques, whereas metal content, % organic carbon, carbon isotopes, carbonate and mud content were determined separately on the same sediments. Although the analysis of estuarine particulates for the latter characteristics is widely known, the usefulness of pyrolysis methods for the characterization of the organic matter fractions has been demonstrated only recently. The fate of sewage sludge in the Boston harbor was traced with thermal distillation-pyrolysis GCMS (3). Colloidal matter in Chesapeake Bay was characterized by stepwise pyrolysis GCMS (4). The fate of organic materials and metals in the river Rhine delta and in the IJsselmeer was investigated by Curie-point pyrolysis mass spectrometry and -GCMS (5). Pyrolysis mass spectrometry was used for the evaluation of the organic matter in the estuaries of the Ems (6) and Gironde (7). These studies demonstrated a good correlation between PMS data, carbon isotopes and other variables.

The pyrolysis methods applied in this study are used as a tool for a general characterization of the organic matter from the sediments. We describe here the results from screening of estuarine and open sea sediment samples by automated pyrolysis low voltage mass spectrometry combined with factor-discriminant analysis. Characteristic mass peaks resulting from this procedure were investigated in more detail by pyrolysis-tandem mass spectrometry and pyrolysis-photoionization GCMS.

Experimental

Sampling and Sample Preparation. Samples were collected by a team from the Netherlands Institute for Sea Research (Texel) and Rijkswaterstaat (the Ministry of Transport and Public Works) by a boxcorer (North Sea stations) and Van Veen grab (Rhine stations). Figure 1 shows a map with the sample locations (for details, see (8)). Sediment samples were taken from the boxcorer or the grab and immediately afterwards resuspended in filtered (0.45 μm) sea or river water; the suspension after settling of the sand fraction was passed over a 50 micron sieve to a Whatman GFF filter with a diameter of 9 cm under gentle suction. After filtration the filtered particulate matter was washed with destilled water to remove sea salts. The filters with their sediments were stored in glass petrie dishes and frozen. Generally five boxcores (coded by letter) were taken in one sampling area (coded by number), whereas in the Rhine estuary usually three separate samples were grabbed in different parts of the river bed. Samples from the North Sea were taken between 22 and 24 November 1983. Samples in the New Waterway (Rhine estuary) were taken on 2 February 1984.

Samples for pyrolysis analysis were taken from the filters and resuspended in water immediately before analysis. Aliquots of these suspensions (about 50 micrograms of dry matter) were transferred to the ferromagnetic sample wires and dried in vacuo. Samples were analyzed immediately afterwards in quadruplicate.

Pyrolysis Mass Spectrometry and Multivariant Data Analysis. The automated pyrolysis mass spectrometer and the multivariant data analysis by factor-discriminant analysis (f.d.a.) procedure used have been

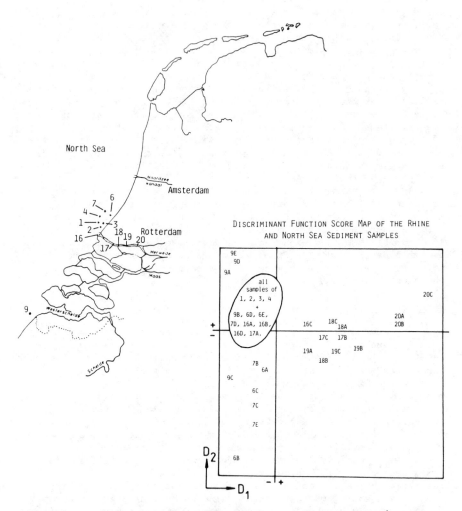

Figure 1. Discriminant function score map of the pyrolysis mass
spectra of the sample sites and their geographical position along
the New Waterway (Rhine estuary) and in the North Sea. The numbers
indicate sampling areas, the letter codes sample sites within an
area.

described recently (9). The Curie-point temperature of the ferromagnetic sample wires was 510°C.

Pyrolysis Tandem Mass Spectrometry. Details of the instrument have
been described (10). Pyrolysis conditions were similar to those used
for the PMS instrument. The pyrolysate was ionized at 70 eV and ions
of interest were selected in the first magnetic sector (resolution
500). These ions were dissociated by collision with helium atoms (pressure 0.9 Pa), accelerated at 15 kV and analyzed in a second magnetic
sector. Ions were detected by simultaneous ion detection using channel
plates, a phosphor screen and a diode array camera. Total analysis
time for each ion is 8 s.

Pyrolysis Photoionization GCMS. Details of the instrument have been
described earlier (11). Curie point pyrolysis was performed in nitrogen which also served as the carrier gas. Pyrolysis occurred directly
in front of the capillary column, which exits in the photoionization
chamber of the mass spectrometer. A 50 m fused silica column coated
with CP-Sil 5-CB (I.D.=0.32 mm, film thickness: 1 micron) was used for
the separation. The oven was programmed from 80-260 C after a period
at room temperature during pyrolysis of the sample. Argon I resonance
photons (11.6 and 11.8 eV) were used for ionization of the GC effluent. Ion source temperature was 150°C. Pressure in the ion source
was 10^{-2} Torr and in the vacuum chamber 10^{-5} Torr. The quadrupole was
scanned at 1 scan/s over 30 to 235 amu.

Results and Discussion

Sampling Plan and Sediment Workup. A number of choices have been
made in design of the sampling program. Samples were taken in late
autumn and winter in order to minimize contributions of phytoplankton
to the bottom sediments. It was not possible to sample all materials
in the same time frame. The open sea samples were taken by boxcoring
to minimize loss of surface sediments during sampling. Generally the
upper five centimeters were used for the isolation of the fraction
smaller than 50 microns. Each sample area in Figure 1 consisted of
five individual sample sites which were separated by several miles.
This replication was chosen for statistical reasons in order to evaluate the reproducibility in the composition of the organic matter of
the sediments in one area. Samples in the river were taken on the
left side, on the right side and in the middle of the bedding for the
same reasons.
    The separation of mud and sand was necessary for the pyrolysis
analysis, which is performed on suspended matter coated on a metal
wire of 0.5 mm, and because sand particles do not stick to these
sample carriers. As the samples are continuously moved about in the
natural environment, a resuspension procedure was thought to be a
minor disturbance of the original sediment. The washing procedure, in
which destilled water is passed over the filter with particulate matter may however result in loss of soluble organic matter together
with the sea salts. This measure has been applied in the earlier pro-
filling studies of estuaries (6,7) with PMS and no negative influence
on the results could be shown. The technical advantage is that the
HCl volatilized from sea salts during pyrolysis analysis is minimized and matrix effects on the organic matter pyrolysis patterns

due to the presence of salts is avoided. It is clear that if differences are found in the composition of the samples despite this workup, the organic matter characteristics must be very stable and strongly connected to the particulate matter fraction.

Screening by Automated Pyrolysis Mass Spectrometry. The total data file after analysis of all samples in quadruplicate consisted of 172 pyrolysis mass spectra. All samples were significantly different from a procedural blank. In the total file only the mud fraction from station 7A (all replicates) was strongly different in pyrolysis mass spectrum from all other spectra. The pyrolysis spectrum of 7A was very similar to the spectrum of cellulose. The cellulose in this sediment may have originated from the outfall sewer of The Hague city, which exits nearby. The carbon isotope δ ^{13}C of $-25.44‰$ (8) points to a nonmarine origin. After removal of the 7A analyses from the data file, factor discriminant analysis was performed again and a discriminant function score map, as shown in Figure 1, was calculated. The distances between the samples in this map are a measure for the differences in composition. Discriminant functions describe a combination of chemical substances, expressed in mass peaks, which are used for an optimal differentiation. The D-scores can be considered as the concentration of these substances. Positive and negative values are obtained, because concentration is expressed relative to the average concentration in all samples. About 50% and 15% of the characteristic variance in the data file was explained in the first and second discriminant functions respectively. Instrument replicates generally fall in the centroid (the station number) on the map.

If we consider only the first D-function, a striking separation between most samples from the river and the marine stations becomes evident. Samples 20A, B and C, collected highest upstream near Rotterdam, were found to have the highest positive D_1 scores, but all stations at 19, 18 and two of the three stations at 17 have positive D_1 scores. Most of the stations from 16 at the mouth of the New Waterway near Hoek of Holland and all open sea stations have negative D_1 values. The stations with negative D_1 scores, are separated by the second discriminant function into a distribution which seems to reflect their geographical position in the sea as well: strongly positive values were found for a number of Flemish Bank stations, strongly negative values were found for most stations north of the Loswal Noord area. A cluster relating to the dredging spoil dumpsite Loswal Noord (1A-E, 2A-E, 3A-E and 4A-E), some stations from north of this area (6D,7D) and some stations from the river mouth at Hoek of Holland (16A,B,D and 17A) and station 9B (Flemish Banks) represent samples which cannot be discriminated further. It is striking that the samples from area 9, 6 and 7 are poorly reproducible which points to relatively strong differences in composition of the bottom sediments. The samples from the Loswal Noord area all plot on top of each other, which implies a common factor in all those samples. It is of interest to note that the dredging area in the mouth of the New Waterway is near station 16, which plots in the cluster of samples from the dredging spoil dump sites.

Chemical Interpretation of the PMS Data. Translation of the mass spectral information into chemical information is tentative unless MSMS or GCMS is used for the identification of mass peaks. Pyrolysis mass spectra and the discriminant function spectra resulting from the numerical analysis of a PMS data file can be interpreted tentatively

by comparison with standards and reference materials (12). On this basis, the zero point spectrum in Figure 2, which represents the mass
peaks common to all spectra, shows information from alkanes and alkenes, lignins, carbohydrates and proteinaceous material, although the
characteristics for the latter substances are not so prominent as in
soil organic matter and humic fractions (12). The mass peaks for HCl
(m/z 36 and 38) are of low intensity which indicates that the washing
procedure has removed most of the interfering salts. The mass peaks
at m/z 34 (H_2S), 48 (CH_3SH) and 64 (SO_2 and/or S_2) point to the existence of sulphur compounds in the mud fraction.

Figure 3 presents the reconstructed mass spectrum of the first
discriminant function which separated the river and marine stations in
the D_1D_2-map of Figure 1. The positive D-function describes the cova-
riant mass peaks with higher intensities with respect to the zero
point spectrum. All sample spectra with such characteristics will have
positive score values. This spectrum is a representation of the characteristics of riverine material. The negative D-function spectrum in
Figure 3 is indicative of the marine characteristics. The D_1^+ spectrum
shows a number of mass peaks indicative for carbohydrates, lignin and
proteinaceous material (12). The mass peak m/z=86 and 100 are uncommon
and a special characteristic of these fluvial samples. It can be speculated to be the molecular ion of (alkyl)thiadiazole (a metal binding
pollutant), however a cyclic ketone, short chain alcohol or unsaturated acid are also possibilities. These mass peaks were chosen for
further study because of their rare occurrence and their high discri-
minating power in the factor-discriminant analysis.

The D_1^- shows a number of characteristics related to sea salts
such as the mass peaks m/z 36 and 38 from HCl generated from chlori-
des, m/z 50 and 52 from CH_3Cl and m/z 142 from CH_3I. It is not clear
how the methyl-chloride and -iodide are formed but they are very common in pyrolysates of marine samples (13). The mass peaks m/z 95, 96,
109 and 110 are from furfural and methylfurfural, which are generated
under sea salt conditions from carbohydrates (14). The mass peaks m/z
67, 81 and 79 are presumably from pyrrole, methylpyrrole and pyridine.

Tandem Mass Spectrometry of Mass 86 and 100. The sample from station
20C near Rotterdam was chosen for a detailed analysis of peaks 86 and
100 because these peaks are relatively abundant in the PMS spectrum of
the sediment (see Figure 5). Figure 4A shows the collision induced
dissociation spectrum of mass 86, obtained by ionization of the pyro-
lyzate at 70 eV. This high voltage was chosen for reasons of increased
sensitivity. CID spectra of unknowns may be compared qualitatively
with electron impact mass spectra taken at 50 eV (15, 16). Based on
this comparison using the EPA/NIH Mass Spectral Data Base, we conclude
that m/z 86 cannot be the molecular ion of thiadiazole, however 2-
butenoic acid is a good possibility. The CID spectrum of this compound
has been studied using the same instrument (17). Its spectrum has relatively abundant fragments at m/z=38 (10.5%), 39 (29.3%), 40 (8.3%),
41 (11.4%), 42 (5.3%), 45 (8.2%) and 68 (2.7%). However, the presence
of a relatively abundant mass 43, 57 and 58 cannot be explained on the
basis of this compound. We assume that other compounds with a molecular ion or fragment ion at 86 are present in the ionized pyrolysate.
The CID spectrum of mass 100 (Figure 4B) shows a series of peaks at
43, 57, 71, which points to a saturated hydrocarbon. A CID peak at 82
(loss of water) and peaks at 39 and 45, point to a small contribution
of an unsaturated acid. This CID spectrum is most likely not from one
compound. These MSMS data are insufficient for a definite identifica-

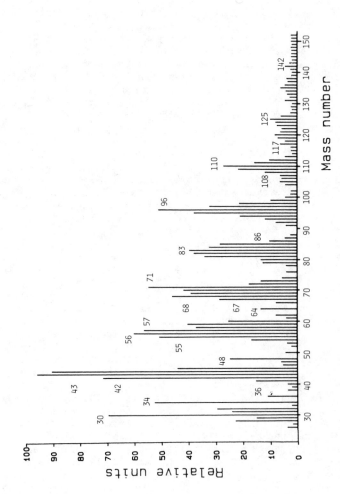

Figure 2. Zero point spectrum of the pyrolysis mass spectral data file of the sample sites shown in the discriminant function score map in Figure 1.

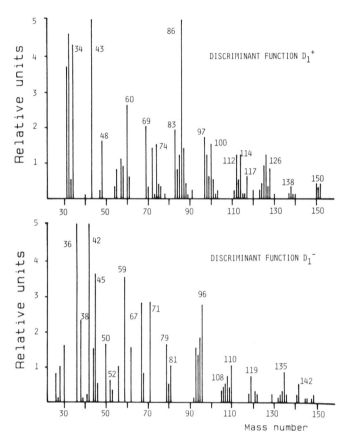

Figure 3.   Reconstructed mass spectra of the first discriminant
function calculated for a file consisting of pyrolysis mass spectra of sediments from sample sites shown in the discriminant function score map in Figure 1.

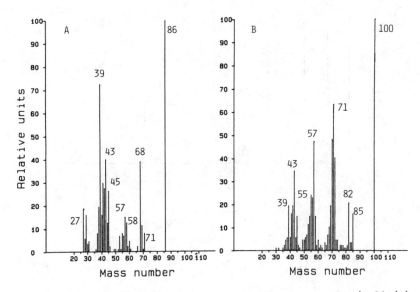

Figure 4. Pyrolysis tandem mass spectra for mass peak m/z 86 (A) and 100 (B) selected from the pyrolysate of sample 20C collected in the New Waterway.

tion of both mass peaks, but clearly indicate it that several compounds of moderate polarity are contributing to these peaks. It should
be possible to separate them by gas chromatography. The technique of
pyrolysis GCPIMS was used for separation and low voltage ionization of
the compounds.

Pyrolysis GC-Photoionization-MS. Genuit and Boon (1) have shown in a
PGCPIMS study of amylose that a cumulative PI spectrum (a summation
of all GCMS scans) ionized with argon I resonance photons is fairly
similar to the low voltage (15 eV) electron impact spectrum. The advantage of this approach compared to the direct introduction of the
whole pyrolysate into the low voltage EI mass spectrometer is that
mass peaks can be searched in the GCMS data file and the complete mass
spectrum of a compound can be retrieved. Figure 5 shows the 15 eV PMS
and the argon I photon PI-PMS of sediment sample 20 C. There is a good
resemblance between both spectra, although alkane fragment ions such
as m/z: 43, 57, 71, 83, 85, 97, 111 and 125 are relatively more intense in the EI spectrum. In both spectra m/z=86 is a relatively intense
mass peak.
    Figure 6 shows the mass chromatograms of m/z=68, 86 and 100 in
the GCMS data file of this sample. The most abundant compound (e) with
m/z=86 and fragment ion at m/z=68 was identified as 2,3-butenoic acid
(a fronting peak in the chromatogram due to its free acid group). An
isomer (cis?) with a shorter retention time (d) has the same spectrum.
The compounds a, b and c were respectively bi-acetyl, 2-methylbutanal
and 3-methylbutanal. The spectrum of biacetyl shows a very intense
fragment ion at m/z 43, whereas the methylbutanals have strong fragment ions at 57 and 58. This explains the CID peaks mentioned above,
which could not be accounted for in the MSMS of peak 86. The mass chromatogram at m/z=100 points to a compound at scan 550, which was identified as a pentenoic acid. The other signals in the m/z=100 trace are
noise. No contributions from alkanes to this mass peak could be detected in this data file, but the response for alkanes in the photoionization mass spectrometer is very poor. Analysis of this sample with
GCMS at 70 eV confirmed the presence of alkanes in the sample (13).
    Studies on the bulk pyrolysis of polyhydroxybutyric acid from
Bacilli and of bacterial polyalkanoates have shown the formation of
2,3-butenoic acid and 2,3-pentenoic acid (18). The presence of
2,3-butenoic acids and pentenoic acid in the pyrolysate of the particulate matter from sample 20C is interpreted as an indication of
polyhydroxy-alkanoates in the sample. These mixed polyesters of hydroxy acids with 4, 5 and sometimes 6 carbon atoms are especially
abundant in activated sludges (19). The occurrence of m/z 86 and 100
as abundant mass peaks in the spectra of the fluvial material and as
very characteristic peaks in the discriminant function spectrum indicates that a significant amount of the mud fraction may consist of
sewage debris. This impression was confirmed by identification of a
number of other pyrolysis products in the data file.
    Figure 7 shows a number of pyrolysis products which were identi-
fied with PPIGCMS. The identity and molecular ions of a number of
major peaks in the TIC have been marked in the chromatogram. The
guaiacol (m/z=124), hydroxystyrene (m/z=120) and methoxyhydroxysty-
rene (m/z=150) are the pyrolysis products of a monocotyledon lignin
(20). Hydroxyacetic aldehyde, hydroxypropanone, furfural and 5-methyl-
furfural are pyrolysis products of carbohydrates (14), but the ab-

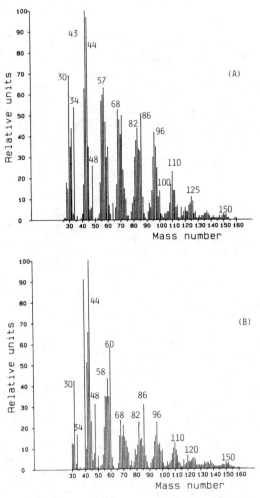

Figure 5. Pyrolysis 15 eV electron impact mass spectrum (A) and pyrolysis argon I photoionization mass spectrum (B) of sample 20C collected in the New Waterway. Pyrolysis wire temperature = 510°C.

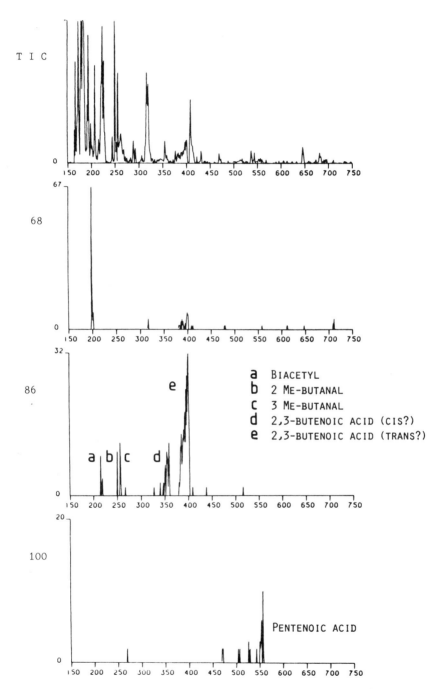

Figure 6. Total ion current and mass chromatogram for m/z 68, 86 and 100 from the py-gc-PI-ms data file of sample 20C.

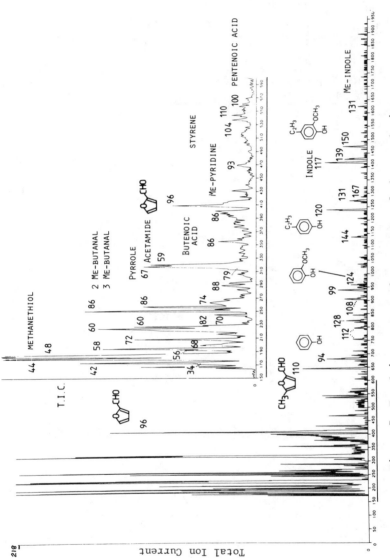

Figure 7. Total ion current chromatogram of the pyrolysis gas chromatography photoionization mass spectrometry data of sample 20C (lower trace). The upper trace (insert) is a magnification of the first part of the chromatogram (scans 150 – 600). The numbers above the peaks in the traces are molecular ions of the compounds. Pyrolysis wire temperature = 510°C.

sence of levoglucosan and related cyclic compounds may indicate that
no intact polysaccharides are present. Methanethiol (m/z=48), methyl-
propanal (m/z=72, scan 205), the methylbutunals (m/z=86), pyrrole
(m/z=67), phenol (m/z=94), p-cresol (m/z=108, scan 905), indole (m/z=
117), methyl-indole (m/z=131), phenylcyanoethane (m/z=131, scan 1270)
and the unknowns at m/z=99 (scan 940) and m/z=167 (scan 1275) are pyrolysis products from proteins (13,21,22). The acetamide (m/z=59) has
been proposed as a marker for bacterial cell walls (23), although it
has been observed as a pyrolysis product of glycine (24). The m/z=144
(scan 1140) is the molecular ion of a free octanoic acid. All these
compounds in the pyrolysate also indicate that the particulate matter
fraction sampled at site 20C must be considered as a sewage sludge
with degraded lignocellulose material from plants and abundant in
bacterial biomass. It is remarkable that the bacterial markers remain
visible in most of the samples from the New Waterway. Their presence
in the most recently dumped dredging spoils in sample area 4 (Loswal
Noord) is indicated in factor discriminant data from a subfile of the
Loswal Noord samples, but remains to be confirmed by GCMS analysis.

The relative abundance of bacterial biomass as well as the reducing sulphur (expressed as methanethiol in the PMS), may play an important role in the binding of metals in this sediment, which has
high concentrations of Zn (900 ppm), Cu (135 ppm), Pb (200 ppm), Cd
(9 ppm) and Hg (4 ppm) (8). The strong adsorption of for example cadmium to suspended matter has been shown (25), as well as the incorporation of various metals into bacterial cell walls (26). Preliminary
results from canonical correlation analysis of metal data, % organic
carbon, mud content and the PMS. data file have shown a strong correlation of Zn, Cu, Pb, Cr, Hg and Cd with the bacterial characteristics found in the riverine samples (27). Understanding of the fate of
this organic matter may help to understand the fate of certain environmental pollutants.

Acknowledgments

This research was financially supported by the Netherlands Organiza-
tion for the Advancement of Pure Research (ZWO), the Foundation for
Fundamental Research on Matter (FOM) and the Rijks Instituut voor de
Zuivering van Afvalwater (RIZA), Lelystad.

Literature cited

1. Salomons, W., Eysink, W., In "Biogeochemical and Hydrodynamic
   Processes Affecting Heavy Metals in Rivers, Lakes and Estuaries".
   Delft Hydraulics Laboratory Publication No. 253: Delft 1981, The
   Netherlands.
2. Critchley, R.F., Wat.Sci.Techn. 1984 16, 83-94.
3. Whelan, J.K., Fitzgerald, M., Tarafa, M.E., Environm.Sci.Techn.
   1983 , 17, 292-298.
4. Sigleo, A.C., Hoering, T.C., Helz, G.R., Geochim.Cosmochim.Acta
   1982 , 46, 1619-
5. Van der Meent, D., Los, A., De Leeuw, J.W., Schenck, P.A., Haverkamp, J., In "Advances in Organic Geochemistry, 1981", Bjorøy,
   M., Ed.; Wiley and Sons: New York, 1983, 336-349.
6. Eisma, D., Boon, J.J., Groenewegen, R., Ittekot, V., Kalf, J.,
   Mook, W.G., In "Transport of Carbon and Minerals in Major World
   Rivers"; Degens, E.T.; Kempe, S.; Soliman, H.; Eds.; SCOPE/UNEP
   Special Publication, Mitt.Geol.Paleont.Inst.Univ.Hamburg 1983,
   55, p.295-314.

7. Eisma, D., Bernard, P., Boon, J.J., Van Grieken, R., Kalf, J.,
    Mook, W.G., Mitt.Geol.Paleont.Inst.Univ.Hamburg, in press.
8. Boon, J.J., "Verkennend Pyrolyse Massaspektrometrisch Finger-
    print Onderzoek naar Organische Stof Karakteristieken in Bodem-
    Slib uit het Rijnmondgebied en in het Kustgebied bij Loswal
    Noord", AMOLF Report 84/177 (FOM nr. 59.714).
9. Boon, J.J., Tom, A., Brandt, B., Eijkel, G.B., Kistemaker, P.G.,
    Notten, F.J.W., Mikx, F.H.M., Anal.Chim.Acta 1984, 163, 193-205.
10. Louter, G.J., Boerboom, A.J.H., Stalmeijer, P.F.M., Tuithof,
    H.H., Kistemaker, J., Int.J.Mass Spectr.Ion Phys. 1980, 33,
    335-338.
11. Genuit, W., Boon, J.J., J.Anal.Appl.Pyrol. 1985, 8, 25-40.
12. Meuzelaar, H.L.C., Haverkamp, J., Hileman, F.D., "Pyrolysis Mass
    Spectrometry of Recent and Fossil Biomaterials", Elsevier, Am-
    sterdam, 1982; pp.293.
13. Boon, unpublished observations.
14. Van der Kaaden, A., Haverkamp, J., Boon, J.J., De Leeuw, J.W.,
    J.Anal.Appl.Pyrol. 1983, 5, 199-220.
15. Cooks, R., "Collision Spectroscopy", Plenum Press, New York
    1978 , p.458.
16. McLafferty, F.W., Bente III, P.F., Kornfeld, R., Tsai, S.-C,
    Howe, I., J.Am.Chem.Soc. 1973, 95, 2120.
17. Dallinga, J.W., Nibbering, N.M.M., Boerboom, A.J.H., J.Chem.Soc.,
    Perkin Trans. II 1984 , 1065-1076.
18. Morikawa, H., Marchessault, R.H., Can.J.Chem. 1981, 59, 2306-
    2313.
19. Wallen, L.L., Rohwedder, W.K., Environm.Sci.Technol. 1974, 8,
    576-579.
20. Martin, F., Saiz-Jimenez, C., Gonzalez-Vila, F.J., Holzforschung
    1979, 33, 210-212.
21. Simmonds, P.G., Appl.Microbiol. 1970, 20, 567-572.
22. Irwin, W.J., J.Anal.Appl.Pyrol. 1979, 1, 3-27/89-122.
23. Hudson, J.R., Morgan, S.L., Fox, A., Anal.Biochem. 1982, 120,
    59-65.
24. Bruchet, M., personal communication.
25. Van der Meent, D., Los, A., De Leeuw, J.W., Schenck, P.A., Salo-
    mons, W., Environ.Technol.Letters 1981, 2, 569-578.
26. Beveridge, T.J., In "Environmental Biogeochemistry and Geomicro-
    biology. Volume 3", Krumbein, W.E., Ed., Ann Arbor Science, Ann
    Arbor 1978, p.975.
27. Hoogerbrugge, R., Eijkel, G.B., Kistemaker, P.G., in preparation.

RECEIVED September 23, 1985

Origin of Organic Matter in North American Basin Cretaceous Black Shales

Rosanne M. Joyce and Edward S. Van Vleet

Department of Marine Science, University of South Florida, St. Petersburg, FL 33701

The expansion of the oxygen minimum zone in the incipient North Atlantic during the Upper Cretaceous enhanced the preservation of sedimentary organic matter. Organic carbon values ranging from 1.7 to 13.7% and the proportions of unbound lipid to total lipid content (>70%) support this premise. A present point of contention is the origin of the organic matter-rich Upper Cretaceous deep-sea sediments. With eustatic sea level transgression and flooding of continental lowlands, transport of terrigenous organic matter into the North American basin may have increased. Analyses of Upper Cretaceous sediments from DSDP Site 603B, lower continental rise east of Cape Hatteras, indicate that the organic matter was continentally derived. $\delta^{13}C$ values of -23.5 to -27.1^{o}/oo, C/N ratios of 32 to 72, and lipid class maxima of unbound alkanes (C_{31} and C_{29}), unbound fatty acids (C_{30} and C_{28}) and bound fatty acids (C_{28}, C_{26} and C_{24}) provide supportive evidence for the terrigenous nature of the organic matter.

The Upper Cretaceous has been described as a period of many important global events. Enhanced sea-floor spreading (up to 18 cm/yr) at many mid-ocean ridge systems led to a reduction in ocean basin volume (1) particularly in the restricted, incipient North Atlantic. Sea level transgressed, surpassing present levels by as much as 350 m (2). As latitudinal temperature gradients diminished, deep-water circulation in the oceans waned. In the North Atlantic basin (3), stagnation and consequent expansion of the oxygen minimum zone to the deep-sea floor may account for the observed preservation of thick organic matter -rich sequences in mid-Cretaceous sediments (4-6). The sluggish deep-water circulation which favored anoxic expansion may have also reduced nutrient recycling to surface waters, decreasing fertility and productivity in the euphotic zone of the open-ocean (7). Paleofertility estimates derived from organic carbon accumulation

0097–6156/86/0305–0091$06.00/0

rates of mid-Cretaceous deep-sea deposits suggest that primary productivity was as much as an order of magnitude lower than productivity measured in today's oceans (8). If these scenarios are correct, increased input of terrestrial organic matter from the prograding continental shelf (9), coupled with reduced sea-surface productivity, would produce an obvious terrestrial influence on the sedimentary organic matter deposited on the continental shelf and slope.

Alternatively, Jenkyns (9) proposed that with the extension of the subaerial continent, terrigenous run-off not only enhanced primary productivity in shelf surface waters by increasing nutrient availability, but also increased the deposition of terrigenous organic matter in the marine environment. In consideration of these scenarios, both marine organic matter and terrigenous organic matter may have been important components of the sediment.

The goals of the present study are to assess the local extent of anoxia on the lower continental rise, east of Cape Hatteras, during specific intervals of the Upper Cretaceous, to evaluate relative contributions to the bulk organic matter by marine and terrigenous sources, and to determine the extent of microbial alteration of specific lipid classes. The above assessments are based on $\delta^{13}C$, organic carbon/organic nitrogen ratios, and analyses of selected lipid classes of Upper Cretaceous black shale and variegated claystone samples from DSDP Site 603B.

Experimental Methods

Sediment was drilled from DSDP Site 603B (lower continental rise, 435 km east of Cape Hatteras, water depth 4643 m, latitude 35°29.71'N, longitude 70°01.71'E) by the D/V Glomar Challenger (Figure 1). Subsurface sampling depths ranged from 1081 to 1146 m (Table I). Based on the evidence of sharp basal contacts and graded silts, it is presumed that the sediments were turbiditically emplaced to Site 603B (10).

Table I.   Site description of DSDP Site 603B lower continental rise, U.S. east coast (10).

Core	29	33	34	34	36
Section (cm)	1(48-53)	CC(1-8)	1(13-116)	2(136-139)	1(39-43)
Depth (m)	1050	1127	1128	1130	1146
Lithology	variegated claystone	-----black carbonaceous claystones-----			
Age	Coniacian-Santonian	Turonian ---------Cenomanian---------			
	84-89 Ma[a]	89-91 Ma		91-98 Ma	

[a]millions of years

Unbound lipid was extracted by refluxing sediment in methanol/toluene/H_2O for three hours following sonication (6).

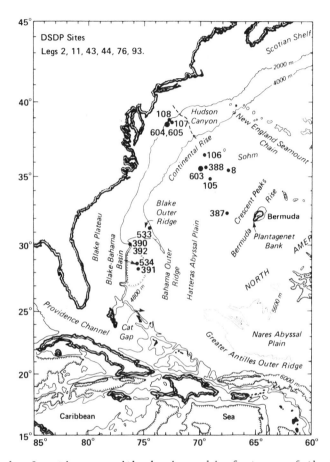

Figure 1. Location map with physiographic features of the North American Basin, and location of DSDP Site 603 and other DSDP drill sites in the North Atlantic Ocean. Bathymetry after Uchupi, 1971 (10).

After partitioning and hexane extraction, the unbound lipid extract
was concentrated to near dryness and diluted to a known volume.
Aliquots of unbound lipid were partitioned employing a modified
serial-elution column chromatography procedure (6, 11) to isolate
aliphatic hydrocarbons and fatty acids. Fatty acids were
methylated (12) and purified by thin-layer chromatography (6).
Bound lipid was extracted by refluxing residual sediment in 0.5N
KOH in methanol/toluene/H_2O and then separated, derivitized and
purified to obtain the bound lipid classes. Aliquots of unbound
and bound total lipid were measured gravimetrically after drying to
constant weight.

Identification and quantitative determinations of lipid
compounds were made by splitless injection capillary gas-liquid
chromatography (GC). Samples were analyzed on a Hewlett-Packard
5880A FID gas chromatograph equipped with a 15 m x 0.2 mm i.d. DB-5
fused silica capillary column, temperature programmed from 90 to
255°C at 4°C/min. using H_2 as a carrier gas. Individual compounds
were identified by comparison of retention times with those of
authentic standards and confirmed by combined high-resolution gas
chromatography-mass spectrometry (6). Quantification of individual
constituents was based on relative percents of integrated peak
areas extrapolated to gravimetrically determined total lipid
fraction weights. Analytical variation of sample replicates
quantified by GC generally ranges from ±10 to 37% (13-14).

Stable carbon isotopes were analyzed on a Finnigan MAT 250
isotope ratio mass spectrometer according to the procedure of Craig
(15). Results are reported relative to the Chicago PDB standard in
terms of $\delta^{13}C$ defined as:

$$\left[\frac{^{13}C/^{12}C_{sample}}{^{13}C/^{12}C_{standard}} - 1 \right] \times 1000$$

(15). Average analytical variability of sample preparation and
stable carbon isotope ratio analysis is ±0.2% (16). A Carlo-Erba
Model 1106 Elemental Analyzer was employed to determine percent
organic carbon and nitrogen in each sediment sample. Analytical
variation of sample analysis is ±20% (17).

Results and Discussion

Total organic carbon (TOC), $\delta^{13}C$, and total lipid data are
summarized in Table II (more detailed results can be found in
Joyce [6]). Total organic carbon was highest in the
Cenomanian-Turonian black shale samples, with ranges from 3.4% to
13.7%. The Coniacian-Santonian variegated claystone had the lowest
TOC (1.7%) in the sample set. TOC in Core 34-1 represents one of
the highest values measured in sediments sampled from the North
American basin. Conditions in the near-shore environment and lower
continental rise following turbiditic emplacement of these
sediments must have favored organic matter preservation since
values >0.6% TOC are not typical of deep-sea sediments (18).
Although an environment entirely depleted in oxygen is not a

requisite for black shale formation (18), the unbioturbated nature of the sediments in the present study suggests that an oxygen-stressed system prevailed during their deposition and burial.

Table II.   Organic geochemical data, DSDP Site 603B lower continental rise, U.S. east coast.

Core	29-1	33-CC	34-1	34-2	36-1
Sub-bottom depth (m)	1050	1127	1128	1130	1146
TOC^a (%)	1.7	3.4	13.7	4.8	3.9
C/N ratio^b	72.0	38.6	43.8	40.0	32.0
δ^{13}C (^{o}/oo)	-24.7	-24.7	-23.6	-23.5	-27.1
Total lipid (mg/g)	0.3	0.7	3.3	1.4	1.9
Lipid in TOC (%)	1.5	2.2	2.4	2.8	4.8
Bound lipid (% of total lipid)	27	7	6	19	9

^{a}total organic carbon; ^{b}organic carbon to organic nitrogen atomic ratio

        Unbound lipids, compounds which are generally more labile than their bound counterparts (19), represented the greatest proportion of total lipid in each sediment sample. More than 80% of the total lipid was unbound in the black shales; Core 34-1 had the highest proportion of unbound lipid (94%) as well as the highest concentration of TOC (13.7%) and total lipid (3.3 mg/g). Conversely, the variegated claystone (Core 29-1) had the lowest proportion of unbound lipid (73%), TOC (1.7%) and total lipid (0.3 mg/g). Lipid concentration and the proportion of unbound to bound lipid may be coupled with the oxidation potential. During the Cenomanin-Turonian, black shale deposition and/or formation probably occurred under conditions of reduced oxidizing potential, and that of the Coniacian-Santonian variegated claystone during elevated oxidation potential conditions.
        The percentage of lipid in TOC increased with depth (1.5 to 4.8%). Dean et al. (20) have related increases in lipid content to enhanced methanogenic bacterial activity, a process that may have produced lighter δ^{13}C values. Work by Pauly (21) has shown that the input of archaebacterial lipids to these sediment samples is very low (<0.1 µg/g). This finding suggests that the increase in lipid/TOC ratio with depth in the present study may not be due to this type of input. The trend in increasing lipid with depth does not coincide with the trend in δ^{13}C values. As discussed above, lipid content in these sediments is probably related to the oxidation potential and not to bacterial biomass.
        The provenance of sedimentary organic matter in the present study appears to be primarily terrigenous. δ^{13}C values of -23.5 to -27.1 ^{o}/oo in the sample set fall within the isotopic range of terrigenous organic matter. Generally, δ^{13}C values ranging from -23 to -33°/oo are found in terrestrial plants utilizing the

Calvin-Benson photosynthetic pathway, a maximum ^{13}C fractionation mechanism. Most warm water marine phytoplankton have $\delta^{13}C$ values between -17 and -22 $^{o}/oo$ ([22]). As global temperatures were thought to have been warm and equitable during the mid-Cretaceous ([3]), warm water phytoplankton probably dominated waters of the North Atlantic and had stable carbon isotope values typical of warm water organisms.

Recently, a point of contention has arisen concerning the use of $\delta^{13}C$ as a source indicator for mid-Cretaceous inputs. Dean et al. ([20]) have suggested that the magnitude of fractionation between ^{13}C and ^{12}C by photosynthetic organisms occupying mid-Cretaceous oceans may not have been the same as in today's oceans. Atmospheric CO_2 levels may have been as much as an order of magnitude greater than present day levels, according to the model developed by Berner et al. ([23]). This condition would shift oceanic bicarbonate to isotopically lighter values. The effects of isotopically lighter dissolved bicarbonate on organic carbon assimilation by phytoplankton have been studied by Wong and Sackett ([24]). It was found that the extent of $^{13}C/^{12}C$ discrimination among species was variable and not all values fell within the terrigenous isotopic range. Although Dean et al. ([20]) proposed that no systematic ^{13}C enrichment occurred in phytoplankton compared to terrestrial plants during the mid-Cretaceous, $\delta^{13}C$ values of -23.5 to -27.1 $^{o}/oo$ observed in the present study are coincident with data from elemental and lipid analyses and indicate that, at least at Site 603B, the Upper Cretaceous organic matter-rich deposits are primarily of continental origin.

Shifts in isotopic signals were observed in the analysis of these Upper Cretaceous sediments. The lightest isotopic value was measured in the oldest sample (603B-36-1; -27.1 $^{o}/oo$). 603B-34-2 and 603B-34-1 were isotopically heavier (-23.5 to -23.6 $^{o}/oo$) and 603B-33-CC and 603B-29-1 were again lighter (-24.7 $^{o}/oo$). A simplistic interpretation of this fluctuation is the variable mixing of marine and terrigenous organic matter. Isotopically lighter values represent smaller contributions by marine organic matter, whereas heavier values indicate a greater marine input.

The trend toward lighter isotopic values from the upper Cenomanian (603B-34-2) to the Santonian (603B-29-1) may also indicate improvement in the oceanic water mass circulation. In oceanic water masses where circulation is diminished, density stratification prevents mixing of ^{13}C enriched surface water with ^{13}C depleted bottom waters. The surface enrichment of ^{13}C produces a 'heavier' phytoplankton community ([25]). Deposition and preservation of phytoplankton detritus would leave a heavier isotopic signal in the sedimentary organic matter. Alternatively, the range of values observed in this study may also represent natural $\delta^{13}C$ variations in species composition.

Elemental analyses provided supportive data for the origin of the sedimentary organic matter. Remineralization of nitrogen in marine organic matter produces C/N ratios between 10 and 15 ([26]). C/N ratios >15 in terrigenous organic matter result from the paucity of nitrogenous compounds in higher plant lignin and cellulose ([27]). C/N ratios of sediment samples from DSDP Site 603B ranged from 32 to 72. These values may represent nitrogen-poor, terrestrially-derived plant detritus. If the sedimentary organic matter had been marine-derived and highly reworked, values closer to 15 would be expected.

In the analysis of selected lipid fractions, it was shown that the unbound non-aromatic hydrocarbons comprised <12% of the total extractable lipid (Table III). Concentrations of the n-alkanes corresponded to those measured in organic matter-rich Cretaceous claystones from the northwestern Atlantic ($\underline{4}$). C_{29} and C_{31}, the most frequently observed alkanes in higher plant waxes[31] ($\underline{28}$), dominated the alkane distributions in the present study (Figure 2). Alkane maxima at C_{31} have been noted in a number of other deep-sea environments including DSDP Site 532B, Walvis Ridge ($\underline{6}$) and Site 386, Northwest Atlantic ($\underline{29}$). C_{29}/C_{17} ratios >>1 were consistent with the continental plant wax input inferred from the above-described alkane distributions (Table III). The short-chain alkanes (C_{17}-C_{21}) are described by an odd-carbon/even-carbon predominance (OEP)($\underline{30}$) of <1. This distribution is typical of continentally-derived inputs. These alkanes are generally less abundant than the longer-chain counterparts and lack odd-carbon preference ($\underline{31}$). The longer-chain compounds (C_{27}-C_{31}) were odd-carbon predominated (OEPs >1.7) and are typical of terrestrial organic matter inputs. Strong odd-carbon preferences in the long-chain alkane distributions indicate early stages of maturation ($\underline{32}$). In a highly evolved mature fraction, longer-chain alkanes would no longer be dominated by odd-carbon compounds. The substantial area of the unresolved complex mixtures may represent branched and cyclic compounds generated by long-term geochemical processes ($\underline{20}$, $\underline{33}$).

Bound aliphatic hydrocarbons comprised <7% of the saponifiable lipid (<1 to 15 μg/g). Unbound/bound ratios of <1 to 67 of this hydrocarbon fraction, with the exception of 603B-29-1, agree with values reported in other studies ($\underline{34}$, $\underline{37}$). Bound alkane distributions were dominated by C_{20} or C_{22} (Figure 3). Even-carbon predominance of alkanes is an anomaly, but distributions similar to those in the present study have been observed in other North Atlantic deep-sea sediments ($\underline{29}$). The source of these compounds, however, is unknown.

Fatty acids comprised 7-29% of the unbound lipid extract (55-370 μg/g; Table IV). 603B-33-CC, 603B-34-1 and 603B-36-1 had fatty acid distributions which maximized at C_{16} and C_{18} and 603B-29-1 and 603B-34-2 at C_{30} and C_{28}, respectively. Dominance of these short-chain acids in the presence of their long-chain counterparts suggests that the short-chain acids were not utilized as readily as the short-chain non-aromatic hydrocarbons. Input of these compounds may have also been subsequent to sediment deposition, possibly through microbial activity (Figure 4). Monoenoic fatty acids were detected in the black shale samples (603B-34-1 and 603B-36-1) but branched fatty acids were observed in all of the sediments analyzed. Alteration or remineralization of the monoenoic compounds, compared to the branched acids, may have occurred more readily.

The bound fatty acid component in the total lipid extract was one to three orders of magnitude less than the unbound fatty acid fraction. This relationship produced unbound/bound ratios

Table III. Gravimetric and gas chromatographic data on unbound and non-aromatic hydrocarbon fractions, DSDP Site 603B lower continental rise, U.S. east coast.

Lipid fractions	Core 29-1[a]	Core 33-CC[b]	Core 34-1[c]	Core 34-2[c]	Core 36-1[c]
Unbound non-aromatic HC[d]					
μg/g dry sediment	4	27	336	44	134
% of unbound lipid	2(3)[e]	4(3)	11(15)	4(5)	8(4)
$<C_{25}$ (%)	2	4	2	2	2
$\geq C_{25}$ (%)	5	5	4	5	4
unidentified (%)	6	44	40	47	20
UCM[f] (%)	87	47	54	46	74
OEP[g] $(C_{17}-C_{21})$	1.0	0.1	0.9	0.4	0.7
OEP $(C_{27}-C_{31})$	8.0	2.2	2.0	1.7	5.8
C_{29}/C_{17}	21	17	33	50	4
pristane/phytane	1.0	0.8	0.4	0.3	0.5
Bound non-aromatic HC					
μg/g dry sediment	5	<1	5	15	3
% of bound lipid	7	<1	3	6	2
unbound/bound ratio	<1	27	67	3	4
$<C_{25}$ (%)	13	16	15	19	7
$\geq C_{25}$ (%)	7	11	2	4	2
unidentified (%)	11	49	54	45	70
UCM[f] (%)	69	24	29	32	22
OEP[g] $(C_{17}-C_{21})$	<0.1	0.1	NC[h]	NC	NC[i]
OEP $(C_{27}-C_{31})$	1.7	0.8	0.3	1.1	0.6[i]
C_{29}/C_{17}	NC	NC	NC	NC	NC

[a]5 replicates; [b]4 replicates; [c]3 replicates; [d]hydrocarbons; [e]standard deviation; [f]unresolved complex mixture; [g]odd-even predominance; [h]not calculated, insufficient data; [i]OEP $(C_{22}-C_{26})$

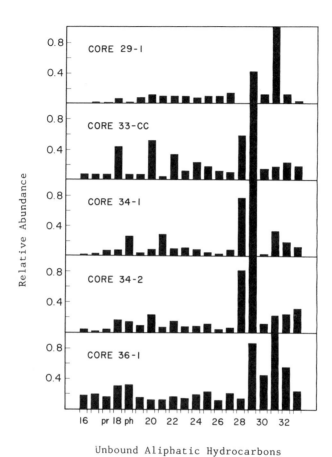

Unbound Aliphatic Hydrocarbons

Figure 2.   Carbon number distributions and relative abundances of unbound aliphatic hydrocarbons normalized to the most abundant hydrocarbon, DSDP Site 603B lower continental rise, U.S. east coast.

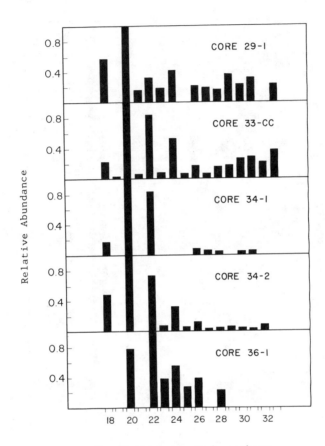

Bound Aliphatic Hydrocarbons

Figure 3. Carbon number distributions and relative abundances of bound aliphatic hydrocarbons normalized to the most abundant hydrocarbon, DSDP Site 603B lower continental rise, U.S. east coast.

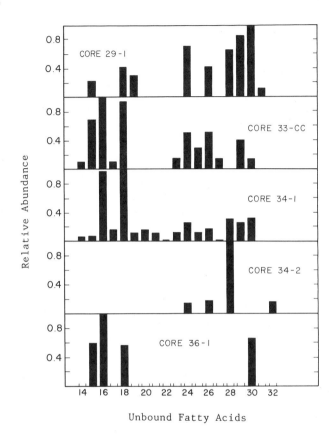

Figure 4. Carbon number distributions and relative abundances of unbound fatty acids normalized to the most abundant fatty acid, DSDP Site 603B lower continental rise, U.S. east coast.

$>>1$. The greater proportion of the unbound fatty acid fraction compared to the bound fraction coincides with similar proportions for total lipid and non-aromatic hydrocarbons. Fatty acids comprised $<1-14\%$ of the bound lipid extract, values equivalent to $<1-10$ µg/g. A summary of gas chromatographic data is presented in Table IV.

Greater relative abundances of compounds $C_{24} - C_{32}$ (with the exception of 603B-34-1) and C_{16} and C_{18} (with the exception of 603B-34-2)(Figure 5), coupled with greater relative percentages of monoenoic and branched fatty acids, suggest that the preservation of compounds in the bound fatty acid fraction was enhanced relative to the those in unbound fraction. Lipophilic humic-like material and/or clay-mineral binding may have reduced the reactivity of these compounds (37). The longer chain acids probably represent components of terrigenous organic matter whereas the short-chain acids are probably components resulting from microbial activity.

Bound and unbound short-chain and long-chain fatty acids were generally predominated by the even-carbon compounds. Even-carbon preference, both in the short-chain and long-chain range, is not uncommmon for fatty acids in ancient sediments. Both the Devonian Antrim shale and the Soudan shale have shown similar even-carbon preferences although the respective n-alkane fractions lacked odd-carbon preference.

Summary

The world-wide distribution of organic matter-rich Upper Cretaceous sediments has prompted a number of organic geochemical investigations to study their origin. Sediments of Upper Cretaceous age sampled from the lower continental rise east of North America (DSDP Site 603B) were characterized by $\delta^{13}C$ values ranging from -23 to -27°/oo and C/N ratios from 32 to 72. Although DSDP Site 603B was, and has been, a deep-sea environment, continental provenance of the organic matter was inferred from the results of TOC and selected lipid analyses. The terrigenous nature of the organic matter in these Upper Cretaceous sediment samples supports previous geological data of it's turbiditic emplacement to Site 603B. Unbound alkanes C_{29} and C_{31}, unbound fatty acids C_{28} and C_{30}, and bound fatty acids C_{24}, C_{26}, and C_{28} distribution maxima indicate that plant waxes were important components in the lipid material. Organic-richness (1.7 to 13.7%) and the relatively large proportions of unbound lipid compared to the bound lipid support the theory that anoxia prevailed in the deep-sea during the time intervals represented by these sediments, particularly during black shale deposition. The limited contribution of marine organic matter in the sediments examined indicates that sea-surface fertility was probably not any higher than in today's oceans.

Table IV. Gravimetric and gas chromatographic data on unbound and bound fatty acid fractions, DSDP Site 603B lower continental rise, U.S. east coast.

Lipid fractions	Core 29-1	Core 33-CC	core 34-1	Core 34-2	Core 36-1
Unbound fatty acids					
μg/g dry sediment	55	122	213	ND[a]	370
% of unbound lipid	29	18	7	ND	22
$<C_{22}$ (%)	2	6	20	0	1
$\geq C_{22}$ (%)	9	5	13	2	<1
antéiso (%)	tr[b]	1	1	0	<1
monoenoic (%)	0	0	tr	0	1
unidentified (%)	45	38	28	40	20
UCM[c] (%)	44	50	38	58	78
EOP $(C_{16}-C_{20})$[d]	NC[e]	2.4	6.4	NC	NC
EOP $(C_{22}-C_{26})$	1.6	2.1	1.7	2.1	NC
C_{26}/C_{16}	NC	0.5	0.7	NC	NC
Bound fatty acids					
μg/g dry sediment	<1	7	10	10	3
% of bound lipid	<1	14	5	4	2
unbound/bound ratio	55	17	21	NC	123
$<C_{22}$ (%)	7	2	14	1	4
$\geq C_{22}$ (%)	22	4	5	9	7
antéiso (%)	<1	1	<1	0	<1
monoenoic (%)	7	3	6	1	2
unidentified (%)	34	38	59	38	46
UCM (%)	30	52	15	51	40
EOP $(C_{16}-C_{20})$	6.8	5.8	22.1	NC	NC
EOP $(C_{22}-C_{26})$	2.2	2.9	2.4	2.5	3.1
C_{26}/C_{16}	0.7	2.2	0.1	6.3	0.4

[a]not detected; [b]trace; [c]unresolved complex mixture; [d]even-odd predominance; [e]not calculated, insufficient data

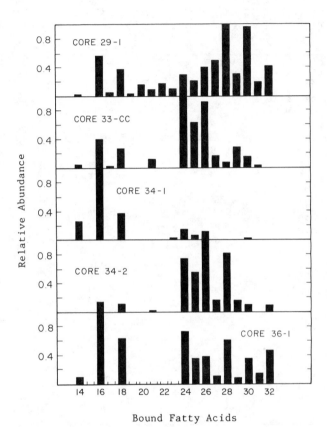

Figure 5.   Carbon number distributions and relative abundances
of bound fatty acids normalized to the most abundant fatty acid,
DSDP Site 603B lower continental rise, U.S. east coast.

Acknowledgments

Our thanks to Dr. Phil Meyers of the University of Michigan for providing the DSDP samples, and to Margarita Conkright for performing the stable carbon isotope analyses.

Literature Cited

1. Hays J.D.; Pitman, W.C. Nature 1973, 246, 18-22.
2. Van Graas, G. In "Advances in Organic Geochemistry, 1981"; Bjoroy, M. et al., Eds.; John Wiley: Chichester, 1981; pp. 471-476.
3. Ryan, W.; Cita, M.B. Mar. Geol. 1977, 23, 197-215.
4. Simoneit, B.R.T. In "Init. Repts. DSDP, 43"; Tucholke B.E.; Vogt P.R. et al., Eds.; U.S. Govt. Printing Office: Washington, D.C., 1979; pp. 643-649.
5. Demaison, G.J.; Moore, G.T. Org. Geochem. 1980, 9-31.
6. Joyce, R.M. M.S. Thesis, University of South Florida, St. Petersburg, 1985.
7. Fischer, A.G.; Arthur, M.A. In "Secular variations in the pelagic realm"; SEPM Spec. Publ. No. 25, 1977; pp. 19-50.
8. Bralower, T.J.; Thierstien, H.R. Geology 1984, 12, 614-618.
9. Jenkyns, H.C. J. Geol. Soc. London, 1980, 137, 171-188.
10. "Initial Core Descriptions, Leg 93", Init. Repts. DSDP, 93: U.S. Govt. Printing Office: Washington, D.C.; in press.
11. Wakeham, S.G. Geochim. Cosmochim. Acta 1982, 46, 2239-2257.
12. Metcalfe, L.D.; Schmitz, A.A.; Pelka, J.R. Anal. Chem 1966, 38, 514-515.
13. Farrington, J.W.; Tripp, B.W. Geochim. Cosmochim. Acta 1977, 41, 1627-1641.
14. Prahl, F.G; Bennett, J.T.; Carpenter, R. Geochim. Cosmochim. Acta 1980, 44, 1967-1976.
15. Craig, H. Geochim. Cosmochim. Acta 1953, 3, 53-92.
16. Mangini, M.E. M.S. Thesis, University of South Florida, St. Petersburg, 1983.
17. DeFlaun, M., personal communication.
18. Katz, B.J.; Elrod, L.W. Geochim. Cosmochim. Acta 1983, 47, 389-396.
19. Aizenshtat, Z.; Baedecker, M.J.; Kaplan, I.R. Geochim. Cosmochim. Acta 1973, 37, 1881-1898.
20. Dean W.E.; Claypool G.E.; Thiede, J. Org. Geochem. 1984, 7, 39-51.
21. Pauly, G., personal communication.
22. Anderson, T.F.; Arthur, M.A. In "Stable Isotopes in Sediment Geology"; Arthur M.A. et al., Eds; SEPM Short Course No. 10: Dallas, 1983; pp. 80-107.
23. Berner, R.A.; Lasaga, A.C.; Garrels R.M. American Journal of Science 1983, 283, 641-683.
24. Wong, W.W.; Sackett, W.M. Geochim. Cosmochim. Acta 1978, 42, 1809-1815.
25. Weissert H.; McKenzie J.; Hochuli P. Geology 1979, 7, 147-151.
26. Muller, P.J. Geochim. Cosmochim. Acta 1977, 41, 765-776.
27. Meyers, P.A.; Leenheer, M.J.; Eadie, B.J.; Maule, S.J. Geochim. Cosmochim. Acta 1984, 48, 443-452.

28.   Tulloch, A.P.   In "Chemistry and Biochemistry of Natural
       Waxes"; Kolattukudy, P.E., Ed.; Elsevier: Amsterdam, 1976; pp.
       235-288.
29.   Erdman, J.G.; Schorno, K.W.   In "Init. Repts. DSDP, 43";
       Tucholke B.E.; Vogt P.R. et al., Eds.; U.S. Govt. Printing
       Office: Washington, D.C., 1979; pp. 651-655.
30.   Scalan, R.S.; Smith, J.E. Geochem. Cosmochim. Acta 1970, 34,
       611-620.
31.   Hunt, J.A. In "Journal of Petroleum Geochemistry and Geology";
       Gilluly J., Ed.; W. H. Freeman: San Francisco, 1979; pp.
       69-149.
32.   Deroo, G.; Herbin, J.P.; Roucache, J.; Tissot, B. In " Init.
       Repts. DSDP, 47B"; Sibuet, J.C., Ryan, W.F.F., et al., Eds.;
       U.S. Govt. Printing Office: Washington, D.C., 1979; pp.
       513-522.
33.   Giger, W.; Schaffner, C.; Wakeham, S.G. Geochim. Cosmochim.
       Acta 1980, 44, 119-129.
34.   Van Vleet, E.S.; Quinn, J.G.   Geochim. Cosmochim. Acta 1979,
       43, 289-303.
35.   Meyers, P.A.; Powaser, J.M.; Dunham, K.W. In "Init. Repts.
       DSDP, 92"; Leinen, M.; Rea, D.K. et al., Eds.; In press.
36.   Van Hoeven, W.; Maxwell, J.R.; Calvin, M. Geochim. Cosmochim.
       Acta 1969, 33, 871-881.
37.   Smith, D.J.; Eglinton, G.; Morris, R.J. Geochim. Cosmochim.
       Acta 1983, 47, 2225-2232.

RECEIVED September 16, 1985

The Biogeochemistry of Chlorophyll

J. William Louda and Earl W. Baker

Organic Geochemistry Group, Florida Atlantic University, Boca Raton, FL 33431

In this paper we attempt a first approximation of
chlorophyll diagenesis, stressing the generation of
DPEP-series geoporphyrins. Samples analyzed include
viable and dead unialgal cultures, sediment traps,
surface to near-surface (0-2m) sediments and long
cores (5-1,000m) obtained from DSDP/IPOD and industry.
Results reveal that chlorophyll(-a) loses Mg and phytol
through the actions of cellular senescence and auto-
trophic recycling in the water column and surface sediments. Following deposition these 'pheo-pigments'
undergo the competing reactions of allomerization,
yielding purpurin and chlorin acids in oxic conditions,
or loss of the 10-carbomethoxy moiety, forming pyro-
phorbides in anoxic settings. These key reactions
appear to fate subsequent diagenesis to either pigment
destruction or fossilization, respectively. DPEP-
series geoporphyrins are thought to be the result of
the stepwise defunctionalization and aromatization of
the pyro-phorbide type precursors. The phenomenology
of chlorophyll geochemistry and tentative identification of several intermediates are described.

The present study is based on the assumption that geoporphyrins
are the diagenetic products of biotic tetrapyrrole pigments.
This we take as a reasonable premise, given the structural complexity of this class of biomarkers (see Figure 1), and was concluded years ago by the late Professor Alfred Treibs (1-5).
However, the development of strong precursor-product relationships in organic geochemistry requires not only statement of
plausible end-members (e.g. chlorophyll-a and DPEP-series geoporphyrins) but must include description of the intervening
reactions and intermediate structures.

These investigations, albeit preliminary, are designed to
ultimately fill the existing gaps in knowledge between biotic

structures (6-8) and the growing data base of product (viz. geo-porphyrins) structures (9-17).

Experimental

Samples. Unialgal cultures of the diatom Synedra sp. were purchased from Carolina Biological Supply. Sediment trap samples and near surface sediments (0-2m) from the Peruvian upwelling region, collected during Cruise 73 Leg 2 of the R/V Knorr, were provided by Woods Hole Oceanographic Institution (18-19). Near-surface sediments from the Guaymas Basin in the Gulf of Califor-nia, collected by the R/V Washington Leg 3-1978, were provided by Scripps Institute of Oceanography (20). Sediments and sapro-pel from Big Soda Lake (Nevada, U.S.A.. 21) and Mangrove Lake (Bermuda. 24) were obtained from the United State Geological Survey. Deeply buried (e.g. 15-1000 m., sub-bottom) marine sediments were obtained from the DSDP/IPOD program, Leg 64 (20-23). An immature marine shale of California (Pliocene/Miocene. 24) was provided by Mobil Research and Development.

Except for viable diatoms, all samples were frozen upon collection and maintained so until extraction.

Solvents, Extraction and Separation of Pigments. All solvents were freshly glass distilled and ethers were freed of peroxides over highly activated (0% H_2O) basic alumina. All procedures were in dim yellow light and extracts/isolates maintained frozen under N_2, whenever possible. Chromatographic separation uti-lized microcrystalline cellulose, Sephadex LH-20, and silica gel in normal and reverse phase modes. Extraction and chromatogra-phy is detailed elsewhere (20).

Purification and Identification of Isolates. In the case of extremely immature samples containing bacteriophytin-a (e.g. Big Soda Lake, Mangrove Lake), interferring carotenols were removed via phase separation into 90% aqueous methanol (25-26).

Low-pressure high-performance liquid-chromatography (LPHLC, Ace-Glass) using 13-24 silica in normal (Whatman #LPS-1) and C-18 reverse phase (Whatman #LRP-1) modes was employed for final pigment purification and co-chromatographic tests. Iso-cractic elution employed methanol/acetone/water (90:5:5, v/v/v) for pheophorbide free acids and methanol/acetone (95:5 v/v) for pheophytins during RP-LPHPLC. The presence of more than trace (ca 0.5- 1.0%) water in the reverse phase mode lead to exceed-ingly long elution times (> 1 hr.) with phytylated pigments. This fact served as a test for the presence of the phytyl ester. Non-polar pigments, pheophytins and decarboxylated species, were purified over normal phase silica with increas-ing percentages of acetone in petroleum ether.

Mass spectrometry was performed on a DuPont #21-491B instru-ment operated at the lowest ionizing voltage (e.g. 4.5-12.0 eV, 40-60 A) possible per sample.

Electronic absorption spectra were recorded on a Perkin-Elmer 575 instrument calibrated with holmium oxide. The absorp-tion spectra of native pigments, sodium borohydride reduction

products (27-28), and the copper chelates of each (20) were compared to numerous authentic chlorophyll derivatives (standards)
in order to classify the chromophore and its auxochromes.

Results and Discussion

Previously we have divided tetrapyrrole diagenesis into early-,
middle-and late-stages (4-5). These divisions encompass the
defunctionalization of dihydroporphyrins (e.g. phorbides), aromatization and chelation, in that order. The geoporphyrins formed
during mid-and late-diagenesis are therefore free-base and metallo-
(e.g. Ni;, V = 0) porphyrins, respectively. Since this study
deals primarily with early-diagenesis, the following section is
presented only to reveal the nature of these initially formed
geoporphyrins.
Products of Diagenesis, Immature Geoporphyrins. The progress of
the tetrapyrrole diagenetic continuum is such that the arbitrarily
defined stages (4-5) can and do overlap. Thus, it is often possible to isolate more than one pigment type (e.g. free-base and
metallo-porphyrins) from the same stratum. Shown in Figure 2 are
the mass spectral histograms, or carbon-number distribution, of
(a) the defunctionalized phorbides (i.e. 7,8-dihyroDPEP-series),
(b) the free-base DPEP series and (c) the nickel DPEP-series
isolated from a Miocene/Pliocene marine shale of California.
While not totally identical, maxima at C31 and the ranges of
pseudohomologs (C27 to C34) reveal their similar origin. Past
studies (4,5,29) have shown that the DPEP- series can be
geochemically generated with C30, C31, or C32 maxima. These
pigments, DPEP-series porphyrins with a limited carbon number
range, represent early diagenetic end products.
    The metalloporphyrins (viz. nickel, vanadyl) characteristic
of catagenesis appear to form via parallel diagenesis with the
exceptions of the degree of alkylation and the immediate organic
environment. That is, carbon-numbers up to C40, C50 and beyond
exist in the vanadyl porphyrins (5,30-33) and these pigments
appear to arise from an inextractable organically bound state
(4,29,34,35). An example of the series and carbon-number distribution of the vanadyl porphyrins from a 'moderately mature'
petroleum is given as Figure 3. The complexity of geoporphyrin
arrays becomes evident upon examination of Figure 3. That is,
since this spectrum was averaged from low voltage (4.5eV) scans
yielding only parent ions, at least twenty-nine compounds (14
DPEP and 15 ETIO), not counting isomers which are known to exist
(13-15,9-12), must be present. Re-examination of Figure 2 reveals
that this spreading of carbon number distributions amongst the
tetrapyrroles probably begins early in the diagenesis of these
pigments and occurs within the free or solvent-extractable species
(e.g.Ni porphyrins), as well as the presumably bound forms thought
to yield the vanadyl pigments. The maturational aspects of metallo-
porphyrins are covered elsewhere (6-7). Phytoplankton cultures
(Synedra, Bacillariophyceae) and water-column detritus (sediment
trap samples) were analyzed in order to describe the pre-deposit-
ional alteration of chlorophyll and therein type the immediate
precursor complement to early diagenesis.

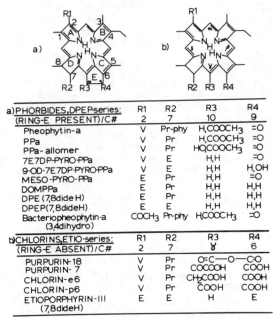

a) PHORBIDES,DPEP-series:	R1	R2	R3	R4
(RING-E PRESENT)/C#	2	7	10	9
Pheophytin-a	V	Pr-phy	H,COOCH$_3$	=O
PPa	V	Pr	H,COOCH$_3$	=O
PPa-allomer	V	Pr	HO,COOCH$_3$	=O
7E7DP-PYRO-PPa	V	E	H,H	=O
9-OD-7E7DP-PYRO-PPa	V	E	H,H	H,OH
MESO-PYRO-PPa	E	Pr	H,H	=O
DOMPPa	E	Pr	H,H	H,H
DPE (7,8dideH)	E	Pr	H,H	H,H
DPEP(7,8dideH)	E	E	H,H	H,H
Bacteriopheophytin-a (3,4dihydro)	COCH$_3$	Pr-phy	H,COOCH$_3$	=O
b) CHLORINS,ETIO-series:	R1	R2	R3	R4
(RING-E ABSENT)/C#	2	7	8	6
PURPURIN-18	V	Pr	O=C—O—C=O	
PURPURIN-7	V	Pr	COCOOH	COOH
CHLORIN-e6	V	Pr	CH$_2$COOH	COOH
CHLORIN-p6	V	Pr	COOH	COOH
ETIOPORPHYRIN-III (7,8dideH)	E	E	H	E

Figure 1. Structures of tetrapyrrole pigments mentioned in text. Code: V=vinyl; E=ethyl; Pr=propionic acid; Phy=phytol (as phytyl ester); DP=despropio-; PD=oxydeoxo; pp-a=pheophorbide-a ; DOMPP-a= deoxomesopyropheophorbide-a; DPE=deoxophylloerythrin; DPEP=deoxo-phylloerythroetioporphyrin (cf. 6,7).

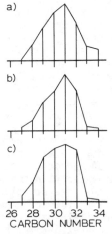

Figure 2. Mass spectral histograms of tetrapyrrole pigments characteristic of mid-/late-diagenesis. (a) free-base 7,8-dihy-dro-DPEP-series; (b) free-base DPEP-series; and (c) nickel DPEP (ETIO- omitted)-series. Sample; Pliocene/Miocene shale of marine origin (cf. 24).

Phytoplankton Cultures. A viable unialgal culture of the diatom
Synedra sp. was split into 2 aliquotes. The first portion
("VIABLE") was extracted immediately and analyzed. The second
part ("DEAD") was purged with nitrogen, sealed and stored in the
dark at room temperature (20-22°C) for 2 months before analysis.
    The electronic spectra of the crude extracts of "VIABLE" and
"DEAD" diatoms, given as Figure 4, reveals the total conversion
of chlorophyll-a to 'pheopigments' through the loss of Mg. That
is, Soret absorption has shifted hypsochromically (429 to 411 nm)
in concert with a bathchromic shift (663 to 667 nm) in the position
of band I ('red') absorption. Chromatographic analyses revealed
that 94+% of the a-series pigments in the "VIABLE" diatoms was
chlorophyll-a. This pigment was below detectable limits (<1%) in
the "DEAD" diatoms. In the latter, a-series pigments consisted of
pheophytin-a (66%), pheophorbide-a (32%) and chlorophyllide-a (2%).
A more rapid destruction of chlorophyll-a series pigments,
relative to chlorophyll-c, was also observed. That is, the
'a'/ 'c' value was found to decrease from 33:1 ("VIABLE") to
7:1 ("DEAD") upon senescence and death.
    Except for the slower demise of chlorophyll-c, the alter-
ation of chlorophyll-a during senescence and death can be attributed to cellular decompartmentalization and resultant action of
cellular acids (Mg-loss) and enzymes (e.g chlorophyllase, phytol
loss, 36). The relative amounts of pheophytin-a versus pheophorbide-a resulting from senescence-death has been shown to be
species specific and is related to chlorophyllase activity (37).
Sediment Trap Samples. The alteration of chlorophyll due to
senescence/death phenomena, described above, yield a certain suite
of pigments. Such arrays, dominated by pheophytin-a, might be
expected to form the precursor complement in sediments deposited
in an environment free of consumers. Such is obviously not the
case in nature. Since the vast majority (e.g. 95-99%) of primary
production serves as fodder in marine food webs (38) we must consider heterotrophic alteration of detrital tetrapyrrole pigments.
    Sediment trap samples from the Peruvian upwelling system
(18-19) were analyzed for chlorophyll derivatives. These samples
were collected at the base of the euphotic zone (FST-16, Z=11 m)
and about 25 m beneath the major pycnocline and zone of remineral-
azation (FST-17, Z=53 m). Both samples were lyophilized (39) and
any intact chlorophyll-a was converted to pheophytin-a. However,
on-board hydrographic data (i.e. fo/fa. cf. 18) allowed back
correction to in situ chlorophyll/pheopigment relationships.
    Table 1 is the compilation of pigment data obtained from
analyses of the sediment trap samples. The data reveal, from the
increase in the relative amounts of pheophytin-a and pheophorbide-a, a large degree of Mg and Mg plus phytol loss, respectively. The deeper sample, upon microscopic examination, was found
to contain a majority of broken, relative to intact, phytoplankton cells (diatom dominated) occurring mainly in fecal pellets.
Thus, the change in the distribution of tetrapyrrole pigments,
occurring with depth in the water column, reveals the combined
effects of senescence/death and predation. The latter is evidenced
by the increase in pheophorbide-a relative to pheophytin-a (37,40).

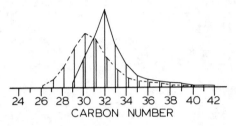

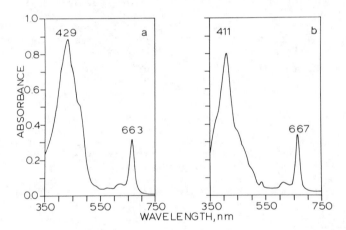

Figure 3. Mass spectral histogram of vanadyl porphyrins isolated from a marine sourced petroleum of 'moderate' maturity (Carboniferous; 3.1%S; 23.1°API). Solid trace = DPEP-, Dashed trace = Etio-series.

Figure 4. Electronic absorption spectra, in acetone, of the crude extracts of the diatom Synedra sp. (a) extracted fresh ("VIABLE") and (b) following dark anoxic storage for 2 months ("DEAD").

Table I.   Tetrapyrrole pigments present in sediment-trap samples from the Peruvian upwelling system.[a]

Pigment	Molar Percentage[b]	
	Z = 11 m	Z = 53 m
Chlorophyll-a	75	20
Pheophytin-a	<2 (77)	26 (46)
Pheophorbide-a	12 (12)	42 (42)
Purpurins	<2 (<2)	2 (2)
Chlorin acids	4 (4)	4 (4)
Chlorophyll-c	4 (4)	6 (6)

a. 15° 09'S x 75° 35'W
b. Back - correct to on-site hydrographic values (18). Parenthesized values are 'as-analyzed.'

Minor amounts of purpurin and chlorin acids, typical of oxidative pigment destruction in marine (5,20,28) and terrestrial (41) ecosystems, were also found. These most likely reside in broken detrital cells exposed to oxygen. Further study is required to clarify destructive pathways.

'Chlorophyll-c, a known component of the diatoms and relatives (6,7,37), was identified in both samples (Table I). The presence of 'chlorophyll-c' in marine destritus is expected, however, to date, its absence in marine sediments collected below the euphotic zone (5,20,28,42) is enigmatic. That is, whether 'cholorophyll-c' is destroyed, complexed into unrecognizable macro-molecular forms or in other ways altered remains to be shown.

The rapid alteration of chlorophyll-a as phytoplankton leave the euphotic zone may well be expected to continue as passage through additional links in the food chain occurs. Thus, the detrital forms of chlorophyll-a, available to sedimentary diagenesis should contain a dominance of pheophorbide-a.

Sedimentary Chlorophyll Derivatives.   In order to unravel the biogeochemistry of chlorophyll, a variety of depositional environments, each representing potentially different fossilized counterparts, require study. Though the present investigation stresses marine sediments, collected well beneath the photic zone in order to divorce detrital from viable photoautotrophic material, photic zone sapropels and peats have also been examined.

Marine Surface Sediments..   (0-2m) sediments beneath the highly productive upwelling regions in the Gulf of California and off the coast of Peru were chromatographically separated (see "Experimental") into (a) pheophytins, (b) pheophorbides (viz. mono carboxylic acids) and (c) chlorin acids (i.e. di-/tri-acids). Table II contains the result of these preliminary analyses and includes the concentration of pigments, present depositional environment and sample depth.

In general, pheophorbides and chlorin acids have become the dominant tetrapyrrole pigments in surface sediments, relative to upper water column detritus. The averaged values for the relation-

Table II. Chlorophyll derivatives in near-surface marine sediments

Sample*	Bottom Waters	Water Depth(m)	Sample Depth(cm)	Conc. µg/g-dry	Chlorophyll Derivatives Molar Percentage		
					Pheophytins	Pheophorbides	Chlorin acids
10G-3	Oxic	1,964	3.5- 5.9	69.5	16	65	19
18G-1	Oxic	2,009	1.0-10.0	51.8	27	21	52
BPA	Anoxic	530	0.0-10.0	52.2	60	33	7
479-3-2	Anoxic	747	1500.0	41.0	22	47	31
BC-5	Oxic	5,380	0.0-2.0	48.3	14	47	39
SC-6	Anoxic	268	0.0-2.0	264.4	36	31	33
SC-4	Oxic/Anoxic	92	0.0-3.0	150.1	17	56	27

*Sample locations

10G (G.C.,SIO): 27°20.2'N x 111°33.6'W
18G (G.C.,SIO): 27°0.9'N x 111°21.5'W
BPA (G.C.,CSM): 23°35.0'N x 107°23.6'W
DSDP/IPOD Leg 64, Site 479 (GC): 27°50.76'N x 111°37.49'W
BC-5 (Peru,WhOI): 15°27.6'S x 75°09.6'W
SC-6 (Peru, WHOI): 15°05.1'S x 75°43.9W
SC-4 (Peru, WHOI): 15°03.1'S x 75°30.3'W

G.C. = Gulf of California, SIO = Scripps Institute of Oceanography,
WHOI = Woods Hole Oceanographic Institution. CSM = Colorado School of Mines.

ship of pheophytins/pheophorbides/chlorin acids, given in Table II,
are found to be 18/47/34 or 39/37/24 for sediments deposited
through oxic or anoxic bottom waters, respectively. The overall
increase of pheophorbides plus chlorin acids, relative to water
column detritus or anoxically deposited sediments, is taken as a
reflection of the prolonged effects of aerobic heterotrophy during
oxic deposition. That is, benthic fauna, as well as zooplankton,
appear able to remove phytol from sedimentary tetrapyrroles.

Overall, these marine sediments appear to contain about 30%
pheophytins, 40% pheophorbides and 30% purpurin plus chlorin acids
at the onset of diagenesis. This, of course, is only a rough
'rule-of-thumb.'

Photic Zone Anoxia. In environments such as meromictic lakes and
peat bogs the oxy- and chemo-clines impinge the photic zone. In
these cases, anoxigenic photosynthetic bacteria (e.g. Thiorhodacea,
Athiorhodacea) contribute not only to overall productivity (5,37,38
but also to the supply of tetrapyrrole pigments in resultant sediments. That bacteriochlorophyll-a derivatives are present in such
sediments, an intuitive conclusion, is shown through the analyses
of Mangrove Lake (Bermuda) sapropel and sediments from Big Soda
Lake (Nevada).

Bacteriopheophytin-a has been isolated from both sets of
samples, as well as from a South Florida freshwater peat accumulation (43). Figure 5 is the electronic absorption spectrum of
bacteriopheophytin-a, and its 'dioxy-dideoxo' derivative obtained
by borohydride reduction. The unique chromophores and resultant
spectra of the bacteriochlorophyll-a derivative makes initial
estimation of the relative amounts of higher plant and bacterial
chlorophylls quite easy. As an example, Figure 5b is the spectrum
of a crude extract of sapropel from Mangrove Lake. In this case,
the ratio of chlorophyll-a to bacteriochlorophyll-a derivatives
as the pheophytins, is approximately 1.3:1. Large amounts of
bacterial carotenoids are also in evidence from their characteristic absorption maxima in the blue (ca.420-500nm) region.

The main point of this simple study is that, in cases such as
peat/coal and aquatic sapropel/'paper shale' accumulations, ultimate geoporphyrin (i.e. DPEP-series) precursors beside chlorophyll-a
need to be considered. In order to mold bacteriochlorophyll-a into
current modified 'Treib's scheme' diagenesis (3-5) only the removal
of the 3,4-dihydro and 2-acetyl features are required. That is, an
aromatization and reduction/dehydration.

Tetrapyrrole Diagenesis; Increasing Sediment Depth, Age and Temperature. Organic-rich diatomaceous oozes recovered within the oxygen
minimum (OMZ) of the northeast slope of the Guaymas Basin (Gulf of
California, DSDP/IPOD Leg 64-Site 479. 23) have been found to
contain a wide range of chlorophyll derivatives which, with depth,
cover the entire period of early diagenesis (20).

Pigments as a Portion of Total Organic Carbon. The value obtained
through the division of tetrapyrrole pigment concentration, in μg-
sediment dry weight, by the percent organic carbon of the host
sediment we have defined as the 'Pigment Yield Index,' or PYI (44).

The plot of PYI versus sub-bottom depth for Site 479, including a box-core sample as a surface sediment reference, is shown in Figure 6. The actual trend, dotted, reveals a series of alternating highs and lows down to about 200 m, sub-bottom. This most likely represents corresponding increases and decreases in productivity and the severity of bottom water anoxia (i.e. paleoenvironment). However, by smoothing the curve we obtain what we suggest is the trend one might find for a hypothetical depositional environment which is constant through time.

Rapid pigment loss occurs within the first 50 meters of burial, after which the rate of loss slows and eventually ceases (Figure 6). During the rapid loss period of early diagenesis it appears that defunctionalization and destructive reactions are competitive as purpurin and chlorin acids, in addition to phorbides, are present. Later and deeper, defunctionalization of the surviving phorbides continues until the pigments attain a higher degree of stability through aromatization, yielding porphyrins of the DPEP-series (cf. Figure 2). Intuitively, incorporation of phorbide and chlorin acids into proto-kerogen may occur during the earlier periods of pigment loss from the bitumen fraction, though little data beyond phenomenological insight (4-5,20,29,34-35) exists to support this supposition.

Individual Tetrapyrrole Characteristic of Early Diagenesis. It should be noted that the 'identifications' of individual pigments given below are only tentative. That is, characterization is based only upon chromatographic mobility, electronic spectroscopy, derivatizations with sodium borohydride and copper and comparisons to authentic pigments treated in like manner. As such, structural-proof is not claimed and resultant geochemical reaction scenarious are given only for the chromophore of each pigment type. That is, this study investigates only diagenetic defunctionalization and offers no insight as to alkyl substituents, beyond inference. Pheophytin-a as an initial isolate, exhibited the electronic spectrum given as Figure 7. Reaction with borohydride was found to yield the typical 9-oxy-deoxo derivative as well (Figure 6, dashed line). However, upon submitting pheophytin-a to LPHPLC, 2 forms of this pigment, neither identical to the authentic compound, were found. That is, as shown in Figure 8, 'pheophytin-a,' from these sediments were found to exist on a more polar and less polar form, relative to the co-injected standard, and both yielded identical electronic spectra (cf. Figure 7). Chromatographic mobility, including comparison to authentic pheophytin-a, and its 'allomer' (i.e. 10-oxy-pheophytin-a) shows that the more mobile form ('PP-a1') may be pyro-pheophytin-a while the more polar form ('PP-a2') is most likely the allomer (i.e. 10-oxy-pheophytin-a 20). In some cases, such as sample 64-481-8-2 (Figure 8), small amounts of "true" pheophytin-a are present.

Meso-Pheophorbides. A compound, or series, exhibiting the electronic spectrum (alt. chromophore) of meso-pyropheophorbide-a has been isolated from Site 479 sediments recovered between 72 and 246 meters (15-31°C 23) sub-bottom (20). Figure 9a is the electronic spectra of the native pigment and its borohydride reduction product. Reaction of each with copper (II) yielded the metallo-

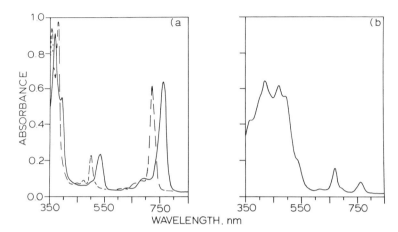

Figure 5. Electronic absorption spectra. (a) bacteriopheo-
phytin-a(solid) and the dioxy-dideoxo derivative (dashed)
obtained following reduction with sodium borohydride; and
(b) the crude extract of Mangrove Lake (Bermuda) sapropel.

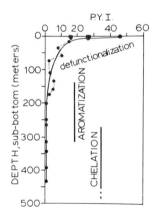

Figure 6. Plot of the 'pigment yield index' (PYI) versus
depth of burial for DSDP/IPOD Site 479 in the northeastern
Guaymas basin slope, Gulf of California (cf. 20).

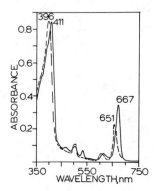

Figure 7. Electronic absorption spectra of pheophytin-a(solid) and the 9-oxy=deoxo derivative obtained following reduction with borohydride, (dashed) Isolated from DSDP sample #64-479-5-3 (<u>20</u>).

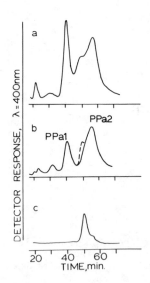

Figure 8. Partial LPHPLC chromatograms of "pheophytins-a." (a) Sample 64-481-8-2, (b) sample 64-479-3-2 (dashed = co-injected authentic pheophytin-a), and (c) partially air-'allomerized' authentic pheophytin-a.

meso-phorbide spectra given as Figure 9b. All of these spectra and
reaction products are consistent with the chromophore of authentic
meso-pyropheophorbide-a. The very non-polar nature of the native
pigment leads us to propose that this is 7-R-7-despropio-meso-
pyropheophorbide-a (R=H,-CH$_3$,-CH$_2$CH$_3$) generated by the
reduction of the 2-vinyl-moiety and the decarboxylation of the
corresponding phorbide acid.

Oxydeoxomesopyropheophorbides. Pigments with the chromophore of
the 9-oxy-deoxomesopheophorbides were isolated from the same strata
as given above for the meso-pyropheophorbides. In these cases, the
Soret(S) and band 1 (I) absorption maxima resemble authentic 9-oxy-
deoxo-mesopyropheophorbide-a (S=396.5,I=646.5nm), its Cu chelate
(S=402.0, I=614.5nm), and the pigment was totally unreactive to
borohydride.

   This compound, tentatively 9-oxy-deoxo-mesopyropheophorbide-a,
is of interest as it offers a clue as to the mode of loss of the
9-ketone moiety inherited from chlorophyll-a. That is, it now
appears that the 9-keto function is reduced to the alcohol which may
then be dehydrated. If this were to occur, the product would contain a highly strained 6,γ-cycloetheno moiety. Most likely,
concerted or subsequent and rapid reduction would follow and result
in the common 6,γ-cycloethano type of isocyclic ring found in geoporphyrins of the DPEP-series.

Oxydeoxo-Vinyl-Phorbides. Pigments exhibiting band I absorption at
651nm, and matching other physiochemical characteristics of authentic 9-oxydeoxo-pyropheophorbide-a, have been reported as being
present in the Gulf of California sediment suite studied herein
(20). Lack of sufficient material has precluded further study on
these isolates.

Deoxomesopyropheophorbide-a and 7,8-Dihydro-DPEP. Sedimentary
bitumen at or near a level of maturity such that free-base
porphyrins are present or dominant also yield tetrapyrroles with
electronic spectra characteristic of deoxomesopyropheophorbide-a
(DOMPP-a). DOMPP-a is isolated, in the greatest amounts, from the
non-polar chromatographic fractions, has an HCL# of ca. 5-7 and,
thus, are decarboxylated (7-R-7-despropio:R=-H,-CH$_3$,-CH$_2$CH$_3$)
analogs. As such, an alternate semi-systematic name would be
7,8-dihydro DPEP.

   Mass spectra of one such isolate (Figure 2) reveals that
diagenetic dealkylation has occurred. That is, at temperatures
below that required for the tetrapyrrole aromatization reaction
(e.g. 20-30°C. 4-5),thermal (viz. catagenetic) dealkylation is
hardly likely. Thus, the presence of carbon numbers below C-32,
the expected product of 'Treibs' Scheme' geochemistry (1,6-7),
most likely reveals methylene-equivalency losses related to
defunctionalization during early diagenesis. The intervention of
sedimentary bacterial infauna is theoretically possible and is
being studied.

   The electronic spectra of the native pigment (S=393,1=640nm),
isolated from a Miocene shale of the Sisquoc formation (California. 24), and numerous diatomaceous oozes from the Gulf of California (20), and its copper chelate (S=395, I=606nm) match
authentic DOMPP-a and CuDOMPP-a, respectively.

Generation of Chlorin and Purpurin Acids; Fating Functions. Sediments of known oxic deposition and/or redeposition have been found
to be relatively enriched in the highly polar purpurin and chlorin
poly-acids (5,20,28,44).
The oxidative scission of the isocyclic ring on the nucleus of
chlorophyll and its derivatives, via allomerization, consists of a
series of well documented reactions (6-7) and leads to such
products as chlorin (e.g. $-e_4, -e_6, -P_6$) and purpurin (e.g.
-7,-9,-18) carboxylic acids (cf. Figure 1).
Sediments from the north rift of the Guaymas Basin (DSDP/IPOD
Leg 64, Site 481.20) yielded 2 pigments of this series. Given as
Figure 10 is the electronic spectra of purpurin-18 and chlorin-
P_6 isolates from this site. Spectra, chromatographic behavior and
reactivity towards borohydride match that of authentic pigments.
As it is known that allomerization is possible only with the
phorbides which contain the 10-carbomethoxy group (45), we have
proposed that competition between the formation of allomers (10-oxy-
phorbides) and the pyro-(10-H-10-decarbomethoxy) phorbides is a
major branch point during very early diagenesis (4-5,20). That
is, the generation of allomerized pigments appears to lead to
subsequent pathways which end in pigment destruction. Mechanisms
of tetrapyrrole destruction in sediments, beyond the purpurin-
chlorin stages, are totally unknown at present. Conversely,
generation of the pyro-phorbides, essentially protecting the
isocyclic ring from oxidation (viz. allomerization), imparts a
first level of stability to these pigments. Subsequent diagenetic
defunctionalization of the pyro-phorbide and aromatization then
leads to the DPEP series, as given in previous sections.
Possible Oxidation of the 2-Vinyl Moiety; 2-Acetyl (or Formyl)
-2-Desvinyl-Pheophorbides. A first indication that, aside from
reduction, the vinyl moiety of chlorophyll derivatives might be
altered stems from the isolation of a pigment we first called
"chlorin 686," after the spectral type and position of band I
absorption. The electronic spectra of the native and borohydride
reduced pigments are given as Figure 11. The hypsochromic shift
of 35.5nm (i.e. 686.5-651.0 nm) in the position of band I absorption is on the same order as that found for the removal of the
auxochrome effects of 2 conjugated carbonyl functions from authentic test pigments (e.g. bacteriopheophytin-a, purpurin-18). These
changes do not however mimic the removal of dicarbonyl conjugation
from such pigments as pheophorbide-b.
Thus, 2 conjugated carbonyl moieties appear to be present and
are most likely located on rings I and V. That is, on "opposite"
side of the macrocycle. The bifurcated nature of the Soret absorption for the native pigment is reminiscent of the 2-acetyl-2-
desvinyl (46) and 2-formyl-2-desvinyl (47) derivatives of pheophytin-a. Should phorbide "686" prove to contain either the 2-acetyl
or 2-formyl moieties this could easily and mistakenly we feel, infer
derivation from bacteriochlorophyll-a or chlorophyll-d, respectively. As this pigments(s) was isolated from deep-sea diatomaceous
oozes, (Z=747m, DSOP/IPOD Site 479. 20,23) significant input of
either purple photosynthetic bacteria or red algae, respectively,
is unlikely. Rather, we feel, that early diagenetic oxidation of

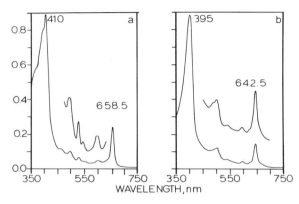

Figure 9. Electronic absorption spectra of (a) decarboxylated mesopyropheophorbide(s)-a and (b) the oxy-deoxo-derivative obtained via borohydride reduction.   Sample 64-479-13-1 (20).

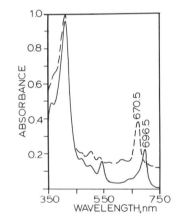

Figure 10. Electronic absorption spectra of purpurin-18 (solid) and chlorin-p6 (dashed)-like compounds isolated from sample 64-481A-2-2 (20).

the 2-vinyl moiety, with potential cleavage to a 2-methyl-2-desvinyl
structure akin to 'Abelsonite' (N⌡-2-methyl-2-desvinyl-DPEP. 16)
formation, is a more likely diagenetic pathway.
Conclusion; Modified "Treibs' Scheme" Diagenesis of Chlorophyll.
The investigations reported here allow a first approximation of the
diagenesis of chlorophyll to be formulated. The scheme presented as
Figure 12 summarizes the proposed reaction sequences, and are, for
now, restricted to sedimentary environment.
     The losses of magnesium and phytol through the combined ef-
fects of cellular senescence and predation (i.e. aerobic heterotrophy) in the water column lead to pheophytin-a and pheophorbide-a
becoming the primary chlorophyll-a derivatives deposited in marine
sedimentary environments. Though it is not known at present, the
heterotrophic processes which cleave phytol more than likely also
affect the 10-carbomethoxy group. Studies are underway to investigate the amounts of pyro-pheophorbides in water column detritus
and surface sediments.
     During very early diagenesis and, to some extent, in the water
column, 2 key competing or 'fating' reactions appear to direct
subsequent alteration, i.e. allomerization versus loss of the 10-
carbomethoxy moiety.
     In the presence of oxygen, allomerization leads to the oxidat-
ive scission of the isocyclic ring. Products found in oxidative
environments include the purpurin and chlorin acids (Figure 11).
Tracing the down-hole fate of these compounds has proved futile, to
date. That is, in cases were oxidation dominates, such as in
sediments low in metabolizable organics (i.e. Philippine Sea.
48-50), tetrapyrrole pigments disappear with depth. Thus, we can
but surmise that the chlorin acids are destroyed and yield un-
identifiable low molecular weight colorless compounds (LMWCC),
Figure 12). It is theoretically possible to derive various ETIO-
series porphyrins from chlorin acids (5-7). However, no data yet
exists to support this alternate pathway.
     The loss of the 10-carbomethoxy moiety from the pheophorbides
generates the pyropheophorbides and prevents allomerization (45).
Based on the down-hole sequencing of pigments at DSDP/IPOD site
479 (20), and past partial sequences (4-5,28-29,44), the generation
of free-base geoporphyrins, at least of the DPEP-series, appears to
follow that given in Figure 12. It should be noted that tentative
identifications have been made only for the decarboxylated species.
However, an analogous series of these pigment types does appear to
be present as the free-acid (viz.7-propionic) analogs. Reduction of
the 2-vinyl and 9-keto moieties has been found to occur in the same
strata and to be unconcerted. Early diagenesis thus yields meso-and
9-oxydeoxo-pyropheophorbides, with or without the carboxylic acid
moeity inherited from chlorophyll. Reduction of either the 9-keto
or 2-vinyl of these derivatives, respectively, followed by dehy-
dration and, ostensibly, rapid reduction yields the corresponding
deoxomesopyropheorbide-a (DOMPPa).
     Following the generation of DOMPP-a, again as either the
carboxylated or decarboxylated forms, increasing thermal stress
leads to aromatization and yields DPE or DPEP, respectively.

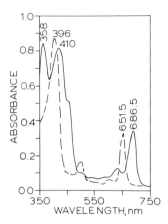

Figure 11. Electronic absorption spectra of a probable dioxo-
phorbide before ('native,' solid trace) and after (dashed trace)
reduction with borohydride. Isolated from DSDP sample 64-479-5-3
(20).

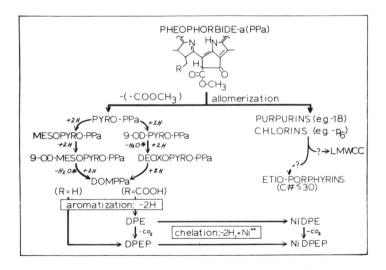

Figure 12. Proposed diagenesis of chlorophyll-a derivatives.
Code: MESO = 2-vinyl reduced to ethyl; PYRO = lacks C# 10 carbomethoxy group; OD = 9-oxydeoxo; asterisk = possible 6,γ-cyclo-
etheno intermediate; LMWCC = low molecular weight colorless compounds (cf. Figure 1).

In conclusion, the early diagenesis of chlorophyll-a derivatives and the generation of the DPEP-series porphyrins roughly parallels the predictions which Professor Treibs made a half a century ago (1). Though certain inroads have been made, much in the way of structural and mechanistic study remains. That is, the generation of dealkylated species and the highly likely incorporation of tetrapyrroles into kerogen during early diagenesis are all but unknown.

Acknowledgments

The authors' are funded by NSF grant #OCE-82-08-107, the support of which is appreciated. The following people and institutions are heartily thanked for samples (see "Experimental" and Table II): The DSDP/IPOD program; B.R.T. Simoneit (U.C.L.A., SIO-Samples); B. Parker and D. Waples (C.S.M.); R. Oremland (U.S.G.S.), P. Hatcher (U.S.G.S.), and R. B. Gagosian and J. F. Farrington (W.H.O.I.). Sampling of The Peruvian upwelling system was funded by NSF grants OCE-77-26084, OCE-079-25352 and OCE-80-18436 to W.H.O.I. (R.B.G.,J.W.F.). Mobil Research and Development Corporation (W. L. Orr) is thanked for shale and petroleum samples and a grant-in-aid. Drs. G. Eglinton and G. Wolff are thanked for their critical reading of the original manuscript. The patience of Ms. Cora Woodman during the revision of this text is appreciated.

Literature Cited

1. Treibs, A. Angew. Chem. 1936, 49, 682-6.
2. Baker, E. W. In "Organic Geochemistry;" Eglinton, A. and Murphy M. T. J., Eds.; Springer-Verlag; Berlin, 1969; pp. 464-97.
3. Baker, E. W.; Palmer, S. E. In "The Porphyrins Dolphin, D., Ed.; Academic Press: New York, 1978; Vol. I, pp: 486-552.
4. Baker, E. W.; Louda, J. W. In "Advances in Organic Geochemistry - 1981," Bjoroy, M. et al ., Eds.; John-Wiley: Chichester, 1983: pp. 401-21.
5. Baker, E. W.; Louda, J. W. In "Biological Markers," Johns, R. B., Pergamon: Oxford, in press.
6. "The Porphyrins," Dolphin, D., Ed.; Academic Press: New York, 1978-1979; Vols. I-VII.
7. "Porphyrins and Metalloporphyrins," Smith, K.M., Ed.; Elsevier: Amsterdam,1975; 910 pp.
8. Lemberg, R.; Barrett, J. "Cytochromes;" Academic Press: London, 580 pp.
9. Quirke, J. M. E.; Eglinton, G.; Maxwell, J. R. J. Am. Chem. Soc., 1979, 101, pp. 7693-7.
10. Quirke, J. M. E.; Shaw, G. J.; Soper, P. D.; Maxwell, J. R. Tetrahedron, 1980, 3G, pp. 3261-7.
11. Quirke, J. M. E.; Maxwell, J. R. Tetrahedron, 1980, 36, pp. 3453-6.
12. Quirke, J. M. E., Maxwell, J. R.; Eglinton, G; Sanders, J. K. M. Tetra. Letts., 1980, 21, pp. 2987-90.

13. Quirke, J. M. E.; Eglinton, G.; Palmer, S. E. Baker, E. W. Chem. Geol., 1982, 35, pp. 69-85.

14. Wolff, G. A.; Murray, M.; Maxwell, J. R.; Hunter, B. K.; Sanders, J. K. M. J. C. S. Chem. Commun., 1983, pp. 922-4.

15. Fookes, C. J. R. J. C. S. Chem. Commun., 1983, pp. 1474-6.

16. Storm, C. B.; Krane, J.; Skjetne, T.; Telnaes, N.; Brantharer, J. F.; Baker, E. W. Science, 1984, 223, pp. 1075-6.

17. Chicarelli, M. I.; Maxwell, J. R. Tetrahedron Letts., 1984, 25, pp. 4701-4.

18. Gagosian, R. B.; Loder, T.; Nigrell: G.; Mlodzinska, Z.; Love, J.; Kogelschatz, J. W. H. O. I. Tech. Rep., 1980, WHOI-80-1, 77 pp.

19. Henrichs, S. M.; Farrington, J. W. Limol. Oceanogr., 1984, 29, pp. 1-19.

20. Baker, E. W.; Louda, J. W. In "Init. Reps. D.S.D.P.," Curray, J. R.; Moore, D. G.; et al., Eds.; U.S. Govt. Printing Office: Washington, D.C., 1982, Vol. 64 - Part II, pp. 789-814.

21. Cloern, J. E.; Cole, B. E.; Oremland, R. S. Limnol. Oceanogr., 1983, 28, pp. 1049-1061.

22. Hatcher, P. G.; Simoneit, B. R. T.; MacKenzie, F. T.; Neuman, A. C.; Thortenson, D. C.; Gerchakov, S. M. Org. Geochem., 1982, 4, pp. 93-112.

23. Curray, J. R.; Moore, D. G.; et al. "Init. Reps. D.S.D.P. LXIV," U.S. Govt. Printing Office: Washington, D.C., Vol. G4-Part I, 507 pp.

24. Baker, E. W.; Louda, J. W.; Orr, W. L. unpublished data.

25. Petracek, F. J.; Zechmeister, L. Analyt. Chem., 1956, 28, pp. 1484-5. s.; Springer-Verlag; Berlin, 1969; pp. 464-97.

26. Krinsky, N.I. Analyt. Biochem., 1983, 6, 283-302. phin,

27. Holt, A. S. Plant Physiol., 1959, 34, pp. 310-314.

28. Baker, E. W.; Louda, J. W. In "Init. Reps. D.S.D.P.LVI-LVII," Langseth, M.; Okado, H.; et al., Eds.; U.S. Govt. Printing Office: Washington, D.C., Vol. 56/57 - Part 2, pp. 1397-1408.

29. Louda, J. W.; Baker, E. W. In "Init. Reps. D.S.D.P. -LXIII," Yeats, R. S.; Haq, B. V.; et al., Eds.; U.S. Govt. Printing Office: Washington, D.C., Vol. 63, pp. 785-818. ss: New York,

30. Baker, E. W. J. Am. Chem. Soc. 1966, 88, 2311-5.

31. Baker, E. W.; Yen, T. F.; Dickie, J. P.; Rhodes, R. E.; Clarke, L. E. J. Am. Chem. Soc. 1967, 89, 3631-9.

32. Didyk, B. M.; Alturk:, Y. I. A.; Pillinger, C. T.; Eglinton, G. Nature 1975, 256, 563-5.

33. Yen, T. F.; Boucher, L. J.; Dickie, J. P.; Tynam, E. C.;em. Vaughn, G. B. J. Inst. Petrol., 1969, 55, pp. 87-99.

34. Baker, E. W.; Palmer, S. E.; Hwang, W. Y. In "Init. Reps.R. D.S.D.P. - XLI," Lancelot, Y.; Seibold, E.; et al., Eds.; U.S. Govt. Printing Office: Washington, D.C., Vol. 41,, pp. 825-37.

35. MacKenzie, A. S.; Ouirke, J. M. E.; Maxwell, J. R. In "Advances in Organic Geochemistry - 1979," Douglas, A. G.; Maxwell, J. R., Eds.; Pergamon: Oxford, 1980, pp. 239-48.

36. Holden, M. In "Chemistry and Biochemistry of Plant Pigments, 2nd. Edit.," Goodwin, T. W., Ed.; Academic Press: London, Vol. 2, 1976, pp. 2-37.

37. Jeffrey, S. W. In "Primary Productivity in the Sea," Falkowski,
     P. G., Ed.; Plenum: New York, 1980, pp. 33-58.
38. Steele, J. H. "The Structure of Marine Ecosystems;"
     Harvard Univ. Press: Cambridge (U.S.A.), 1974, 128 pp.
39. Gagosian, R. B. personal communication.
40. Currie, R. I. Nature, 1962, 193, pp. 956-7.
41. Aronoff, S. Adv. Food Res. 1953, 4, pp. 133-84.
42. Orr, W. L.; Emery, K. O.; Grady, J. R. Bull. Am. Assoc. Petrol.
     Geol., 1958, 42, pp. 925-958.
43. Palmer, S. E.; Charney, L. S.; Baker, E. W.; Louda, J. W.
     Geochim. Cosmochim. Acta, 1982, 46, pp. 1233-41.
44. Louda, J. W.; Palmer, S. E.; Baker, E. W. In "Init. Reps.
     D.S.D.P.-LVI-LVII," Langseth, M.; Okado, H.; et al., Eds.;
     U.S. Govt. Printing Office: Washington, D.C., Vol. 56/57 -
45. Pennington, F. C.; Strain, H. H.; Svec, W. A.; Katz, J. J.
     J. Am. Chem. Soc.., 1967, 89, pp. 3875.80.
46. Smith, J. R. L.; Calvin, M. J. Am. Chem. Soc., 1966, 88,
     pp. 4500-6.
47. Holt, A. S.; Morley, H. V. Can. J. Chem., 1959, 37,
     pp. 507-14.
48. Baker, E. W.; Louda, J. W. In "Init. Reps. D.S.D.P.-LVIII,"
     DeVries-Klein, G.; Kobayashi, K.; et al., U.S. Govt.
     Printing Office: Washington, D. C., Vol. 58, 1980, pp. 737-9.
49. Pennington, F. C.; Strain, H. H.; Svec, W. A.; Katz, J. J.
     J. Am. Chem. Soc.., 1967, 89, pp. 3875.80. D.S.D.P.LVI-LVII,"
50. Smith, J. R. L.; Calvin, M. J. Am. Chem. Soc., 1966, 88,ng
     pp. 4500-6.
51. Holt, A. S.; Morley, H. V. Can. J. Chem., 1959, 37,   III,"
     pp. 507-14. ; Haq, B. V.; et al., Eds.; U.S. Govt. Printing
52. Storm, C. B.; Krane, J.; Skjetne, T.; Telnaes, N.;
     Brantharer, J. F.; Baker, E. W. Science, 1984, 223,
     pp. 1075-6. ; Yen, T. F.; Dickie, J. P.; Rhodes, R. E.;
53. Baker, E. W.; Louda, J. W. In "Init. Reps. D.S.D.P.-LVIII,"
     deVries-Klein, G.; Kobayashi, K.; et al., Eds.; U.S. Govt.n,
     Printing Office: Washington, D.C., Vol. 58, 1980, pp. 737-9.
54. Baker, E. W.; Louda, J. W. In "Init. Reps. D.S.D.P.-LX,"
     Hussong, D.; Uyeda, S.; et al., Eds.; U.S. Govt. Printing
     Office: Washington, D.C., Vol. 60, 1981, pp. 497-500. ps.
55. Baker, E. W.; Louda, J. W. In "Init. Reps. D.S.D.P.-LXI,"
     Larson, R. L.; Schlanger, S.; et al, Eds.; U.S. Govt.
     Printing Office: Washington, D.C., Vol. 61, 1980, pp. 619-20.

RECEIVED January 15, 1986

HUMIC SUBSTANCES

8

Structural Analysis of Aquatic Humic Substances by NMR Spectroscopy

Andrew H. Gillam[1] and Michael A. Wilson[2]

[1] Institute of Offshore Engineering, Heriot–Watt University, Research Park, Riccarton, Edinburgh, EH14 4AS, Scotland
[2] CSIRO Division of Fossil Fuels, P.O. Box 136, North Ryde, NSW 2113, Australia

Recently developed techniques in NMR spectroscopy have been applied to the analysis of marine and estuarine organic material. For example, the advantages of dipolar dephasing NMR spectroscopy in elucidating the types of aliphatic and aromatic structural groups in these materials are demonstrated. It is shown that methoxy and amino acid carbon of similar chemical shift can be distinguished. Problems in quantifying the different functional groups in marine and estuarine organic material by NMR are discussed and specific examples are given in which nonquantitative data might be expected.

It was not so long ago that water chemists were content to measure the amount of dissolved organic carbon in fresh, estuarine or marine waters. At best, all that was known of chemical structure was the concentration of trace, albeit important organic compounds. All that has now changed and the chemical structure of all the dissolved organic substances (humic substances) is being investigated at a detailed level.

Although conventional functional group analysis and Fourier transform infra-red spectroscopy are providing useful information, new revelations concerning the chemical structure of these ubiquitous materials have been largely due to developments in 1H- and 13C-NMR spectroscopy. Although some questions need to be answered concerning quantitation, major advances have been made in determining the aromaticity (fraction of carbon which is aromatic) and carbohydrate content of these substances by NMR.

Recently, 'second generation' (dipolar dephasing, two dimensional NMR) 13C-NMR experiments have begun to appear in the coal science literature which are particularly applicable to humic substances. In this paper, some applications of these techniques to the study of the structure of organic materials from aquatic systems are demonstrated. These studies concentrate on: 1) obtaining additional

0097–6156/86/0305–0128$06.00/0

information on structure, particularly functional groups and 2)
establishing limits on quantitation.

Experimental

The location and sampling methods of samples used in this work have
been described elsewhere (1-5).
1H- and 13C-NMR solution spectra were determined on a Jeol FX90Q
spectrometer.   The sample (~20 mg) was added to 0.5 cm³ of
deuterium oxide.   A few drops of 1 mol dm⁻³ sodium deuteroxide were
added until the sample dissolved.
1H-spectra were obtained at 89.99 MHz under homogated decoupling
conditions in which the HOD peak produced by adventitious water
impurities and proton exchange was irradiated.   The radiofrequency
(RF) level of the HOD irradiation was optimized for each sample,
i.e. a sufficient RF level to reduce the HOD peak to zero was
applied but not enough to distort the intensity of the humic material
in the vicinity of the HOD peak.   Spectra were determined using a
45⁰ pulse and 8K data were acquired using 1500 Hz spectral width.
Acquisition time was 2.727 s, with a pulse delay of 2.0 s.   Approximately 1000 scans were collected for adequate signal to noise ratio.
Chemical shifts are quoted with respect to internal tetramethylsilane
(TMS) but were measured with respect to external TMS.   A capillary
inserted into the sample tube was used as a reference.   The values
so obtained were corrected by measuring the chemical shifts of
n-butanoic acid with respect to both internal and external TMS.
Solution 13C-NMR spectra were determined at 22.5 MHz.   8K data
were collected using a 7000 Hz spectral width.   Acquisition time was
0.584 s.   Pulse delays of 1.0 to 5.0 s were used with, and without,
inverse gated decoupling.   Up to 60,000 scans were collected.
13C-cross polarization magic angle spinning (CP-MAS) spectra
were obtained on a Bruker CXP-100 instrument.   A rotor consisting of
a barrel of boron nitride and a base of Kel-F was used.   Rotor speed
was ~3.8 kHz.   Recycle time was varied from 0.3 to 1 s.   A variety
of contact times from 0.5 to 3ms were employed.   The Hartmann-Hahn
condition was set using a sample of hexamethylbenzene.   Chemical
shifts were measured with respect to external hexamethylbenzene (by
storing the hexamethylbenzene spectrum in another computer memory
block) but are quoted with respect to TMS.   It is assumed that the
chemical shifts of hexamethylbenzene with respect to TMS are the same
in solution as in the solid state.
Dipolar dephasing experiments were performed by inserting a
delay before data acquisition in which the decoupler was gated off.
A 180⁰ refocussing pulse along the spin locked coordinate was
inserted midway in the delay period.   Full details are given elsewhere (6).
In general, data was worked up using line broadening factors of
50 Hz or greater for 13C or 1Hz or less for 1H.   This is
particularly important for solution 13C-spectra where signal to noise
is poor.

Solution versus solid state NMR

In principle, solution or solid state NMR can be used to study the

structure of dissolved organic carbon.   Both methods have advantages
and disadvantages.   The main advantages of solution NMR are:

1. some degree of extra resolution might be expected because of
    conformational freedom of the organic macromolecules;
2. quantitative aspects are more fully understood than for solid
    state techniques.

Advantages of the solid state techniques are:

1. increased sensitivity due to signal enhancement techniques, rapid
    relaxation, and the fact that a higher concentration of nuclei
    are between the poles of the magnet;
2. no ambiguity concerning degradation by the solvents used for
    solution NMR.

1H-NMR

To obtain a solid state NMR spectrum it is necessary to remove interactions from nuclei closely bonded to the nucleus under investigation.   For 13C-nuclei this is done through a combination of magic
angle spinning and high power proton decoupling techniques.   In the
latter case, the protons are irradiated close to their resonant
frequency.   To obtain a high resolution 1H-spectrum of a solid,
high power decoupling techniques cannot be employed.   Moreover,
using magic angle spinning alone, the required spinning speeds are
unobtainable.   Thus, spectra obtained with current technology are
unsatisfactory in that they do not give high resolution information.
Although multiple pulse techniques may solve this problem (7), at
present it is necessary to obtain 1H-NMR spectra of estuarine and
marine organic matter by solution techniques.   This presents other
difficulties mainly due to the high concentration of adventitious
water.   For example, even when deuterated solvents are used, only a
broad featureless spectrum of water is obtained unless additional
techniques are employed.

One method that has been successful is to continuously gate and
irradiate the water peak so that these molecules cannot relax and
hence, give an NMR signal (Figure 1).   Figure 1 also shows spectra
of marine, sea loch and terrestrial humic substances.   The main
application of 1H-NMR is to estimate the proton aromaticity of the
samples by integrating the 6-8 ppm region against the rest of the
spectrum, and secondly, to measure the amount of branching in the
aliphatic structures.   The peak at < 0.9 ppm arises from methyl
groups at the end of alkyl chains and can be integrated against the
other aliphatic protons.   It should be appreciated that protons
which exchange with the solvent are not observed.

13C-NMR

Typical 13C-spectra for humic acids from a range of sources are
shown in Figures 2 and 3.   Five major regions of resonance can be
recognised.   These are 0-50 ppm (alkyl carbon), 50-108 ppm (carbohydrate, alcohol, ether and the COOH α-carbon of amino acid), 108-
160 ppm (aromatic carbon), 160-200 ppm (carboxylic, ester, amide
carbon) and 200-220 ppm (ketonic or aldehydic carbon).

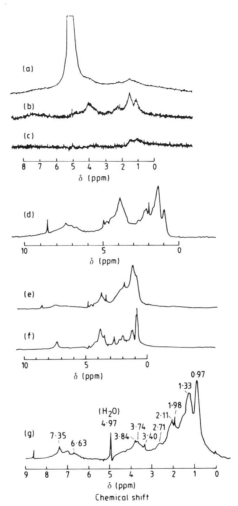

Figure 1. ^{1}H-NMR spectrum of humic acid (242 FIDs): (a) without irradiation of water peak; (b) with optimum irradiation of water peak; (c) with over irradiation of water peak; (d) terrestrial humic acid (Levin) (refn 4); (e) dissolved marine humic acid (refn 1); (f) intracellular algal material (refn 1); (g) sea loch sedimentary material. (Reproduced with permission from reference 4. Copyright 1983 Blackwell Scientific Publications.)

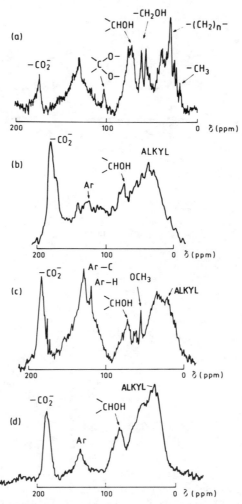

Figure 2. Typical ^{13}C-NMR spectra of humic substances from various sources. (a) terrestrial (Levin)(solution) (refn 5); (b) terrestrial (Patua) (solution); (c) freshwater aquatic (solution) (refn 2); (d) dissolved marine (solid state) (refn 1)

Recent developments in high resolution 13C-solid state NMR spectroscopy have demonstrated that additional information can be obtained if the proton decoupler is turned off for a short period between irradiating the nuclei in the sample under investigation and observing their subsequent behaviour (8). If the correct period of time is left, (dipolar dephasing period), spectra can be obtained from only methyl and non-protonated carbon. However, during the dipolar dephasing period the signal intensities of these resonances are also attenuated, albeit at different rates from tertiary and secondary carbons. Hence for quantitative analysis of structural groups in complex materials such as dissolved organic matter in marine and estuarine waters, knowledge of the decay constants for different structural groups is essential.

To this end, the dipolar dephasing behaviour of a wide range of organic compounds has been established (9-12). Various types of carbons experience a wide range of 13C-1H interactions. Carbons weakly coupled to protons, e.g. quarternary carbons follow a single exponential law given by

$$I_B = I_B^0 \exp\left(-t_1 / T_2'\right)$$

(1)

where I_B^0 is the signal intensity at zero time and T_2' is the exponential decay constant for the signal intensity. When the carbons are strongly coupled to protons, e.g. methine and methylene carbons, the signal decay is modulated by the strong 13C-1H coupling and the overall decay of the signal in the short time limit is better described by the equation

$$I_A = I_A^0 \exp\left(-t_1^2 / 2T_2'^2\right)$$

(2)

Values of T_2' for various carbon types are listed in Table I. In principle, it should be possible to identify a number of structural groups in aquatic organics by their decay constants.

Nevertheless, for a complex material such as the organic matter in marine and estuarine waters, each resonance is broad and arises from a number of structural groups. For example, in the aromatic region the broad signal is due to CH and non-protonated carbon. In this case signal decay will be given by the sum of equations (1) and (2) as (3).

$$I = I_A + I_B = I_A^0 \exp\left(-t_1^2 / 2T_{2A}'^2\right) + I_B^0 \exp\left(-t_1 / T_{2B}'\right)$$

(3)

However, this equation can be computer fitted to give time constants T_{2A}', T_{2B}' and I_B^0 and I_A^0. For the aromatic resonance, $I_B^0/(I_A^0 + I_B^0)$ is a measure of the fraction of aromatic carbon which is non-protonated and the time constants tell us which structural groups are present.

Typical dipolar dephased spectra of a sedimentary sea loch humic acid are shown in Figure 4 as a function of dephasing time. Comparison of Figure 4 with Figure 3 clearly shows substantially more alkyl and O-alkyl carbon is protonated than aromatic carbon since the signal from these two types of carbon are attenuated at long (40 μs)

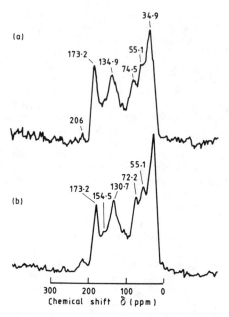

Figure 3. ^{13}C-NMR (solid state) spectra of sea loch sedimentary humic substances. (a) fulvic acid; (b) humic acid.

Table I.   Typical T_2' Values for Various Carbon Types

Carbon Type	T_2' (μs)
CH$_2$ (acyclic)	12–29
(alicyclic)	15–32
CH	15–32
CH$_3$–R	50–121
Methoxy	49
Non-protonated sp^3	252
Non-protonated sp^2-aryl	75–218
Non-protonated sp^2-COOH	122–126
Protonated sp^2-aryl	15–22

dephasing times. Moreover T_2' values (20 μs and 26 μs) show that the resonances at 30 and 75 ppm arise from protonated carbons. The resonance at 155 ppm has a long T_2 value (> 120 μs) and is more prominent in the dipolar dephased spectrum (Figure 4, t_1 = 40 μs) showing that these carbons are non-protonated, i.e. oxygenated. This signal is due to the presence of phenols or aryl ether functionalities. An approximate estimate of various protonated structures, using equation (3), shows that 85% of the aromatic carbon in the humic acid is non-protonated. The corresponding figure for the fulvic acid sample is 90%.

The decay constant (T_2' = 26 μs) for the resonance at 55 ppm is considerably shorter than expected for methoxy groups but similar to that expected for CH carbons in amino acids. Hence the results suggest a substantial contribution of amino acid carbon to the sample. In contrast, we have observed relatively long decay constants (T_2' = 60 μs) for resonances at similar chemical shifts in terrestrial organic materials which can therefore be assigned to methoxy groups.

Table II illustrates the types of structures which may be distinguished from each other by dipolar dephasing experiments on humic substances. Clearly, methine and methyl, protonated aromatic and non-protonated aromatic, ketone and aldehyde, ketal and acetal carbons and also protonated olefinic and non-protonated olefinic carbon can be distinguished. Examples of the use of the method (6, 13), are shown in Figure 5.

Quantitation

The question of how quantitatively NMR measurements on macromolecules such as those comprising organic matter in coals, sediments and soils can be interpreted, is a matter of continuing debate. Similar considerations are applicable to studies on aquatic organic samples.

Solution NMR

There is general agreement about the problems of measurement by solution NMR. To determine whether data is quantitative, it is necessary to determine the magnitude of the spin-spin (T_2) and spin-lattice (T_1) relaxation times of the sample. The spin-lattice relaxation time constants of carbon in organic substances can be as short as 0.1 msec or as long as 1,000 sec. Unfortunately, short pulse delays of 1 sec or less must be employed when obtaining 13C-spectra of humic substances because large numbers of transients (scans) must be collected with signal averaging to obtain reasonable signal to noise ratios. Thus, it is not always possible to allow nuclei to fully relax between pulses and this can cause problems when attempting to estimate the proportions of various carbon types in humic extracts by NMR. The use of short pulse delays may lead to an over estimation of rapidly relaxing functional groups and an under estimation of slowly relaxing functional groups.

Because a large number of transients are needed to obtain even a nonquantitative 13C-spectrum of humic substances in solution, the time needed for measurement of relaxation times can be prohibitive.

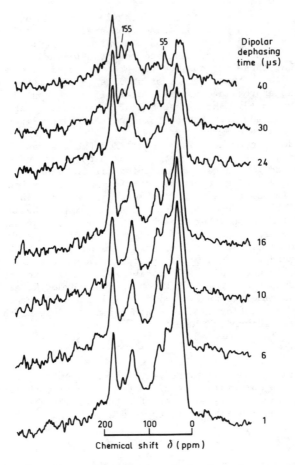

Figure 4. Effect of varying dipolar dephasing time on signal intensity from sea loch sedimentary humic acid.

Table II.   Differentiation of Functional Groups in Organic Material
in Terrestrial and Aquatic Environments
by Dipolar Dephasing

Chemical Shift	Structure
0-30 ppm	methyl groups from methine and methylene
50-60 ppm	amino acid CH carbon from methoxy
100-108 ppm	acetal from ketal
110-160 ppm	protonated from non-protonated aromatics, protonated from non-protonated olefins
200-220 ppm	aldehyde from ketone

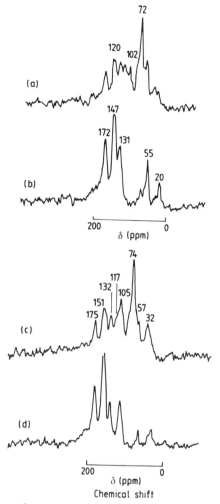

Figure 5.   Effect of dipolar dephasing on signal intensities from various carbons in (a), (b) peat and (c), (d) decayed pine leaves.   The methoxy (55 ppm) and methyl (20 ppm) signals in conventional spectrum (a) are retained after 40 μs dipolar dephasing (spectrum b).   Also acetal resonance (102 ppm) (spectrum a) is lost after dipolar dephasing in (spectrum b).   In decayed pine leaf spectra, ketal resonance (105 ppm) in conventional spectrum (c) is retained after 40 μs dipolar dephasing, spectrum (d). (Reproduced with permission from reference 6. Copyright 1983 Pergamon Press.)

However, a number of authors have been concerned whether the area of
the resonances in 13C-solution NMR spectra ascribed to aromatic plus
olefinic carbon (f_a), can be integrated quantitatively.   Hence,
experiments have been carried out in which the pulse delay between
transients is varied (1,2,5,14).   The effect of pulse delay on
f_a appears to be quite small, which suggests that most carbons in the
sample relax at similar rates.   Nevertheless, it is clear that
signal intensity increases rapidly with increasing pulse delay and at
pulse delays ($\leqslant$ 1 sec) normally used to obtain 13C-spectra of humic
substances, nuclei have not fully relaxed and carboxyl carbon is
underestimated.

In samples relaxing through nuclear-nuclear dipole-dipole
mechanisms, another source of error in quantitative 13C-NMR is the
nuclear Overhauser effect.   This arises in double resonance
experiments when one nucleus (in this case ^{1}H) is irradiated to
simplify the spectrum while another (in this case ^{13}C) is observed.
The signal from the observed species will differ in intensity from
that of the signal from a single resonance experiment.   Nuclear
Overhauser enhancements ($\mathcal{n}$) have been measured on humic substances
(2, 14).   The enhancements, while not large, are significant and
suggest that substantial numbers of humic substances can relax
through nuclear-nuclear dipole-dipole mechanisms.   Different
functional groups in molecules relax by different mechanisms, so they
will experience different N.O.E.'s.   Indeed 13C experiments on humic
substances obtained so far indicate that N.O.E.'s for carboxyl
carbons are different so that irradiating protons causes problems
(1,2,5,14).   For soil humic materials this has little effect on
aromaticity measurements but in aquatic materials the carboxyl
content can constitute nearly one quarter of the total carbon (2).
Hence to obtain spectra of these samples, it is necessary to turn off
the proton irradiation during the delay period between pulses to
reduce N.O.E.   This is commonly termed inverse gated decoupling.

Further complications in quantifying data can occur because of
unusually short relaxation times.   In the close proximity of paramagnetic ions, relaxation can be extremely fast.   In effect, if T_2
is extremely short, the nucleus will not be observed.   One way of
avoiding this possible source of error is to demineralise humic
substances.   Nevertheless, the intrinsic organic free radicals may
assist relaxation of certain structures in humic substances.   If the
radicals are sufficiently delocalised, this may affect quantitation.

Solid state NMR

Solid state 13C-NMR spectra of humic substances are usually obtained
by cross polarization techniques (15).   In this method, relaxation
occurs at almost the proton T_1 rate, rather than the carbon T_1 rate
which can be slower by a factor of 100.   Hence, in principle, the
problems of long carbon T_1's in solution are overcome by obtaining the
spectrum in the solid state.   However, solid state T_1's are
intrinsically much longer than in solution and hence it is possible
that some molecules are not observed because now the proton T_1's are
too long.   These species are normally small polycyclic aromatics
(12), although, fortunately, when these materials are coordinated
together as polymers, $T_1(H)$ is reduced.   It can be expected that

$T_1(H)$ problems in quantitation can be restricted to low molecular weight material ($<$ 1000).

There are several other problems associated with quantitation using cross polarization techniques. In determining the aromaticity of humic substances, cross polarization is achieved by applying r.f. magnetic fields at the resonant frequency of both 13C and 1H nuclei. By adjusting their relative intensities, contact between the two populations is established and cross polarization occurs between protons and carbons. However, the decay of polarization of the population of nuclei occurs at a rate determined by a time constant $T_{1\rho}(H)$ which is independent of the rate of cross polarization.

During the cross polarization process magnetization is transferred from protons to carbons according to the strength of the C-H dipole interaction, at a rate governed by a second time constant T_{CH}. If $T_{1\rho}(H)$ is smaller than the time needed for the most weakly coupled carbons to cross polarize, these signals will be reduced in intensity and in the extreme case, become lost in the background noise. Several laboratories have turned to model compound studies to elucidate the magnitude of this problem (16-18). The rate of cross polarization depends on the proximity of protons to carbons and on molecular motion. Hence carbons remote from protons may not be observed. Alemany et al. (17,18) have suggested that a serious reduction in intensity is obtained when carbons are four bonds removed from protons. It is difficult to imagine such structures constituting more than a minor proportion of the carbon in humic substances.

Of more concern for humic substances than the effect of long T_{CH}'s is the variable nature of $T_{1\rho}(H)$. Paramagnetic ions are of particular concern since they may assist relaxation and greatly shorten $T_{1\rho}(H)$'s in their immediate vicinity.

We have measured average ($<>$) $T_{1\rho}(H)$ and T_{CH}'s of humic materials (6) through cross polarization experiments. Whereas the $<T_{CH}>$'s for aliphatic and aromatic carbons are of the order of 50 and 300 μs respectively, $< T_{1\rho}(H)>$'s are much greater and range between 3.5 and 7 ms. Although encouraging, these results do not demonstrate that quantitative data are obtained. In rigid solids, polarization diffusion between protons usually operates to average $T_{1\rho}(H)$ values. However, in a complex mixture like humic substances, a number of $T_{1\rho}(H)$'s may exist. Indeed we have been able to demonstrate lack of spin diffusion in some humic substances (6).

A method of checking whether all protons in a sample have $T_{1\rho}(H)$'s long enough for carbons to cross polarize, is to measure the $T_{1\rho}(H)$'s independently of carbon and directly through protons. Data of this sort is presented in Table III. The shortest $T_{1\rho}(H)$ can be as small as 0.15 ms. Clearly these protons will not be able to transfer polarization quantitatively to carbons. Whether these protons are associated with organic as well as inorganic matter has, however, yet to be established. This is important because it is only the organically bound protons which may influence aromaticity measurements.

One way to check whether the cross polarization method is responsible for introducing errors in aromaticity measurements, is to perform experiments using single pulse excitation in much the same way as solution NMR measurements are made. This method is time

consuming and unsuitable for routine measurements because long
delays are needed between pulses in solids to ensure complete
relaxation of 13C nuclei. There is also a considerably lower signal
to noise ratio compared with cross polarization experiments of equal
number of transients. Single pulse experiments have been reported
by Hatcher (19) to be consistent with cross polarization experiments.
Moreover, data obtained by solution and solid state NMR is similar
(1,3,19).

Table III. Relaxation Data for Some Humic Materials

| Time Constant (ms) | Substrate[a] | | |
	Snares Island Peat	Maungatua Soil	Decayed Pine Litter
$T_{1\rho}H$ through protons[b]	13	10	13
	0.15	1.1	0.63
% fast decaying component	88	60	50
$T_{1\rho}H$ through carbon	6.3	4.4	6.9
T_1H through protons[b]	510	96	380
	0	2.5	46
% fast decaying component	100	56	35
T_1H through carbon	112	10	43

a) Samples dried at $100^{\circ}C$ before study
b) Results obrained by Dr J. Kalman

Conclusions

It is clear that caution should be taken when quantitatively inter-
preting 13C-spectra of humic substances, including aquatic humic
substances, unless an in depth study of relaxation is undertaken.
Unfortunately these studies are not routine. It has been
established that certain structures may cause problems in quantitation. These are:

a) in solution:
   1. molecules with carbons having long T_1's. These are
       invariably non-protonated carbons;
   2. molecules with carbons with very short T_2's. These are
       those carbons bonded to paramagnetic ions or intrinsic free
       electrons.
b) in the solid state:
   1. molecules with carbons with long T_{CH}'s and long $T_1(H)$'s,
       e.g. polycyclic hydrocarbons which have carbons remote from
       protons;
   2. molecules with protons with short $T_{1\rho}(H)$'s. These are
       molecules with protons bonded to, or close to, paramagnetic
       ions.

If the proviso's above are met, dipolar dephasing methods provide useful quantitative information on methoxy, aldehyde, ketone and protonated and non-protonated aromatic and olefinic carbon content.

Literature Cited

1. Wilson, M.A.; Gillam, A.H.; Collin P.J. Chem. Geol. 1983, 40, 187.
2. Wilson, M.A.; Barron, P.F.; Gillam, A.H. Geochim Cosmochim. Acta. 1981, 45, 1743.
3. Gillam, A.H.; Wilson, M.A. Org. Geochem. 1985, 8, 15.
4. Wilson, M.A.; Collin, P.J.; Tate, K.R. J. Soil Sci. 1983, 34, 297.
5. Newman, R.H.; Tate, K.R.; Barron, P.F.; Wilson, M.A. J. Soil Sci. 1980, 31, 623.
6. Wilson, M.A.; Pugmire, R.J.; Grant, D.M. Org. Geochem. 1983, 5, 121.
7. Gerstein, B.C. Phil. Trans. R. Soc. London A. 1981, 299, 521.
8. Opella, S.J.; Frey, M.H. J. Amer. Chem. Soc. 1979, 101, 5854.
9. Alemany, L.B.; Grant, D.M.; Alger, T.D.; Pugmire, R.J. J. Amer. Chem. Soc. 1983, 105, 6697.
10. Wilson, M.A.; Pugmire, R.J. Trends Anal. Chem. 1984, 3, 144-147.
11. Murphy, P.D.; Cassady, T.J.; Gerstein, B.C. Fuel 1982, 61, 1230.
12. Wilson, M.A.; Vassallo, A.M.; Collin, P.J.; Rottendorf H. Anal. Chem. 1984, 56, 433.
13. Wilson, M.A.; Vassallo, A.M.; Russell, N.J. Org. Geochem. 1984, 7, 161.
14. Newman, R.J.; Tate, K.R. J. Soil Sci. 1984, 35, 47.
15. Pines, A.; Gibby, M.G.; Waugh, J.S. J. Chem. Phys. 1973, 59, 569.
16. Wemmer, D.W.; Pines, A.; Whitehurst, D.D. Phil. Trans. R. Soc. London 1981, A300, 15.
17. Alemany, L.B.; Grant, D.M.; Pugmire, R.J.; Alger, T.D.; Zilm, K.W. J. Amer. Chem. Soc. 1983, 105, 2133.
18. Alemany, L.B.; Grant, D.M.; Pugmire, R.J.; Alger, T.D.; Zilm, K.W. J. Amer. Chem. Soc. 1983, 105, 2142.
19. Hatcher, P.G.; Breger, I.A.; Dennis, L.W.; Maciel, G.E. In: "Aquatic and Terrestrial Humic Materials"; R.F. Christman; E.T. Gjessing, Eds., Ann Arbor, Michigan, 1983, pp.46-48.

RECEIVED November 19, 1985

9

Structural Interrelationships among Humic Substances in Marine and Estuarine Sediments
As Delineated by Cross-Polarization/Magic Angle Spinning ¹³C NMR

Patrick G. Hatcher and William H. Orem

U.S. Geological Survey, Reston, VA 22092

Nuclear magnetic resonance studies of the structural composition of humic substances in marine and estuarine sediments have provided a wealth of information regarding the mode of formation for these macromolecular organic substances. The NMR data show that humic acids are highly aliphatic in nature. The aliphatic structures are thought to be derived from macromolecular residues of algae and other micro-organisms and have a high degree of branching and cross-linking. Fulvic acids, the most soluble of the humic substances, are generally unlike their less soluble counterparts, humic acids and humin, in that they appear to be mostly composed of carbohydrates and/or polyuronic acids. The interrelationships among fulvic acids, humic acids, and humin in a variety of marine and estuarine sediments suggests that humin is the parent material from which humic and fulvic acids are formed. The pathway for this formation appears to be oxidative, either chemical or biological oxidation. Fulvic acids are metabolic products of bacterial degradation of plant remains, whereas humic acids appear to be oxidized structural equivalents of the macromolecular insoluble humin in sediments receiving most of their contibutions from algal or microbial biomass. This humin probably originates from primary macromolecular structures in algae and/or bacteria and is concentrated in sediments by a process of selective preservation during early diagenesis. Humin from some coastal marine and estuarine sediments appears to contain a significant proportion of refractory, coal-like materials. In these sediments, humic acids bear no structural relationship to the humin.

The use of nuclear magnetic resonance spectroscopy for structural studies of humic substances has become commonplace since some of the early work of Vila and others (1), Stuermer and Payne (2), Wilson and Goh (3), and Hatcher and others (4). These early studies relied on solution ^1H and ^{13}C NMR to derive structural information on humic substances, primarily humic and fulvic acids, that are soluble in a suitable solvent, usually 0.5 N NaOD. The development of a technique that was capable of examining solids, solid-state ^{13}C NMR using cross polarization with magic-angle spinning (CPMAS), provided an opportunity to not only examine soluble humic substances but also the insoluble humin (5-8) and whole soil (9-11). Such a capability has for the first time allowed for direct chemical structural comparisons between the various humic fractions in marine and estuarine sediments. Some of these comparisons were made earlier for marine sediments (6). This paper provides new data for estuarine sediments and attempts to provide a more complete analysis of the structural interrelationships.

Solid-state ^{13}C NMR is the method of choice in this study for a number of reasons. First, the method is not limited by solubility, allowing intercomparisons among all humic isolates including the insoluble humin. We felt it important that humin be comparatively examined because our previous studies (7, 8) have suggested that this humic fraction may be the precursor from which humic acids originate. Second, the CPMAS technique is more sensitive and provides a spectrum we feel is more representative of carbon structures than solution ^{13}C NMR. Finally, the structural information obtained by NMR is far more useful for intercomparison purposes than that obtained from many other organic geochemical methods such as infrared spectroscopy, pyrolysis/gas chromatography/mass spectrometry (Py/GC/MS), elemental analysis, and others. The spectral data provides a striking visual presentation that enables rapid structural intercomparisons among humic fractions. Though the data cannot provide detailed discrimination at the molecular level like Py/GC/MS and similar techniques, major structural differences are readily discernable with NMR. When used in combination with other organic geochemical techniques, ^{13}C NMR becomes a powerful tool for chemical structural determinations.

Methods

Humic and fulvic acids as well as humin were isolated from the samples described in Table I by standard methods (6). In short, humic and fulvic acids are extracted with 0.5 N NaOH under N_2. Humic acids are protonated on an ion exchange resin, precipitated by acidifying to pH 2, separated by centrifugation, and lyophilyzed. The soluble fulvic acids are concentrated by ultrafiltration and lyophilyzed. Humin, the residue after treatment with NaOH, is treated with concentrated HCl:HF to remove a large portion of the mineral matter and hydrolyzable substances such as proteins and polysaccharides.

Dried humic substances were analyzed by placing them in a bullet-type rotor machined from Kel-F. CPMAS ^{13}C NMR spectra were obtained on a Chemagnetics CMC 100S/200L spectrometer operating at a field strength of 2.35 Tesla. Approximately 10,000 – 50,000 scans were obtained with a 1 s delay and 1 ms contact time. Spinning speeds of 3 to 3.5 kHz were achieved to minimize spinning sidebands.

Table I. Sample locations and descriptions

Sample Name	Location and Description

Mangrove Lake

Samples collected in Mangrove Lake, Bermuda, a small marine lake located on the northeastern corner of the island. A core (5.5m in depth) of the organic-rich sapropel at the bottom of the lake was obtained.

Potomac River (fluvial)

Sample of humic acid was obtained from G. Diachenko (U.S. Department of Agriculture). Sediment was collected at Point of Rocks, Maryland, in the Potomac River.

Potomac River (estuarine)

Samples were collected at the mouth of the Potomac River near Piney Point, Maryland. A 1-m core was obtained.

New York Bight

Samples were collected on the Continental Shelf, 10 nautical miles southeast of Rockaway Point, New York.

Hudson Canyon

Samples were collected at the head of the Hudson Canyon in the New York Bight 100 nautical miles east of Rockaway Point, New York, in 266m of water.

Walvis Bay

Samples were collected on the Continental Shelf, Walvis Bay, Namibia (South West Africa) in 846m of water. Surface sediment was collected on Cruise AII-93-3 of the Atlantis II in 1975 by J.W. Farrington (Woods Hole Oceanographic Institution).

Georgia soil

Surficial soil sample was collected from a fallow agricultural field near Kingsland, Georgia.

Results

Fulvic acids. Marine sedimentary humic substances soluble in base and acid (fulvic acids) have previously been examined by ^{1}H and ^{13}C NMR (12). The dominant structural components were postulated to be polysaccharide – like substances, probably polyuronic acids. Solid-state ^{13}C NMR spectra of fulvic acids isolated from a number of marine and estuarine sediments are shown in Figure 1. Major peaks at 72 and 106 ppm betray the overwhelming presence of polysaccharide – like substances, and, as shown by Hatcher and others (12), the moderate peak for carboxyl or amide carbon at 175 ppm suggests that these polysaccharides are more like polyuronides. Aromatic carbons (110 to 160 ppm) are decidedly minor components. Aliphatic carbons (0–50 ppm) are also minor components. ^{1}H NMR spectra shown by Hatcher and others (12) indicate that these aliphatic structures are highly branched.

It is noteworthy that fulvic acids from aerobic and anaerobic sediments and from offshore marine and estuarine sediments all have similar ^{13}C NMR spectra. The abundance of polysaccharides and structural similarity in this fraction precludes attempts to correlate sources of sedimentary organic matter on the basis of structural differences. The uniform structural character also suggests that the process leading to the production of fulvic acids is ubiquitous. If polyuronides are the likely structural components of fulvic acids, then it is likely that these components derive from algal or microbial remains known to be enriched in polyuronides.

The classical definition of fulvic acids is not very specific. Many biochemical substances such as proteins, sugars, and fatty acids would fall under this classification. These substances, in many instances, can hardly be considered "humic" in nature. But, if one uses the classical definition of fulvic acids these substances are included. No doubt the observations made above that polysaccharide-like substances constitute the major components of sedimentary fulvic acids is partly attributable to the fact that the operational definition classes polyuronides as fulvic acids. It is not the intent of this paper to discuss the merits of using the classical operational definition for fulvic acids as opposed to one's perception of what true fulvic acids are. We know far too little about the composition of humic isolates and about their origin to begin discussions of whether they are humified or not. We therefore chose to use operational definition with recognition that well defined structural entities can sometimes be a part of what is isolated.

Humic acids. CPMAS [13] NMR spectra of representative marine and estuarine humic acids are shown in Figure 2. Solution ^{1}H and ^{13}C NMR and solid state ^{13}C NMR spectra of marine sedimentary humic acids have previously been described by Hatcher and others (4) and Dereppe and others (13). These spectra showed that marine sedimentary humic acids are predominantly composed of paraffinic structures that have a relatively high degree of branching, compared to long-chain alkyl structures. Aromatic structures are generally depleted in marine humic acids whose source is predominantly from algal or microbial detritus.

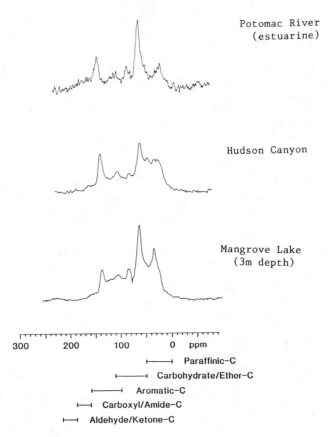

Figure 1. Representative CPMAS ^{13}C NMR spectra of fulvic acids
from marine and estuarine sediments.

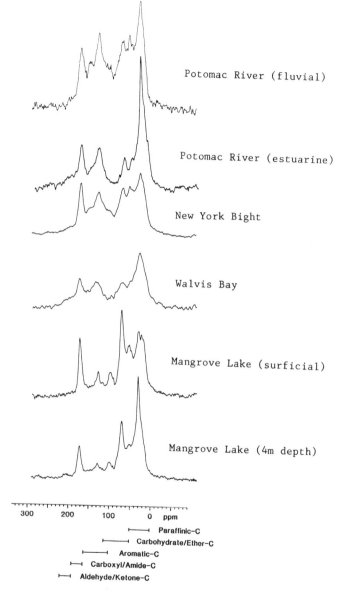

Figure 2. Representative CPMAS ^{13}C NMR spectra of humic acid from marine and estuarine sediments.

The spectrum of humic acid from an algal sapropel from Mangrove
Lake (4m depth) shown in Figure 2 is characteristic of such marine
humic acids. Note the strong signals for paraffinic,
carbohydrate/ether, and carboxyl/amide carbons at 30, 72, and 175
ppm, respectively. Carbohydrates have been shown to be present in
humic acid isolates, probably existing as uronic-acid-like
carbohydrates (4). The spectrum of humic acids from surficial
sediments in Mangrove Lake (Figure 2) shows a greater proportion of
carbohydrate signals, probably representing biological degradation
products from early diagenetic reactions. Hatcher and others (7) and
Spiker and Hatcher (14) showed that early diagenesis leads to the
degradation and removal of carbohydrates in Mangrove Lake sapropel.
Presumably, these degraded or partially degraded carbohydrates could
become incorporated in humic acid isolates, especially if they
contain carboxyl functional groups. Yields of humic acids from
Mangrove Lake sediments are low (2%); thus it is likely that
carbohydrates or uronic acid-like carbohydrates, usually a major
fraction of the total sapropel in surficial layers, are being
incorporated into humic acid isolates. The carbohydrates are most
likely incorporated in humic acids from Mangrove Lake because the
sapropel is in its initial stages of diagenesis and, as such, still
contains sizeable quantities of carbohydrates (7). At depth, where
the sapropel has been diagenetically altered further, carbohydrates
are minor components of humic acids, whereas the paraffinic
structures, alluded to above, are dominant. Associated with the
paraffinic structures are carboxyl/amide groups (175 ppm) and
alcoholic/etheric groups other than carbohydrates (70 ppm). Note
also that a small peak is observed at about 50 ppm. Peaks in this
region are usually assigned to methoxyl or amino groups. Because
methoxyl carbons in humic acids are usually associated with aromatic
structures (from lignin-like substances) and because contributions of
aromatic, lignin-derived components are lacking in Mangrove Lake, the
peak at 50 ppm is probably that of amino groups. The elemental data
which indicate approximately 4 to 5 percent nitrogen in these humic
acids (7) are in accord with this assignment.
    Solid-state ^{13}C NMR spectra of humic acids from other marine and
estuarine sediments (Figure 2) show peaks in similar regions as those
noted for the Mangrove Lake humic acids, but the relative intensities
vary considerably. Most marine and estuarine humic acids contain few
carbohydrate-like structures, as the peaks for carbohydrates at 72
and 106 ppm are minor. This probably is because these sediments are
more advanced diagenetically than those from Mangrove Lake. The
yields of humic acids are greater than those from Mangrove Lake (6)
and it is likely that less extractable carbohydrate-like material
from undegraded plant detritus has been incorporated.
    The content of aromatic carbon varies considerably in marine and
estuarine humic acids, but is, in all cases, greater than that of
Mangrove Lake humic acids. This is probably a reflection of the
greater contribution of vascular plant-derived material which can be
expected to provide lignin-like components rich in aromatic
structures. Note that the humic acids from fluvial sediments of the
Potomac River are the most aromatic. Peaks at 150 and 55 ppm are
characteristic of oxygen-substituted aromatic carbons typically
associated with lignin of vascular plants. Humic acids from New York

Bight sediments (Figure 2) also contain considerable quantities of lignin-derived structures, as expected, because large quantities of sewage sludge and dredge spoils are being dumped at the site. Sediments from Walvis Bay, a coastal upwelling zone off the west coast of Africa, and estuarine muds from the lower Potomac River contain humic acids that show lesser proportions of aromatic structures than those mentioned above. Their NMR spectra (Figure 2) very nearly approximate those of marine algal or microbial detritus (Mangrove Lake sapropel). The carbohydrate signals (72 ppm) are less, however, probably reflecting the fact that a greater degree of decomposition of algal biomass has occurred in these sediments. Like humic acids from Mangrove Lake, paraffinic structures are dominant.

In summary, solid-state ^{13}C NMR spectra of humic acids from marine and estuarine sediments reveal some diagenetic structural changes. In recently deposited and well preserved marine sediments such as those from Mangrove Lake, Bermuda, carbohydrates and paraffinic structures constitute the major structural entities. Burial or increasing degree of decomposition, leads to the diminution of carbohydrates, whereupon paraffinic structures become dominant. In marine sediments derived from algal or microbial sources, that are more exposed to decomposition, paraffinic structures predominate in humic acids. Such is the case for sediments from Walvis Bay and the lower Potomac River estuary. It is important to note that these latter sediments contain humic acids that have a predominantly algal or microbial signature even though substantial contributions of terrestrial materials are expected due to their proximity to sources of such contributions. The high algal productivity and organic rich sediment accumulation rates for the lower Potomac, as well as the stable isotopic compositions of the sedimentary organic matter (E.C. Spiker, personal communication), are consistent with a finding that algal detritus is the major contributor to sedimentary carbon. Thus the NMR data for humic acids are consistent with this conclusion.

As one examines humic acids from sediments where large terrestrial or vascular plant inputs are expected, the CPMAS ^{13}C NMR spectra show higher proportions of aromatic carbons and notable peaks for lignin-like contributions at 55 and 150 ppm. Such distinctions could possibly be used to estimate the relative contribution of vascular plant residues to the sediments.

Humin. Humin, the fraction of humic material that is insoluble in alkali, has often been referred to as kerogen, protokerogen, or stable residue in the geochemical literature(15-17). Because it exists as a residue and is admixed with inorganic components of sediments, which generally constitute the sizeable portion of the total weight, it has been necessary to concentrate the humin prior to analysis. The most common means of achieving this is by removal of mineral matter. Usually, the sediment residue from alkali extraction is treated with concentrated HCl mixed in a 1:1 v/v ratio with 48% aqueous HF. Carbohydrates and proteins are selectively hydrolyzed in the process, but these substances cannot strictly be called humic in nature. Their removal from the sediment allows us to examine refractory organic matter or stable residue. Of course, we must concern ourselves with the effect of such treatment on the humic

material. Sediments from Mangrove Lake, Bermuda provide ideal
samples for testing the effect of acid treatment because the carbon
content of the whole sapropel is high enough to obtain CPMAS ^{13}C NMR
directly without concentration. Figure 3 shows the NMR spectra of
algal sapropel near the sediment surface (0.2 m) and at depth (2.9m),
humin obtained after alkali extraction, and humin after HF/HCl
treatment. Note that very little difference exists between spectra
of the whole sapropel and the humin after alkali-extraction.
Carbohydrates (peaks at 72 and 106 ppm) are major contributors to the
sapropel in the surface intervals and diminish in concentration with
depth. Hatcher and others (7) have described the diagenetic trends
as being attributable to degradation and loss of carbohydrates with
selective preservation of the insoluble, macromolecular paraffinic
substances, which are the dominant components of humin. When treated
with HF/HCl, the humin is altered, primarily by loss of carbohydrates. Peak intensities for non-carbohydrate moieties (paraffinic
structures) remain relatively unchanged, to the extent that can be
detected by NMR. Thus, humin treated with HF/HCl is primarily
altered by removal of carbohydrates. This is essentially the same
process that is the result of early diagenetic transformations
described above (7). Because carbohydrates cannot be considered
"humic" in nature (18), the HF/HCl removal of carbohydrates is useful
in allowing intercomparisons among the various humic fractions of
sediments.

     Solid-state ^{13}C NMR spectra of humin, isolated by the HF/HCl
treatment, from various representative marine and estuarine sediments
are shown in Figure 4. Humin from Mangrove Lake was described as
being essentially composed of paraffinic structures (30 ppm)
containing carboxyl/amide (175 ppm) and etheric carbon (70 ppm)
functional groups (7). The low aromaticity (7%) is typical of humin
from algal and microbial sources. The peak at 70 ppm can be
attributed to ether or alcohol-like structures because the compounds
that are usually characteristic of this resonance, the carbohydrates,
have most likely been removed by the acid treatment. Spectra of
humin from other samples examined here also show an intense peak at
30 ppm for paraffinic structures, but the relative proportion of
aromatic carbon (130 ppm) increases significantly in some samples.
Those having strong contributions from algal/microbial sources,
namely sediments from Walvis Bay and the Hudson Canyon, show a
greater proportion of paraffinic structures. The increased amounts
of aromatic carbons when compared to humin from Mangrove Lake
probably reflects increased contributions from vascular plants.

     Humin isolates from sediments of the New York Bight and Potomac
River estuary have spectra that are notably different in that
aromatic carbons are the dominant components. The spectra resemble
that of humin isolated in the same manner from an aerobic soil from
southern Georgia (Figure 5). However, unlike the humin from soil
which shows a significant peak for carboxyl carbon (175 ppm), spectra
of humin from the New York Bight and the Potomac River do not display
a discreet peak at 175 ppm and appear to be depleted of
carboxyl/amide groups. Elemental data for these humins (19) are
consistent with the NMR results. Atomic H/C ratios of less than 0.8
are not typical of humic material but more like those of highly
aromatic coal or coal-like products. The NMR spectra also resemble

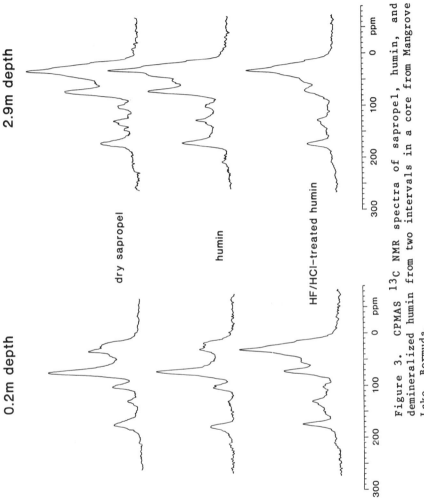

Figure 3. CPMAS 13C NMR spectra of sapropel, humin, and demineralized humin from two intervals in a core from Mangrove Lake, Bermuda.

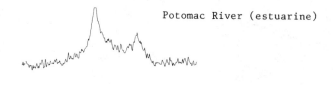

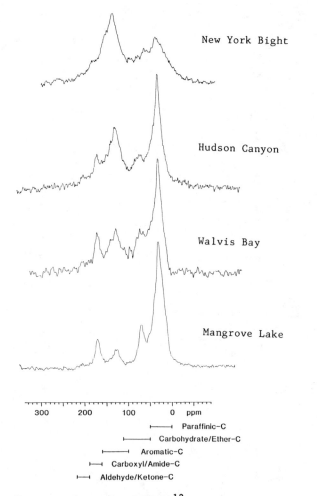

Figure 4.   Representative CPMAS ^{13}C NMR spectra of demineralized humin from marine and estuarine sediments.

humin-HF/HCl

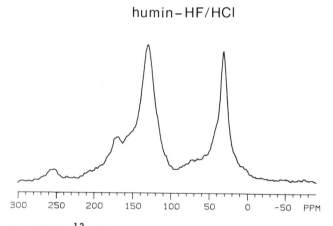

Figure 5. CPMAS ^{13}C NMR spectrum of demineralized humin from the Georgia soil.

those of coal, with little or no carboxyl peaks and broad, unresolved
peaks for mostly aromatic carbons (20). The presence of coal and
coal-like materials in estuarine sediments is not unexpected. The
presence of coal has been noted in sediments of the Chesapeake Bay
(21). Storm runoff and sewage from New York City, which are both
collected and dumped in the New York Bight, are also likely to
contain coal-derived substances. Certainly, dredge spoils from the
Hudson-Raritan estuary, which are also dumped in the New York Bight,
are likely to contain coal-like substances as well.

The possibility exists that carboxyl functional groups and the
chemical composition of the humin from the two samples in question
have been affected by the HF/HCl treatment. Decarboxylation and
hydrolysis of proteinaceous substances by strong acid treatment is
possible and cyclization of carbohydrates, if present, to form
aromatic groups could also take place. This seems unlikely,
considering that a similar treatment of Mangrove Lake sapropel, that
is rich in carbohydrates and carboxyl functional groups, did not
produce such components. We believe that the highly aromatic
structures are indigenous and possibly indicative of contributions of
coal-like material. Based on the relatively low yield of humin in
these two sediments, the contribution of coal-like material may be
relatively minor.

Discussion

The solid-state ^{13}C NMR spectra of humin from algal or microbially
derived sediments are similar to those of corresponding humic acids.
Aside from the presence of carbohydrates in humic acids from the
Mangrove Lake sapropel, the spectra of humins are almost identical to
those of humic acids, suggesting a close structural relationship and
possibly a close genetic relationship. Humic acids generally contain
more oxygen in their elemental analysis (6) and we would suspect that
this would be refelected in the presence of more carboxyl,
ether/alcohol, and carbonyl groups. In comparing spectra of humic
acids and humin from Walvis Bay sediments, the increased contents of
such functional groups in humic acids are subtle but noticeable as
mostly increased relative intensities at 175 and 190 ppm. Because of
the close structural similarities, we believe that humic acids are
structural equivalents of humin that have been oxidized, resulting in
the introduction of oxygen-functional groups (i.e. carboxyl groups).
The increased oxygen functionality allows these structures to be more
readily extracted by dilute base. When acidified, the carboxyl
groups are protonated, thereby reducing solubility such that the
oxidized remains precipitate as humic acids. In estuarine and
coastal sediments strong structural relationships between humic acids
and humin are not observed (see Figures 2 and 4). This is most
likely attributable to the fact that highly refractory, coal-like
components dominate the humin residue. These refractory materials
probably do not produce humic acids as readily as modern plant
remains upon oxidation. However, modern plant-derived materials
within the sediment, whose detection by NMR may be masked by the
broad peaks of coal-like substances, are probably responsible for the

production of humic acids. Thus, the humic acids in these sediments have structural features that are like those expected from modern source materials but the humin is probably not representative of source materials derived from modern carbon.

CONCLUSIONS

Solid-state ^{13}C NMR provides a visual presentation of the chemical structural composition of humic isolates that allows for direct structural intercomparisons among fulvic acids, humic acids, and the insoluble humin in marine and estuarine sediments. Though the structural detail provided by this technique is no more than a "broad brush" examination, such an approach is useful from the standpoint that gross structural interrelationships can provide clues to the origin of humic substances. Previous studies involving the use of solid-state ^{13}C NMR in combination with other organic geochemical and stable isotopic analyses have led to the suggestion that insoluble macromolecular humin is an original component of algal/microbial biomass in marine sediments and that this material, rich in paraffinic structures, is selectively preserved during early diagenesis (7). Early diagenesis basically involves degradation and loss of microbially labile substances such as carbohydrates, proteins, and lipids. With this frame of reference we examined structural features of marine and estuarine humic substances by CPMAS ^{13}C NMR and conclude the following:
1. Fulvic acids, the most soluble humic fraction, are predominantly polysaccharides in all marine and estuarine sediments examined. Uronic acid-like polysaccharides are the most likely entities. Fulvic acids may be the degradation products that evolve during diagenetic alteration of sedimentary plant residues. We envision that microbial degradation renders plant polysaccharides soluble, entraining them into a fulvic acid fraction by definition. Ultimately, this material is degraded further to CO_2, CH_4, and other low molecular weight organic compounds.
2. Humic acids of marine and estuarine sediments are characterized by major amounts of paraffinic structures that previous studies have shown to be highly branched and to contain significant quantities of carboxyl/amide and alcohol/ether carbon. Some humic acids, namely those from well preserved sapropelic marine sediments show significant quantities of carbohydrate-like structures incorporated. This, no doubt, is a reflection of the solubility characteristics of polysaccharides which may have some carboxyl functionalities (uronic acid groups).
3. Humin varies widely in composition. Sediments derived from algal/microbial biomass have humin with paraffinic structures resembling those of corresponding humic acids. Estuarine or coastal marine sediments examined in this study have humin with highly aromatic structures which resemble coal-like materials rather than modern plant residues. In these latter sediments no structural correspondence exists between humin and humic acids which appear to more nearly reflect the nature of modern plant

residues being incorporated in sedimentary organic matter.
Consequently, a certain amount of caution must be exercized when
using aromaticity and possibly other structural characteristics
as source discriminants.
4. Finally, the structural interrelationships among humic
fractions in marine and estuarine sediments suggest that the
pathway for humification is one of degradation rather than
condensation as proposed by others (15, 22). Insoluble humin, an
original component of sedimentary detritus, is degraded to
smaller molecules which become extractable and thereby classed as
humic acids. Fulvic acids are ultimately formed by intense
degradation; but degradation products of labile macromolecules of
sedimentary detritus such as carbohydrates also become
incorporated and, in fact, dominate the fulvic acid fraction.

LITERATURE CITED

1. Vila, F.J.G.; Lentz, H.; Ludemann, H.D. Biochem. Biophys.
    Res. Commun. 1976, 72, 1063-9.
2. Stuermer, D.H.; Payne, J.R. Geochim. Cosmochim. Acta, 1976,
    40, 1109-14.
3. Wilson, M.A.; Goh, K.H. J. Soil Sci., 1977, 28, 645-52.
4. Hatcher, P.G.; Rowan, R.; Mattingly, M.A. Org. Geochem.,
    1980, 2, 77-85.
5. Hatcher, P.G.; VanderHart, D.L.; Earl, W.L. Org. Geochem.,
    1980, 2, 87-92.
6. Hatcher, P.G.; Breger, I.A.; Dennis, L.W.; Maciel, G.E. In
    "Aquatic and Terrestrial Humic Materials", Christman, R.F.;
    Gjessing, E.T.; Eds.; Ann Arbor Science: Michigan, 1983;
    Chap. 3.
7. Hatcher, P.G.; Spiker, E.S.; Szeverenyi, N.M.; Maciel, G.E.
    Nature 1983, 305, 498-501.
8. Hatcher, P.G.; Breger, I.A.; Maciel, G.E.; Szeverenyi, N.M.
    In "Humic Substances in Soil, Sediment, and Water"; Aiken, G.;
    McKnight, D.; Wershaw, R.; MacCarthy, P. Eds., John Wiley: New
    York, 1985; Chap. 11.
9. Wilson, M.A.; Pugmire, R.J.; Zilm, K.M.; Goh, K.M.; Heng, S.;
    Grant, D.M. Nature 1981, 294, 648-50.
10. Wilson, M.A.; Pugmire, R.J.; Grant, D.M. Org. Geochem. 1983,
    5, 121-9.
11. Preston, C.M.; Ripmeester, J.A. Can. J. Spectrosc. 1982, 27,
    99-105.
12. Hatcher, P.G.; Breger, I.A.; Mattingly, M.A. Nature 1980,
    285, 560-2.
13. Dereppe, J.M.; Moreaux, C.; Debyser, Y. Org. Geochem. 1980,
    2, 117-124.
14. Spiker, E.C.; Hatcher, P.G. Org. Geochem. 1984, 5, 283-90.
15. Huc, A.Y.; Durand, B. Fuel 1977, 56, 73-80.
16. Stuermer, D.H.; Kaplan, I.R.; Peters, K.E. Geochim.
    Cosmochim. Acta 1978, 42, 989-97.
17. Pelet, R. In Advances in Organic Geochemistry, 1981, Bjoroy,
    M., Ed., John Wiley: New York, 1983; pp.241-250.

18. Stevenson, F.J. Humus Chemistry, Genesis, Composition, Reactions, John Wiley: New York, 1982, Chap. 2.
19. Hatcher, P.G. Ph.D. Dissertation, The University of Maryland, 1980.
20. Miknis, F.P; Sullivan, M.; Bartuska, V.J.; Maciel, G.E. Org. Geochem. 1981, 3, 19-28.
21. Goldberg, E.D.; Hodge, V.; Koide, M.; Griffin, J.; Gamble, E.; Bricker,O.P.; Matisoff, G.; Holdren, G.R. Jr.; Braun, R. Geochim. Cosmochim. Acta 1978, 42, 1413-25.
22. Nissenbaum, A.; Kaplan, I.R. Limnol. Oceanogr. 1972, 17, 570-2.

RECEIVED September 23, 1985

10

Early Diagenesis of Organic Carbon in Sediments from the Peruvian Upwelling Zone

W. T. Cooper, A. S. Heiman, and R. R. Yates

Department of Chemistry, Florida State University, Tallahassee, FL 32306-3006

The diagenesis of organic matter in recent sediments from the Peruvian Upwelling Zone has been studied by solid state ^{13}C NMR spectroscopy of the intact sedimentary organic carbon and by conventional GC and GC-MS analyses of sterol biomarkers in lipid extracts of the sediments. The distribution of sterols in surficial sediments from two sites on the continental margin indicate that the distinction between marine and terrestrial sources of organic carbon in aquatic sediments is not as well defined as previously thought. Both sites exhibit large relative abundances of twenty-eight (C-28) and thirty (C-30) carbon atom sterols, suggesting a marine source. The C-27 and C-29 abundances do not reflect this expected marine source, however, confirming recent observations that sterol distribution patterns in nature are extremely complex.

Magic angle spinning ^{13}C NMR spectra with variable cross polarization contact times were obtained on the intact, non-extracted sediments. The time-dependent spectra reveal subtle differences in organic carbon with depth; differences not observed in single contact experiments. Dipolar-dephased spectra of these same sediments indicate the presence of substantial amounts of substituted aromatic/olefinic carbons which are rapidly altered with depth.

The transformations of organic matter in young, recently deposited aquatic sediments play a unique role in the biogeochemical cycle of organic carbon. During sedimentation and residence in the upper sediment horizons, organic matter is subject to a number of chemical, biological and physical processes which alter it in various ways. While the nature and extent of these transformations depend on both the type and amount of organic material and the depositional environment at the time of burial, it is clear that the composition of sedimentary organic matter is significantly altered, and its fate during subsequent geochemical/geophysical evolution is largely dependent on these initial transformations.

0097-6156/86/0305-0158$06.00/0
© 1986 American Chemical Society

The process by which biopolymers, the remnants of living organisms buried in sediments, are degraded and rearranged into insoluble geopolymers is usually referred to as diagenesis. One theory (1) holds that diagenesis includes microbial degradation of biological macromolecules into smaller components; condensation of these small, highly functionalized compounds into geopolymers such as humic acids, fulvic acids and less functionalized humin residues; and insolubilization of these condensed structures via elimination of hydrophilic functional groups to form insoluble kerogen. While other scenarios have been proposed for the formation of kerogen (2), it is generally considered the main source of petroleum and is the primary organic material found in ancient sediments.

Degradation occurs rapidly in the overall evolution of organic matter; usually in the first 1-10 m interval of a sediment. Insolubilization is much slower, normally occurring in the 10-100 m range. At this point, temperature and pressure become important, and the second phase of the evolution of organic matter commences; catagenesis, thermal alteration and the resulting formation of oil and gas.

In this paper we present the initial results of studies of early diagenesis of organic carbon in the upper horizons (0-30 cm) of marine sediments. Studies of this sort are complicated by the inherent complexity of sedimentary carbon and the presence of large geopolymers. These geopolymers are in general insoluble and extractable only by extreme treatments (3). These treatments are such that there is real doubt about whether the extract is truly representative of the organic material as it exists in the sedimentary environment (4).

Our approach is significantly different. We have attempted to obtain information we feel is representative of the in-situ character of sedimentary organic matter. This approach has involved two distinct phases of sediment evaluation:

> –analysis of biomarkers, or geochemical fossils, in the lipid fraction of the sedimentary organic matter, and

> –solid state NMR studies of intact sediments, where the organic fraction is viewed in-situ.

Lipid Biomarkers

Lipids incorporated into aquatic sediments behave somewhat differently than most other forms of organic debris. By lipid, we are using the geochemist's definition; those compounds extractable with organic solvents. Containing less functional groups and thus more hydrophobic, lipids generally are much less reactive and persist in discrete chemical forms longer than proteins, polysaccharides, etc.. Indeed, a number of different lipid molecules of known biological origin are routinely found in petroleum, coal and ancient sediments. Molecules which persist in sediments over geologic time spans in an unaltered or slightly altered state and which can be attributed directly or indirectly to living organisms are frequently referred to as "biomarkers", "molecular fossils", or "geochemical fossils".

A number of geochemical fossils have been identified and their presence in sedimentary organic matter utilized for various

geochemical purposes (5). These include: n-alkanes, 2-methyl (iso)
and 3-methyl (anteiso) branched alkanes; n-alkanoic acids; acyclic
isoprenoid-type molecules; steroids; and a variety of cyclic triterpenoids.
    One of the most valuable groups of geochemical fossils is the
steroidal alcohols, or sterols. Regulators of metabolic processes
and major components of eucaryotic cell membranes, sterol structures
appear to be strongly related to the organisms that produce them.
Marine organisms produce primarily twenty-seven carbon atom sterols
(C-27) (6-8), while terrestrial plants produce primarily twenty-nine
carbon atom sterols (C-29) (9). This specificity has been used by a
number of investigators to characterize the ecological history of
water columns. Gaskell and Eglinton (10), for example, attributed
variations in the C-27/C-29 ratio with depth in a recent lacustrine
sediment to a change in the lipid source of the organic carbon in
the sediment.
    In addition to this source specificity, sterols undergo a
variety of reactions which are of particular value in studies of
early diagenesis. These reactions are not only indicative of the
sedimentary environment, but also result in products which are
similar enough to precursor molecules to be easily related to them.
One of the most intensely investigated reactions has been the hydro-
genation which converts Δ^5, Δ^7 and Δ^{24} unsaturated sterols (stenols)
into their saturated analogues (stanols) (11-13). Other reactions
identified include bacterial dealkylation of the sterol side chain
(14), and aromatization to form phenanthrene homologues (15).
Sedimentary reactions of sterols which have been identified or
suggested are summarized in Figure 1.

NMR Studies of Sedimentary Organic Matter

Recent developments in ^{13}C nuclear magnetic resonance spectroscopy
have made it possible to probe the chemical environment of carbon
atoms in rigid and semi-rigid systems. Using the techniques of
cross polarization (CP) (16) and magic angle spinning (MAS) (17),
the primary obstacles to obtaining NMR spectra of solids with reasonable line widths (chemical shift anisotropy, anisotropic dipole-
dipole interactions (dipole coupling), and long spin-lattice relaxation times) can now be circumvented. Since its introduction by
Schaefer and Stejskal (17), the ^{13}C NMR technique combining cross
polarization and magic angle spinning (CP-MAS-NMR) has been applied
to a wide variety of insoluble organic complexes such as fossil
fuels, polymers, coal, kerogen and humic/fulvic acids (18). These
applications represent what Maciel (18) describes as the "first
generation" of NMR solids applications.
    It is the "second generation" of solid state techniques, in
which the onset of relaxation of magnetization is observed in the
time domain, which offer promise for the detailed, in-situ investigation of early diagenetic processes. For example, Hagaman and
Woody (19) showed that the time dependence of cross polarization
could be used to improve the resolution of complex spectra of coal,
since different carbon types cross polarize at different rates.
Wilson (20-21) has used dipolar dephasing to distinguish carbon
types in soils and resins based on relaxation rates.
    We have used both time dependent cross polarization and dipolar

dephasing techniques to view carbon atoms in recent marine sediments. Variations in cross polarization contact times have been used to improve the resolution of adjacent resonances which would otherwise overlap. Dipolar dephasing experiments were employed to distinguish protonated and nonprotonated carbon atoms, particularly in the aromatic region.

Experimental

Sampling. Sediment samples from the coast of Peru were collected during Cruise 23-06 of the R/V Robert Conrad, June - July, 1982. Sampling locations are indicated in Figure 2. A specially designed box corer (22) was used in order to obtain undisturbed surface sediments (<50 cm depth). Subcores were taken on board, sectioned over 2 cm intervals, and each subsection immediately frozen. The frozen samples were packaged in dry ice and shipped to the FSU chemistry department, where they were stored in a $-10^{\circ}C$ freezer until analyzed.
    Sediments from two locations have been analyzed for this study. These are sites BX-3 and BX-6 of Figure 2. Important characteristics of these sites are listed in Table I.

Table I. Characteristics of Surficial Sediments, Peruvian Coast

	BX-6	BX-3
Latitude	12° 05.0'	15° 16.9'
Longitude	77° 39.5'	75° 23.9'
Water Depth (m)	183	387
% Organic Carbon in Surficial Sediment	16	6
Sedimentation Rate[a] (mm/y)	2.3	1.6

a. from ^{210}Pb activity.

Sterol Analyses. The overall analytical procedure for each 2 cm subcore section is schematically represented in Figure 3. After homogenization and lyophilization, approximately 1 g of each sediment was mixed with 37.5 ml of chloroform, 75 ml of methanol and 30 ml of a buffered aqueous solution (pH=7). The sediment-solvent slurry was then sonicated for 3-5 minutes and the extract decanted to a separatory funnel. Seventy five ml of water was then added to the funnel, resulting in separate aqueous and organic phases. The chloroform layer was then removed and the aqueous phase washed five times with chloroform. The chloroform fractions were then combined and the volume reduced to 10 ml under nitrogen at $37^{\circ}C$. The entire extraction was repeated until the chloroform phase was visually colorless (6-9 extractions).
    Extracted lipids were then separated into neutral, glyco-, and phospholipid fractions by silica gel column chromatography. Neutral lipids were eluted with 10 ml of chloroform per g of silica gel. Glycolipids were eluted with 20 ml/g of acetone, followed by phospholipids with 20 ml/g of methanol. The neutral lipid fraction

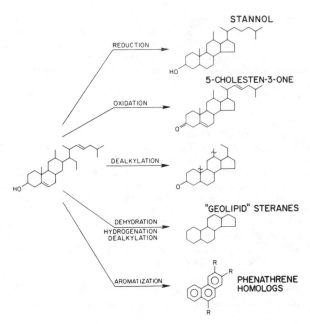

Figure 1. Sedimentary reactions of sterols.

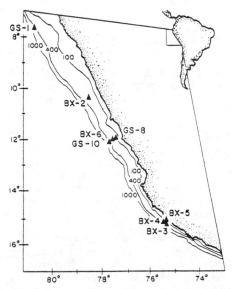

Figure 2. Cruise track of R/V Conrad, June – July, 1982.
Samples for this study obtained from sites labelled BX-6 and BX-3.

was then concentrated and fractionated by TLC into n-alkane, sterane, sterol, sterone and fatty acid subfractions using a 55/45 hexane/chloroform solvent. The sterol fraction was further purified by TLC with 45/55 hexane/chloroform. The sterol spot was scraped from the plate and sterols were extracted with hexane/chloroform.

After concentration under nitrogen at $37^\circ C$, sterols were derivatized to silyl ethers with BSTFA (Pierce Chemical Co.). The derivatized sterols were then diluted to a known volume in chloroform containing 100 picomoles of hexadecane, cholestane, cholestanol, cholesterol, stigmasterol and sitosterol as internal standards. The mixture was then analyzed with a Hewlett-Packard 5880A GC equipped with a flame ionization detector. Separations were achieved with a Hewlett-Packard 50m x .322 mm ID crosslinked methyl silicone fused silica capillary column. The column temperature was held initially at 50 $^\circ C$ for 1 min, programmed at 10 $^\circ C$/min to $160^\circ C$, then 4 $^\circ C$/min to a final temperature of $300^\circ C$. Samples were injected via the splitless method (23) with the injector at $375^\circ C$. The splitter was turned on at 30 sec. The FID was held at $350^\circ C$.

Sterol identifications were made by comparing relative retention times of unknowns with those of authentic standards obtained from Applied Science Labs, Inc. and Suppelco, Inc. Identifications were confirmed by gas chromatography/mass spectrometry using a Hewlett-Packard 5890 GC/MS.

NMR Experiments. Cross polarization/magic angle ^{13}C-NMR spectra were obtained with a Bruker WP-200SY spectrometer equipped with an IBM solids control accessory, operating at 50.325 MHz for carbon and externally tuned with t-butylbenzene. Samples were spun (air driven) at the magic angle at 3 KHz.

A variety of pulse sequence programs were employed. A conventional cross polarization, single contact program was used to obtain spectra of intact sediments. Contact times were varied from 200-3000 μsec with a three sec recycle time. For dephasing delay experiments, a 50 μsec delay was inserted prior to data collection. This delay consisted of two 25 μsec intervals separated by a 10 μsec, 180° refocusing pulse. Data was collected in 2K of memory, exponentially multiplied with 50 Hz of line broadening, and expanded to 8 K prior to Fourier transformation. All spectra are the result of 5000 accumulations.

Results and Discussion

Sterols. Distributions of free sterols in the surficial sediments (0-2 cm) at both sampling stations are displayed as a function of carbon number in Figure 4. BX-6 sediments exhibit a total C-27/C-29 ratio greater than unity as expected, since this is an indication of a marine source of organic carbon in sediments. The ratio of C-27/C-29 sterols in the BX-3 sediments, however, is indicative of significant terrestrial input, in spite of the fact that BX-3 sediments are at greater depth. This situation has been observed by other workers (24). However, it has also been noted that certain marine organisms produce C-29 sterols (25). These and other results suggest that use of sterol compositions to determine biological sources of organic carbon in aquatic sediments may not be completely

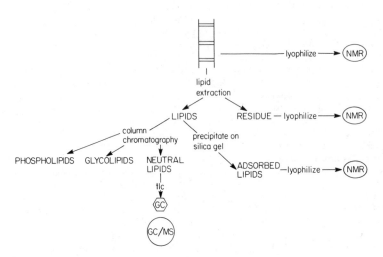

Figure 3.   General analytical scheme for analysis of sediment lipids by extraction and GC/MS, and intact organic carbon by solid state NMR.

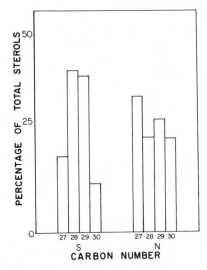

Figure 4.   Histogram depicting relative proportions of sedimentary sterols by carbon number.   S refers to BX-3, N refers to BX-6.

valid. Huang and Meinschein (9), for example, maintain that the relative abundance of C-27, C-28 and C-29 sterols can be used to determine the proportions of terrigenous and autochthonous organic materials in marine and lascustrine sediments, since C-27 and C-28 sterols are most abundant in plankton and marine invertebrates (marine source) whereas C-29 and C-27 sterols are more prominent in higher plants and animals (terrigenous source).

Also noteworthy in Figure 4 are the large relative abundances of C-28 and C-30 sterols at both sites. The C-28 sterols are most commonly associated with marine diatoms and bacteria (26,27). The C-28 abundance is not surprising, since the Peruvian coast is known to contain large amounts of diatoms. The C-30 abundance is due primarily to dinosterol, a 4-methyl sterol characteristic of dino-flagellate input (28). de Leeuw et al. (29) also observed large relative abundances of 4-methyl sterols in a Black Sea surface sediment and concluded that free, living dinoflagellates were the major contributors of organic matter to this sediment. The overall sterol pattern observed by de Leeuw was quite different from that observed by Lee et al. (30) in a similar Black sea sediment, however, indicating that the nature of phytoplankton living in the water column varies at different geographical sites and these variations are reflected in the organic matter contained in the underlying sediments.

Active biological reworking of organic carbon at the sediment-water interface is also supported by the data of Figure 5a. The most obvious feature of Figure 5 is the pronounced reduction in monounsaturated sterols (stenols) and diunsaturated sterols (stenol-dienes) below the upper (0-2 cm) horizon. This rapid depletion is probably due to biodegradation. It also appears that stenols are decreasing more rapidly with depth than saturated sterols (stanols); that is, there is not a corresponding increase in stanols associated with the decrease in stenols. This suggests that not only simple biogenic hydrogenation reactions are occurring, but also reactions which alter the basic hydrocarbon skeleton – possibly microbially mediated dealkylations.

NMR Studies of Intact Sediments. Figure 6 contains spectra of intact BX-6 core material from three separate core depths. The sedimentation rates listed in Table I indicate that this interval represents approximately 100 years of sedimentary history. While the spectra contain interesting features, such as the sharpening of certain resonances in the carboxyl region (165 - 180 ppm) and the loss of some resonances in the aromatic region (135 - 160 ppm) with depth, it is difficult to draw definitive conclusions regarding subtle diagenetic changes. These data are valuable, however, in the gross characterization of sedimentary organic matter. Table II contains the relative areas of peaks from selected regions of the spectra. Within the experimental uncertainty of such an analysis (approximately 10%), there appears to be little or no change in the gross composition of organic matter with depth at either site. One interesting result in the data of Table II is the suggestion that surficial sediments from BX-3 contain less methyl and methylene carbon than the corresponding sediments from BX-6. This is in contrast to the data of Figure 4, in which C-29 sterols are relatively

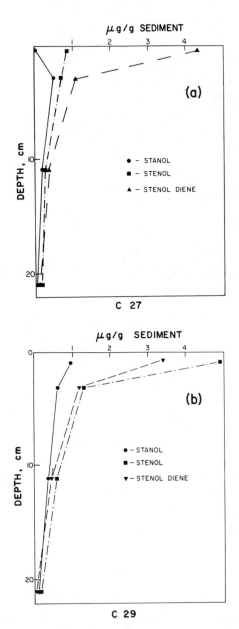

Figure 5. (a) Plot of total C-27 stenol and stanol concentrations with sediment depth, BX-3. (b) Plot of total C-29 stenol
and stanol concentrations with sediment depth, BX-3.

more pronounced in BX-3 surficial sediments than in BX-6. This further suggests that extractable lipids may not always be representative of the total organic carbon in sediments.

Table II. Intensities of Resonances from Selected Spectral Regions with Core Depth, Intact Sediments.

Chemical Shift (ppm)	Depth(cm)	Relative Integrated Peak Areas %					
		BX-6			BX-3		
		0-2	10-12	20-22	0-2	10-12	20-22
0-50		45	49	41	38	40	42
60-90		16	13	13	13	14	12
120-150		8	10	13	13	16	14
165-190		6	4	7	5	6	3

More definitive conclusions about subtle changes with depth can be drawn from spectra which exploit the time dependence of carbon magnetization. Figure 7a contains plots of cross polarization contact times vs. peak intensities for a number of different resonances. Clearly, different types of carbon atoms relax at different rates in these sediments. This is a completely expected result based on previous studies of carbon atoms in model compounds (31) and other geochemical matrices such as coal resins (32).

Figure 8 demonstrates how this time dependence can be exploited to improve spectral resolution and delineate subtle diagenetic changes in bulk organic matter. At short contact times (200 and 800 μsec), carboxyl and/or amide carbons are only partially magnetized (Figure 7b), and only the most prominent appear in the spectra. This improves resolution of the peaks at 172 and 175 ppm relative to the 1000 μsec spectra. The importance of this is apparent when comparing the 800 μsec spectra of Figure 8a (10-12 cm) and 8b (20-22 cm). In 8a, the 175 peak is more prominent, albeit slightly, whereas in 8b, the 172 peak dominates.

It is important to note that the experiments which generate the data of Figure 7 are a necessary part of any NMR studies of intact sedimentary carbon. Before characterizations can be made, it is first necessary to establish that there exist no selective relaxation pathways for any particular carbon type. Otherwise, quantitative comparisons such as those of Table II could not be made, since carbon types with fast relaxation would be underestimated. For example, Figure 7a indicates that spectra obtained with 800 μsec contact times (Table II) may underestimate the methyl and methylene content of the organic carbon in these sediments. Figure 7 further indicates that magnetic relaxation in these intact sediments is very rapid relative to previously studied carbonaceous materials. In general, 1000 μsec contact times have been used for studies of coal, kerogen and extracted humic/fulvic acids. We presume contact time experiments were done to maximize the magnetization of all carbon types. However, the optimum contact times indicated in Figure 7 are much shorter than 1000 μsec.

Further information about diagenetic alteration of sedimentary organic matter can be obtained by exploiting the time dependence of magnetic relaxation. For example, it has been shown by several workers that nonprotonated carbon atoms relax at much slower rates

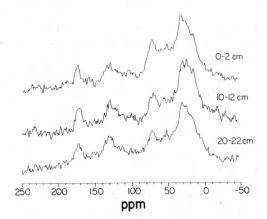

Figure 6. NMR spectra of intact core material, BX-6.

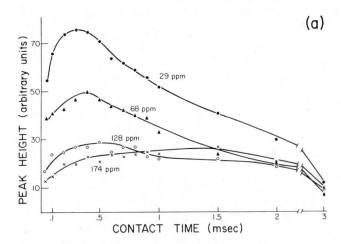

Figure 7. (a) Peak intensity plotted as function of cross
polarization contact time, selected resonances.

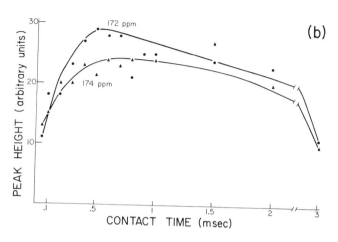

Figure 7. (b) Peak intensity plotted as a function of cross polarization contact time, carboxyl/amide resonances.

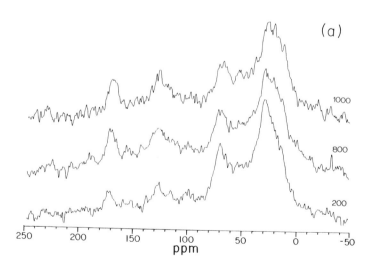

Figure 8. (a) NMR spectra obtained with varying contact times, 10-12 cm, BX-3.

(b)

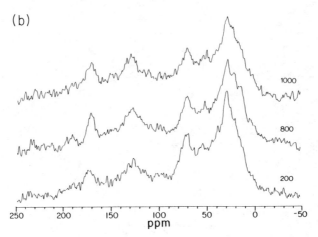

Figure 8. (b) NMR spectra obtained with varying contact times,
20-22 cm, BX-3.

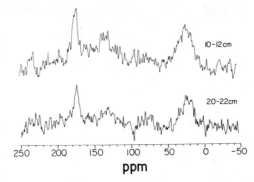

Figure 9. Dipolar dephased spectra of intact sediments, BX-3.

than protonated carbons (20). By delaying the onset of data
acquisition after cross polarization has ceased (dipolar dephasing),
this effect can be used to discriminate between protonated and non-
protonated carbon types. Figure 9 contains the dipolar dephased
spectra of Figure 8. These spectra were obtained with a 50 μsec
dephasing delay inserted prior to data acquisition. The most
prominent feature of Figure 9 is the greatly enhanced carboxyl
signal of both spectra. However, there is also a substantial signal
in the 130-145 ppm region of these spectra, indicating that most of
the aryl/olefinic carbons are non-protonated. The aryl/olefinic
spectra of the upper core section (10-12 cm) also appear more
intense than the corresponding peaks in the lower section. This
would indicate diagenetic alteration involving some loss of
aromatic/olefinic functional groups, provided that the signal-to-
noise ratios are sufficiently large so that the differences in the
spectra are real and the presence of paramagnetics are not affecting
relaxation rates to an appreciable degree. These questions are
currently being investigated in order to determine whether the
diagenetic pathway suggested is actually occurring.

Summary

In this paper we have demonstrated the utility and potential of time-resolved NMR spectroscopy for studying early diagenetic alteration of sedimentary organic matter. Conventional analyses such as the identification and quantification of lipid biomarkers give complimentary information. We anticipate that, as NMR techniques become more widely used in studies of sedimentary organic matter, the complimentary nature of conventional and spectroscopic results will become more apparent, leading to a better understanding of the chemical processes involved in early diagenesis.

Acknowledgments

This work was supported by a grant from the Petroleum Research Fund, administered by the American Chemical Society. Sediment samples were obtained by Professors P. Froelich, W. Burnett and M. Andreae of the Florida State University Oceanography Department. The assistance of D. Rosanske and T. Gedris of the FSU Magnetic Resonance Lab is greatly appreciated, as are the helpful discussions with P. Nichols and J. W. de Leeuw concerning sterol distributions.

Literature Cited

1. Tissot, B. P., Welte, D. H. "Petroleum Formation and Occurrence"; Springer - Verlag: New York, 1984; pp. 69-92.
2. Hatcher, P. G.; Spiker, E. C.; Szeverenyi, N. M.; Maciel, G. E. Nature, 1983, 305, 498.
3. Scnitzer, M.; Khan, S. U. "Humic Substances in the Environment"; Marcel Dekker: New York, 1972, p. 327.
4. Eglinton, G. In "Organic Geochemistry, Methods and Results"; Murphy, M. T. J.; Eglinton, G. Eds; Springer-Verlag: New York, 1966; Chap. 2.
5. Blumer, M. Pure Appl. Chem. 1973, 34, 591.
6. Huang, W. Y.; Meinschein, W. G. Geochem. Cosmochem. Acta 1978, 42, 1391.
7. Attaway, D. H.; Parker, P. L. Science 1970, 169, 674.
8. Gagosian, R.B. Geochem. Cosmochem. Acta 1975, 39, 1443.
9. Huang, W. Y.; Meinschein, W. G. Geochem. Cosmochem. Acta 1979, 43, 739.
10. Gaskell, S. J.; Eglinton, G.; In "Advances in Organic Geochemistry"; Tissot, B.; Bienner, F., Eds; Editions Technip, 1973; pp. 936-976.
11. Gaskell, S. J.; Eglinton, G. Nature 1975, 254, 209.
12. Nishimura, M.; Kayoma, T. Geochem. Cosmochem. Acta 1977, 41, 379.
13. Nishimura, M. Geochem. Cosmochem. Acta 1978, 42, 349.
14. Mallory, F. B.; Connor, R. L. Lipids 1971, 6, 149.
15. Wakeham, S. G.; Schaffner, C.; Giger, W. Geochem. Cosmochem. Acta 1980, 44, 403.
16. Pines, A.; Gibby, M. G.; Waugh, J. S. J. Chem. Phys. 1973, 59, 569.
17. Schaefer, J.; Stejskal, E. O. J. Am. Chem. Soc. 1976, 98, 1031.
18. Maciel, G. E. Science 1984, 226, 282.

19. Hagaman, E. W.; Woody, M. C. Fuel 1982, 61, 53.
20. Wilson, M. A.; Pugmire, R. J.; Grant, D. M. Org. Geochem.
    1983, 5, 121.
21. Wilson, M. A.; Collin, P. J.; Vassallo, A. M.; Russell, N. J.
    Org. Geochem. 1984, 7, 161.
22. Soutar, A.; Johnson, S.; Fisher, K.; Dymond, J. EOS 1981, 62,
    45.
23. Grob, K.; Krob, K., Jr. J. Chromatogr. 1974, 94, 53.
24. Lee, C.; Farrington, J. W.; Gagosian, R. B. Geochem. Cosmochem.
    Acta 1979, 43, 35.
25. Wardroper, A. M. K.; Maxwell, J. R.; Morris, R. J. Steroids
    1978, 32, 203.
26. Rubenstein, I.; Goad, L. J. Phytochem. 1974, 13, 485.
27. Orcutt, D. M.; Patterson, G. W. Comp. Biochem. Physiol. 1975,
    50B, 579.
28. Boon, J. J.; Rijpstra, W. I. C.; de Lange, F; de Leeuw, J. W.;
    Yoshioka, M.; Shimizu, Y. Nature 1979, 277, 125.
29. de Leeuw, J. W.; Rijpstra, W. I. C.; Schenk, P. A.; Volkman,
    J. K. Geochem. Cosmochem. Acta 1983, 47, 455.
30. Lee, C.; Gagosian, R. B.; Farrington, J. W. Org. Geochem. 1980,
    2, 103.
31. Alemany, L. B.; Grant, D. M.; Pugmire, R. J.; Alger, T. D.;
    Zilm, K. W. J. Am. Chem. Soc. 1983, 105, 2142.
32. Wilson, M. A.; Collin, P. J.; Vassallo, A. M.; Russell, N. J.
    Org. Geochem. 1984, 7, 161.

RECEIVED September 16, 1985

ORGANIC POLLUTANTS

11

The Biogeochemistry of Polychlorinated Biphenyls in the Acushnet River Estuary, Massachusetts

John W. Farrington, Alan C. Davis, Bruce J. Brownawell, Bruce W. Tripp,
C. Hovey Clifford, and Joaquim B. Livramento

Chemistry Department and Coastal Research Center, Woods Hole Oceanographic
Institution, Woods Hole, MA 02543

Analyses of PCBs in sediment, water and organisms by
high resolution gas chromatographic quantitative
determination of individual chlorobiphenyls revealed
marked compositional differences in segments of the
Acushnet River estuary ecosystem. Dramatic differences in chlorobiphenyl compositions in tissues of
lobster, crab and fish compared to compositions in
industrial PCB mixtures released to the environment
suggest that public health criteria based on PCB
industrial mixture determinations should be revised.
Predictions of bioconcentration based on K_{ow} of
chlorobiphenyls are of limited accuracy for several
chlorobiphenyls for which metabolism or membrane
transfer selectivity are apparently the major deter-
minants in some organisms. The importance of understanding the biogeochemistry of individual chlorobiphenyls as toxic compounds and as model compounds
for studies of biogeochemistry of organic matter in
aquatic ecosystems is discussed briefly.

An understanding of the biogeochemistry of organic matter in the
contemporary marine environment is important in order to facilitate
a greater understanding of: i) the C, N, S and P cycle; ii) interactions between organic compounds and biota, e.g. chemotaxis; iii)
interpretations of molecular paleontology in ancient sediments and
petroleum formation; iv) the inputs, fates and effects of pollutants as has been set forth in several reviews (1-6).

Our study reported herein primarily addresses this latter issue
although we think that xenobiotic compounds can be used as valuable
tracers of processes influencing most naturally occurring organic
compounds. This is analogous to the use of anthropogenic releases
of radioactive elements in the studies of the biogeochemistry of
metals (7, 8). Polychlorinated biphenyls are useful compounds in
this regard because of the wide range of solubilities and reactivi-
ties among individual chlorobiphenyls (9-12). In addition, there
are continuing concerns about the adverse impacts of environmental
burdens of PCBs with respect to human health and the viability of

valuable living natural resources even though releases to the
environment have been markedly reduced (12-14). This is the
result, in part, of the preferential accumulation of PCBs in parts
of contemporary environments such as landfills, toxic waste disposal areas, and aquatic sediments near prior effluent discharges,
and the probability of release of PCBs back to other components of
contiguous ecosystems long after the initial input to the environment (12, 15). For example, in estuaries there is a close dynamic
coupling between the atmosphere/water/sediment/biota at relatively
short time scales.

   Most of the previous research on the biogeochemistry of PCBs
has focused on the measurement of types of industrial mixtures
(e.g. Aroclor 1016, 1242, 1254) by packed column gas chromatography, although it was recognized that several of the individial
chlorobiphenyls had different types and intensities of biological
effects (12, 14). Recent advances in analytical methodology, particularly glass capillary gas chromatography with electron capture
detection (16, 17), and the more general availability of standards
of individual chlorobiphenyls via impressive synthesis and verification analyses (e.g. 18) have made it feasible to undertake in-
depth studies of the biogeochemistry of individual chlorobiphenyls
(10, 17-20).

   We report here on the distributions of several chlorobiphenyls
in samples of water, sediment and biota of the Acushnet River
Estuary - New Bedford Harbor, Buzzards Bay, Massachusetts, U.S.A.
Our general objective is to gain information of generic utility in
addition to providing specific data and interpretations of assistance to remedial action at this Superfund site. Our specific
objectives in this paper are to: i) document the composition of
individual chlorobiphenyls in biota normally harvested by commercial and recreational fishermen and discuss factors which could
lead to the observed distributions and potential implications for
public health standards for PCBs in fish; and ii) to investigate,
in a preliminary manner, the adherence of bioconcentration of PCBs
to predictions based on equilibrium assumptions and octanol/water
(K_{ow}) partition coefficients (21, 22).

   The study site is shown in Figure 1 which also presents a
generalized composite of PCB concentrations in surface sediments
based on analyses of hundreds of surface sediment samples by
several different laboratories (23). Concentrations range from
over 1 part per thousand (10^{-3} g/g dry weight) in segments of
the inner harbor sediments to generally less than 1 x 10^{-6} g/g
dry weight (ppm) in segments of the outer harbor area (Figure 1).
The commercial harvesting of lobsters in the harbor area and of
certain fish and bivalves in segments of the harbor is banned by
the Massachusetts Department of Public Health because of PCB concentrations in excess of the 5 ppm wet weight edible tissue guide-
lines (24). Warnings have been posted to inform recreational and
subsistence fishermen about the PCB pollution in these same areas.
Descriptions of the history and severity of PCB contamination in
the area leading to designation as a U.S. EPA Superfund site are
available (24, 25).

Sampling and Methods

The dates and types of samples are given in Table I and locations
of stations in Figure 1. Water samples were obtained using glass
sampling devices (26, 30). Sediment samples were obtained by
coring using a box core and careful sectioning (10). Biota were
obtained by hand collection techniques or by net hauls, and
precautions were taken to avoid contamination during sampling and
dissection (27).

Water samples were stored in precleaned glass carboys and
returned to the laboratory where they were filtered and extracted
within 18 hours of sampling. Filters were Soxhlet extracted with
hexane/acetone (1:1) for 24 hours and again with fresh hexane/ace-
tone for an additional 24 hours. Water and pore water samples were
extracted three times with CH_2Cl_2 in a separatory funnel. Extracts
were dried over Na_2SO_4, concentrated to near dryness, and hexane
was added with further concentration by rotary evaporation under
vacuum until hexane replaced the CH_2Cl_2.

Sediments were Soxhlet extracted with acetone/hexane and the
extract was concentrated (10). Biota samples were extracted by
aqueous KOH digestion followed by extraction of non-saponifiable
lipids into ethyl ether (28).

Column chromatography of the lipid or non-saponifiable lipid
extracts on alumina over silica gel columns, or silica gel columns,
to partially isolate polychlorinated biphenyls from other classes
of compounds used procedures described previously (10, 28, 29).

Chlorobiphenyls were quantified by high resolution capillary
chromatography using response curves generated for a standard of
each chlorobiphenyl. A 0.32 mm i.d. x 30 M SE-52 column (J & W
Scientific Company) installed in a Carlo Erba Model 2150 GC
equipped with a split/splitless injector and Ni-63 electron capture detector, interfaced with a Columbia Scientific Instruments
Supergrator 3 electronic integrator and a 30 m DB5 fused silica
column (J & W Scientific Company) installed in a Hewlett-Packard
Model 5840 GC equipped with Ni-63 electron capture detector and
splitless injector were used for analyses. A Finnigan 4510 quadrupole mass spectrometer interfaced with a Carlo Erba 4160 gas
chromatograph (0.32 mm i.d. x 25 M DB-5 bonded fused silica column
- J & W Scientific), and interfaced with a Finnigan INCOS 2300
data system and standard EI/CI ion source and PPNICI accessory,
was employed for GC-MS analyses to confirm that the compounds
under study were chlorobiphenyls.

Duplicate analyses of homogenate samples agree within ± 20%
based on replicate analyses of tissue homogenates of Mytilus
edulis. All data are corrected for recovery of internal standards
(chlorobiphenyls number 29 and 143), added at the time of extraction. Average recoveries were 80-95% for the different types of
samples.

Results and Discussion

Figure 2 presents representative glass capillary gas chromatograms
for two industrial Aroclor mixtures used in the area. Aroclor 1242
is very similar to Aroclor 1016 (not shown) which was the predominant mixture used, but for which we had no standard. Figures 3 and

Table I. Sampling Data (see Figure 1 for station locations).

Station	Date (year, month, day)	Sample Type*
67	830901	Ast. 67- 0.25 m^2 Sandia Hessler MKIII Box Core.
	790709	Pseudopleuronectes americanus (black back, winter flounder) 15-49 cm length, 3 each, fillets edible tissue homogenized; Lephopsetta maculata (sand flounder) 20-28 cm length, 3 each, fillets of edible tissue homogenized; Homarus americanus (lobsters) 2 each, tail and claw muscle tissue, and viscera analyzed for each individual; Neopanope taxons (green crab) 16 each, whole crabs homogenized.
233 approx. 1 mile south of Station 92 - not shown	800617	Homarus americanus (one each), tail and claw muscle tissue and viscera analyzed.
74	820922	Ast. 65 - 14-17 L composite of 2 L samples each hour during ebb and flood tides sampled with glass stoppered 2 L glass flask (29).
92	810724	Ast. 49 - 20 L sample obtained with glass Bodman type sampler (25).
48	780311	Clarks Point - Mercenaria mercenaria (hard shell clam, quahog) - pooled tissue samples, 2-10 individuals.
Fort Phoenix (intertidal area near hurricane barrier)	810504	Mytilus edulis (blue mussel) - homogenized pooled tissue, from 20-30 individuals.
Cleveland Ledge Light (eastern Buzzards Bay Station 227-not shown)	780320	Aequipecten irradians (scallops) homogenized pooled tissue from 10-15 individuals.

*Ast. means R/V Asterias Cruise No.

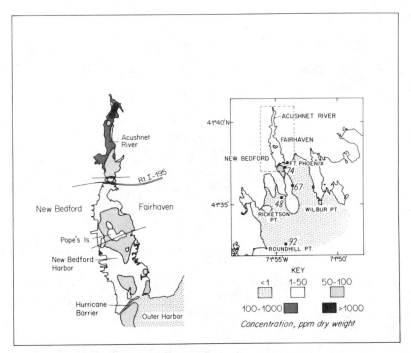

Figure 1. Study site location, station locations and contours of PCB concentrations in surface sediments.

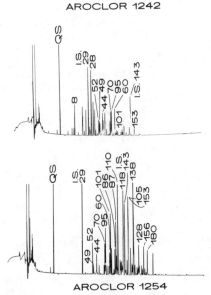

Figure 2. High resolution gas chromatograms of Aroclor 1242 and Aroclor 1254. QS refers to quantitation standard, IS refers to internal standard, and numbers are IUPAC numbers for chlorobiphenyls.

4 present glass capillary gas chromatograms of PCBs in lobster (Homarus americanus), a small crab (Neopanope taxons), surface sediments and tissue of Mytilus edulis (the common blue mussel). Table II gives the IUPAC numbers for chlorobiphenyls and corresponding chlorine substitution patterns as a key to the peaks identified in the gas chromatograms with IUPAC numbers. Figure 5 gives examples of packed column gas chromatograms of some of the same samples shown in Figures 2-4. The increased information content about PCBs in glass capillary gas chromagography is illustrated by comparison of Figure 5 with Figures 2-4.

The composition of the chlorobiphenyl mixtures present in the surface sediment, particulate matter in the water column, filtrate from water column samples, and mussels reflect a combination of Aroclor 1242 and 1254 mixtures of chlorobiphenyls although there are distinct differences in each sample type. The chlorobiphenyl composition of water column samples (gas chromatograms not shown) resembles that of the mussels and surface sediments, although further measurements of a larger set of samples may reveal small but significant differences in composition.

Examination of the gas chromatograms of PCBs in the samples of lobster and crab reveals the marked contrast in composition of PCBs in these types of biota samples and the composition of PCBs in water, sediment, bivalves and Aroclor mixtures. For example, the chlorobiphenyl composition of the lobsters are dominated by IUPAC chlorobiphenyl (CB) numbers 118, 153 and 138 while numbers 118 and 153 dominate the composition in the crab (Figure 2). Several more chlorobiphenyls, e.g. 28, 52, 44, 70, 95, 101, 110, are present in the mussel and sediment samples in addition to 118, 153 and 138 (Figure 2). PCBs in flesh of flounder species (L. maculata and P. americanus) were intermediate in composition between the lobsters and the mussels based upon examination of the gas chromatograms (not shown), i.e. 118, 153 and 138 predominate but not as much as in the lobsters. The possible reasons for these differences will be discussed below.

Table III contains data for concentrations of several individual chlorobiphenyls dissolved in the water or associated with particulate matter in the water column, and in surface sediments and pore waters. Concentrations of total PCBs estimated from these data are quite high for water, sediments and biota (2, 24). There are few quantitative data for individual chlorobiphenyls in water samples and this prevents extensive comparisons with other data of this type. The concentrations we report are within a factor of two to four of individual chlorobiphenyl data for a few water samples from the upper Hudson River (31). We have discussed water column and sediment data in more detail in two other papers (10, 30).

Concentrations of individual chlorobiphenyls in selected biota samples are presented in Table IV. The data for viscera and combined tail and claw muscle tissue from two different lobsters collected at the same location and one lobster collected at a second location provides an example of the variation in concentrations to be found for biota samples when comparing individuals. Concentrations in the viscera are expected to be higher than in muscle tissue because of the lipid rich nature of the viscera and the lipophilicity of the chlorobiphenyls.

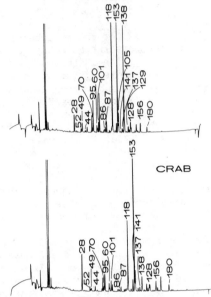

Figure 3. High resolution gas chromatograms of PCBs in lobster
tail and claw mussel tissue and whole green crab (N. taxons).
Also see legend Figure 2.

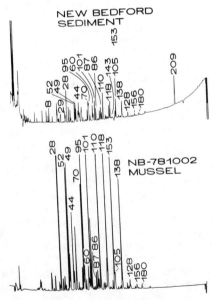

Figure 4. High resolution gas chromatograms of PCBs in surface
sediment and mussel (M. edulis) tissue. Also see legend Figure 2.

Table II.   Individual Chlorobiphenyls

IUPAC No.	Chlorine Substitution
28	2,4,4'
29	2,4,5
44	2,2',3,5'
49	2,2',4,5'
52	2,2',5,5'
60	2,3,4,4'
70	2,3',4',5
86	2,2',3,4,5
87	2,2',3,4,5'
95	2,2',3,5',6
101	2,2',4,5,5'
105	2,3,3',4,4'
110	2,3,3',4',6
118	2,3',4,4',5
128	2,2',3,3',4,4'
129	2,2',3,3',4,5
137	2,2',3,4,4'5
138	2,2',3,4,4',5'
143	2,2',3,4,5,6'
153	2,2',4,4',5,5'
156	2,3,3',4,4',5
180	2,2',3,4,4',5,5'

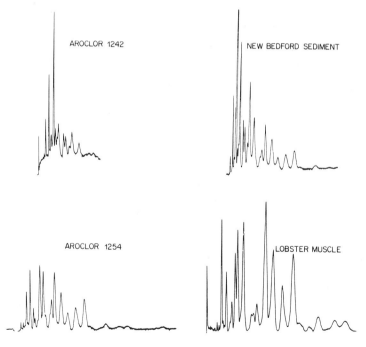

Figure 5.   Packed column gas chromatograms of PCBs in representative samples from the study site.

Table IIIa. Concentrations of Chlorobiphenyls of Water and Particulates (x 10^{-12} g/g) and Sediment (x 10^{-9} g/g)

Sample	Station No.	28	52	49	IUPAC No. 44	70	95	101	87
Ast. 67 3-5 cm Sediment	67	1200	360	400	390	590	1700	650	270
Ast. 67 3-5 cm Pore Water	67	160	94	100	89	120	450	210	69
Ast. 65 Surface Water	74	9.5	2.1	2.4	2.1	1.8	5.7	1.4	.43
Ast. 65 Bottom Water	74	7.1	1.7	1.9	1.5	1.3	4.6	1.3	.35
Ast. 65 Surface Particulates	74	3.0	1.1	1.4	1.0	1.3	3.7	1.9	.58
Ast. 65 Bottom Particulates	74	1.8	0.75	.87	.71	.98	2.5	1.3	.44
Ast. 49 Surface Water	92	.18	.14	.14	.19	.12	.44	.29	.07
Ast. 49 Bottom Water	92	12	3.8	3.4	3.7	3.2	8.7	2.1	.66

Table IIIb. Concentrations of Chlorobiphenyls of Water and Particulates (x 10^{-12} g/g) and Sediment (x 10^{-9} g/g)

Sample	Station No.	IUPAC No. 60	153	141	138	128	156	180	194
Ast. 67 3-5 cm Sediment	67	300	540	58	610	130	90	75	0
Ast. 67 3-5 cm Pore Water	67	54	210	21	250	41	32	35	0
Ast. 65 Surface Water	74	.94	.46	.02	.47	.08	.02	.02	.02
Ast. 65 Bottom Water	74	.76	.37	.02	.50	.07	.04	.04	.02
Ast. 65 Surface Particulates	74	.67	1.3	.14	1.8	.30	.18	.19	.05
Ast. 65 Bottom Particulates	74	.42	.99	.13	1.9	.36	.24	.20	.05
Ast. 49 Surface Water	92	.06	.16	.02	.14	.04	.02	.02	.02
Ast. 49 Bottom Water	92	1.8	.68	.11	.76	.16	.13	.13	.02

Table IVa.   Concentrations of Chlorobiphenyls of Selected Organisms
(x 10^{-9} g/g wet wt.)

Sample	Station Location or No.	IUPAC No.							
		28	52	49	44	70	95	101	87
Bivalves									
M. edulis (mussel)	Ft. Phoenix	52	56	70	64	130	340	180	57
M. mercenaria (clam)	48	22	13	15	15	15	36	22	8.6
A. irradians (scallop)	Cleve. Ledge	2.7	1.1	12	9.3	1.5	5.7	4.9	0.3

Table IVb.   Concentrations of Chlorobiphenyls of Selected Organisms
(x 10^{-9} g/g wet wt.)

Sample	Station Location or No.	IUPAC No.								Percent Water
		60	153	141	138	128	156	180	194	
Bivalves										
M. edulis (mussel)	Ft. Phoenix	52	140	4.1	130	21	14	5.6	0.0	86
M. mercenaria (clam)	48	7.3	21	3.0	26	4.0	3.6	3.6	0.4	88
A. irradians (scallop)	Cleve. Ledge	0.7	5.2	0.4	5.8	0.7	0.2	0.4	0.0	84

Table IVc. Concentrations of Chlorobiphenyls of Selected Organisms (x 10^{-9} g/g wet wt.)

Sample	Station No.	IUPAC No.							
		28	52	49	44	70	95	101	87
Crustacea									
N. texons (crab)	67	63	9.5	4.2	3.0	13	99	36	9.9
H. americanus									
Lob A Musc.	67	130	19	18	13	29	320	72	16
Lob A Visc.	67	2200	590	520	260	780	11000	2800	580
Lob B Musc.	67	71	6.3	3.2	2.7	9	160	18	5.8
Lob B Visc.	67	840	150	74	58	220	3600	760	160
Lob C Musc.	67	0.8	0	0	0	0	10	2.4	1.0
Lob C Visc.	67	360	36	36	41	58	1600	220	52
Ground Fish									
L. maculata (flounder)	67	180	290	270	240	300	1100	640	250
P. americanus (flounder)	67	21	10	21	11	83	200	150	84

Table IVd. Concentrations of Chlorobiphenyls of Selected Organisms
(x 10^{-9} g/g wet wt.)

Sample	Station No.	60	153	141	138	IUPAC No. 128	156	180	194	Percent Water
Crustacea										
N. texons (crab)	67	22	220	6.0	54	15	25	25	1.5	N.C.*
H. americanus										
Lob A Musc.	67	50	260	6.2	250	48	44	31	2.5	82
Lob A Visc.	67	1300	13000	270	11000	2000	1800	1500	110	N.C.
Lob B Musc.	67	18	110	1.4	91	21	17	11	0.9	N.C.
Lob B Visc.	67	480	5500	65	4800	1000	890	800	59	N.C.
Lob C Musc.	67	1.0	29	0.3	26	3.9	3.7	3.7	0.4	N.C.
Lob C Visc.	67	150	2300	22	2200	440	310	290	34	N.C.
Ground Fish										
L. maculata (flounder)	67	150	770	72	750	120	100	84	0	82
P. americanus (flounder)	67	29	370	34	330	77	63	62	0	78

*N.C. = not calculated or measured.

The visual inspection of gas chromatograms (Figures 2-4) provides only a qualitative impression of the relative differences in
composition of PCBs in the biota and their habitat. Therefore, we
have calculated a parameter to provide a more quantitative means
for evaluating the differences in composition. We have chosen a
chlorobiphenyl, IUPAC No. 153, which is present in appreciable
quantities in the industrial mixture of Aroclor 1254 and was identified as one of the major components in the samples of biota from
the study area, and calculated the following ratio:

$$\frac{\dfrac{[\text{chlorobiphenyl i}]}{[\text{IUPAC No. 153}]} \quad \text{sample}}{\dfrac{[\text{chlorobiphenyl i}]}{[\text{IUPAC No. 153}]} \quad \text{PCB standard Aroclor 1254}} \tag{1}$$

where i = any specific chlorobiphenyl.

Calculated ratios for several chlorobiphenyls are given in
Tables V and VI. Values greater than 1.0 indicate that the chlorobiphenyl is enriched in that segment of the ecosystem being sampled, relative to No. 153. Values less than 1 indicate the opposite and values close to 1 indicate that the chlorobiphenyl has a
biogeochemical behavior close to that of No. 153. There are groupings of chlorobiphenyls in all three categories (> 1, $\approx$ 1, < 1),
(Tables V and VI).

The processes which change the chlorobiphenyl composition of
the Aroclor type mixtures once they are discharged to the estuary
are: i) volatilization – the lesser chlorinated biphenyls, e.g.
tri- and tetrachlorobiphenyls, would be partitioned to the atmosphere to a greater extent than the more chlorinated penta-, hexa-,
and heptachlorobiphenyls (12); ii) sorption – the complicated partitioning interactions between particulate matter, colloids, water
in the water column, and surface sediments, can have a marked influence on compositions of chlorobiphenyl mixtures (10, 30); iii)
microbial degradation – evidence to date suggests that lesser
chlorinated biphenyls would be more rapidly degraded than the more
chlorinated biphenyls (12); iv) selective uptake and metabolism by
marine biota – the influence of this on compositions of chlorobiphenyls in our biota samples is discussed in the next several
paragraphs.

The hypothesis has been advanced that changes in relative concentrations of lipid type compounds, when comparing aquatic biota
and their habitat, can be explained in large part by an estimate
of their tendency to partition into tissues which has been related
to octanol/water partition coefficients – K_{ow}s (21, 22). Table
VII presents tabulated data for K_{ow} and water to biota bioaccumulation concentration factors calculated from data in Tables III
and IV. Representative data from Table VII are plotted in Figure
6 in the manner of Mackay (21) and Chiou (22), who have reviewed
data on bioaccumulation of neutral hydrophobic compounds in aquatic
biota. The solid line is the expected distribution of data based
on Chiou's review (22) of predictability for equilibrium situations. Our data is different in an absolute sense than the data
used by Mackay and Chiou, because they used concentrations in biota

Table Va.   Concentration of Chlorobiphenyls Relative to No. 153

Sample	52	IUPAC No. 49	44	70	95	101	87
Ast. 67 3-5 cm Sediment	1.5	6.2	2.5	2.7	2.2	1.2	.85
Ast. 67 3-5 cm Pore Water	1.0	4.0	1.5	1.4	1.5	1.0	.55
Ast. 65 Water Surface	7.0	25	9.3	7.1	6.2	2.2	1.2
Ast. 65 Water Bottom	7.1	25	8.5	6.4	6.4	2.6	1.2
Ast. 65 Part. Surface	1.4	5.5	1.6	1.9	1.5	1.1	.57
Ast. 65 Part. Bottom	1.2	4.3	1.5	1.8	1.3	.93	.55
Ast. 49 Water Surface	2.1	7.3	4.1	1.8	2.0	1.9	.81
Ast. 49 Water Bottom	12	41	19	12	9.1	3.1	1.6

Table Vb.   Concentration of Chlorobiphenyls Relative to No. 153

Sample	60	IUPAC No. 141	138	128	156	180
Ast. 67 3-5 cm Sediment	8.9	.77	1.2	1.5	1.1	1.4
Ast. 67 3-5 cm Pore Water	4.1	.70	1.2	1.2	.95	1.7
Ast. 65 Water Surface	24	.33	.70	.70	.29	.48
Ast. 65 Water Bottom	24	.42	.92	.76	.48	.80
Ast. 65 Part. Surface	6.2	.58	.99	.97	.63	1.1
Ast. 65 Part. Bottom	5.0	.68	1.3	1.5	1.1	1.5
Ast. 49 Water Surface	5.7	1.1	.93	1.8	.56	.84
Ast. 49 Water Bottom	3.9	1.1	1.1	1.4	1.2	1.9

Table VIa. Concentration of Chlorobiphenyls Relative to No. 153

Sample	52	49	IUPAC No. 44	70	95	101	87
Bivalves							
M. edulis	0.93	4.2	1.6	2.2	1.8	1.3	.69
M. mercenaria	1.4	6.0	2.5	1.7	1.2	1.0	.69
A. irradians	.48	19	6.2	.71	.79	.94	.10

Table VIb. Concentration of Chlorobiphenyls Relative to No. 153

Sample	60	141	IUPAC No. 138	128	156	180
Bivalves						
M. edulis	6.1	.21	.97	.94	.63	.40
M. mercenaria	5.7	1.0	1.3	1.2	1.1	1.7
A. irradians	2.1	.57	1.1	.81	.25	.80

Table VIc. Concentration of Chlorobiphenyls Relative to No. 153

Sample	52	49	IUPAC No. 44	70	95	101	87
Crustacea							
N. texons	.09	.16	.05	.15	.33	.17	.08
H. americanus							
Lob A Musc.	.16	.58	.17	.27	.86	.28	.10
Lob A Visc.	.09	.33	.07	.15	.57	.21	.07
Lob B Musc.	.14	.24	.09	.20	1.1	.17	.08
Lob B Visc.	.07	.11	.04	.10	.47	.14	.04
Lob C Musc.	.05	0	0	.05	.24	.08	.04
Lob C Visc.	.05	.13	.06	.07	.51	.10	.03
Fish							
L. maculata	.86	2.9	1.1	.95	1.1	.84	.54
P. Americanus	.07	.47	.10	.56	.40	.40	.39

Table VId. Concentration of Chlorobiphenyls Relative to
No. 153

Sample	60	141	IUPAC No. 138	128	156	180
Crustacea						
N. texons	1.6	.21	.25	.44	.75	1.2
H. americanus						
Lob A Musc.	3.1	.14	.97	1.2	1.1	1.2
Lob A Visc.	1.6	.07	.86	.94	.88	1.1
Lob B Musc.	2.7	.07	.86	1.3	1.0	1.0
Lob B Visc.	1.5	.07	.89	1.2	1.0	1.4
Lob C Musc.	.48	.07	.91	.81	.81	1.3
Lob C Visc.	1.1	.07	.96	1.2	.81	1.3
Fish						
L. maculata	3.2	.64	.99	1.0	.81	1.1
P. americanus	1.3	.64	.91	1.3	1.1	1.7

Table VII.   Octanol-Water Partition Coefficients and Bioaccumulation Concentration Factors on a Wet Wt. Basis of Selected Chlorobiphenyls

Log BCF

IUPAC No.	Log K_{ow}*	(M. edulis) Mussel	(M. mercenaria) Clam	(H. americanus) Lobster A Muscle	Viscera	Lobster B Muscle	Viscera	(P. americanus) Flounder
28	5.69	3.74	3.66	4.14	5.37	3.87	4.95	3.35
52	6.09	4.42	4.06	3.95	5.44	3.47	4.86	3.67
49	6.22	4.47	4.08	3.88	5.35	3.13	4.50	3.95
44	5.81	4.48	4.11	3.80	5.09	3.11	4.44	3.72
70	6.23	4.84	4.17	4.20	5.63	3.69	5.07	4.66
95	6.55	4.78	4.07	4.75	6.28	4.44	5.81	4.56
101	7.07	5.10	4.41	4.70	6.28	4.10	5.72	5.01
87	6.37	5.12	4.54	4.57	6.13	4.13	5.56	5.29
60	5.84	4.74	4.16	4.72	6.15	4.28	5.71	4.49
153	7.75	5.48	4.83	5.75	7.46	5.37	7.08	5.90
138	6.96	5.42	4.81	5.78	7.39	5.42	7.12	5.85
128	6.96	5.42	4.81	5.78	7.39	5.42	7.12	5.33

*Brownawell and Farrington, 1985 (10) tabulated from several studies.

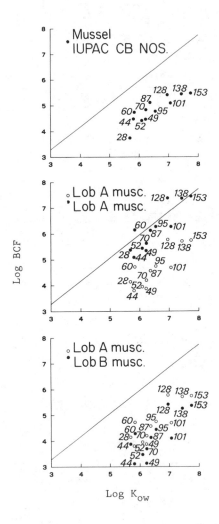

Figure 6. Plot of log K_{ow} vs. log BCF. Numbers are IUPAC numbers for chlorobiphenyls. Solid line is expected plot from Chiou (22) using his equation log (BCF) = 0.893 log K_{ow} + 0.607.

normalized to lipid concentrations. However, we have not done this for three reasons. First, we do not have lipid concentration data for these samples; unfortunately. Second, methods of lipid determination are operational in practice, and standardized or even common methodology for "total" lipid measurements are not commonly used. Third, incorporation of lipid values does not alter the expected linear correlation in a log-log plot and thus the lipid data are not essential to our interpretations and discussion which follow.

We have included examples for M. edulis and H. americanus muscle and viscera from an individual organism and H. americanus muscle from duplicate samples at the same station and sampling time to indicate the range of variability for these types of plots. The viscera data plot closer to the region of the theoretical line because of the lipid rich nature of viscera relative to muscle and whole mussel tissue. The data for the mussel are arranged in a more linear fashion or more tightly grouped around a linear plot than the lobster data. Our data agree with those reviewed by MacKay (21) and Chiou (22) in that K_{ow} provides an estimate of bioconcentration for neutral hydrophobic molecules.

There appear to be some significant departures from linearity in the log BCF-log K_{ow} plot, but we cannot be certain because of the limited set of data. Also, we must caution that our assessment is only preliminary in this regard because we have assumed for purposes of this paper that our data are representative of the field situation over a period of time sufficient for "equilibrium" to be approximated closely. There are very few water measurements of this type available and certainly more are needed to test these assumptions of equilibrium. The few measurements we have made over the last few years have indicated that fluctuations of a factor of 2 to 4 may occur in total PCBs in the water column in this area. Storm conditions may be an exception to this, but no data are available. Biota samples in this study were not obtained immediately after a storm. Repeated measurements of total PCBs in lobsters from this area by Massachusetts agency laboratories using packed column gas chromatography (23, 32), and less voluminous (unpublished) data from our laboratory by both packed column and glass capillary gas chromatography, suggest that concentrations of PCBs and more importantly relative amounts of chlorobiphenyls, do not fluctuate temporally by a factor of more than 4 in a given area, when averaging over a number of organisms. Data for the mussel M. edulis at a station near Ft. Phoenix (Figure 1) on the hurricane barrier, remained within a factor of 3 for the period 1978 to 1981 (25), encompassing the sampling period for most of the samples being considered.

There can be errors in estimation or measurement of K_{ow} values that could cause some departure from linearity in these plots by as much as factors of 2 to 4 (less than a log unit or factor of 10), which is much less than some of the departures from linearity for a few chlorobiphenyls. Chiou (22) has noted that non-steady state exposure concentrations, short exposure times, storage of compounds in non-lipid portions of tissues, and metabolism by organisms will lead to discrepancies between predicted and measured BCF-K_{ow} relationships.

We think that factors other than non-equilibrium conditions explain much of the departure from prediction of BCFs using K_{ow}'s. There is evidence to suggest selective metabolism of chlorobiphenyls in certain species (14) and it has been shown that some species of fish and crustacea contain mixed function oxidase activity that is probably capable of metabolizing certain biphenyls at different rates (33). Furthermore, the mixed function oxidase activity in selected species of fish and crustacea tested has been induced to higher levels of activity by exposure to xenobiotics while, thus far, only low levels of activity can be demonstrated in bivalves (33). This is interesting as it is consistent with our finding that the mixture of chlorobiphenyls in bivalves more closely approximates a mixture of Aroclors than does the mixture of chlorobiphenyls in lobster, crab and flounder.

Duinker and co-workers (20) noted results similar to ours for comparison of PCBs in Crangon sp. (a shrimp) and PCBs in its habitat. Ballschmitter and co-workers have published detailed analyses of chlorinated hydrocarbon pesticides and PCBs in various individual fish from different habitats, usually well removed from the immediate vicinity of point source inputs (34). The glass capillary gas chromatographic patterns they report are generally similar to those we have found for fish from our study site in that they depart significantly from an Aroclor 1254 or 1242 type mixture. There are some differences in detail among species in our study and in their studies, that are as yet unexplained, but could be related to specificity of uptake or metabolism by biota. For example, there is the marked difference of the presence of chlorobiphenyl No. 138 in the lobster and its absence in appreciable amounts in the small green crab (Figure 3), even though both are crustacea and inhabit the same benthic region.

This may be due to specificity of metabolism of certain chlorobiphenyls by species specific enzyme systems.

There are as yet no conclusions nor even a consensus hypothesis as to which types of chlorine substitution patterns govern ease of metabolism by enzymes. Some researchers favor the hypothesis that chlorine substitution at the 4,4' or combinations at the 3,5; 3'5' positions of the biphenyl molecule block ease of enzymatic epoxidation of easily accessible vicinal carbons (14, 35). Other researchers suggest that 2,2' or 6,6' substitutions or some combination reduce coplanarity of the biphenyl rings thereby causing a steric hinderance to enzymatic activity or transfer across membranes (14, 36). Our data on individual chlorobiphenyls in biota does not yet encompass a wide enough range of chlorobiphenyl structures to test these hypotheses which must be tested rigorously with isotopically labeled chlorobiphenyls in carefully controlled experiments in any event.

General Discussion

It should have been obvious from first principles and certainly reinforced by the advent of several sets of data by glass capillary gas chromatography, that packed column gas chromatographic analyses do not provide adequate information about marked compositional differences between species of biota and between species and their habitat. More detailed compositional information is important for understanding factors controlling the biogeochemical cycle of PCBs in the environment; more specifically aquatic eco- systems including estuaries. Our data have demonstrated this for the case of a severely polluted coastal estuarine area, both for the data discussed herein and for pore water, sediment and water column data presented and discussed elsewhere (10, 30).

Parameters such as solubilities and K_{ow}'s provide a first order predictive capability concerning bioconcentration in biota, but departures from predicted distributions due to kinetic factors involved with uptake and release and metabolic transformations have a marked influence. While equilibrium considerations are a good starting point, it is necessary to move beyond these to dynamical considerations to provide better general knowledge for such ques- tions as risk assessment in waste disposal to the ocean, and clean- up of severely polluted estuarine coastal areas. There has been some progress, but more research concerned with kinetic factors and dynamics in biogeochemical cycles of pollutant organics is needed.

The most important message contained in our glass capillary gas chromatographic analyses of PCBs, and those of others, is that public health standards are probably outdated. The edible portions of fish and lobsters in samples from our study area contain mix- tures of PCBs markedly different in composition compared to indus- trial Aroclor mixtures used in most experiments assessing adverse effects on animals. The 2 to 5 ppm (wet weight) total PCBs guide- line may be overprotecting or underprotecting public health. This dual edge sword problem could cut either way and needs evaluation in the public health research sector. Some attention has been focused on this issue in the European community (37). Since it has been established that individual chlorobiphenyls can have a range of potency in regard to biological activity (14), it is important to understand the biogeochemical cycles of individual chlorobiphenyls as well as the bulk mixture, for which there is a significant amount of information in regard to first order environ- mental behavior (12, 14).

Furthermore, as we emphasized in the introduction, PCBs can serve as model compounds for studying several aspects of processes active on a wide range of organic compounds. The full potential of this approach can be realized only via experiments and field programs involving determinations and interpretations of a range of types of individual chlorobiphenyls.

Acknowledgments

We wish to thank Capt. A. D. Colburn for assistance in sampling, Dr. N. M. Frew for GCMS analyses, and Ms. Nancy Hayward for assis- tance. Financial support was provided by the Andrew W. Mellon Foundation and the Mobil Foundation to the Coastal Research Center, Woods Hole Oceanographic Institution; the Education Office of the Woods Hole Oceanographic Institution/Massachusetts Institute of Technology Joint Program in Oceanography; and the U.S. Dept. of Commerce, NOAA, National Sea Grant Program under grants NA 83 AA-D-00049 (R/P-13) and NA 84 AA-D-00033 (R/P-17). This is Con- tribution No. 6043 from the Woods Hole Oceanographic Institution.

Literature Cited

1. Duursma, E. K.; Dawson, R., Eds. "Marine Organic Chemistry";
    Elsevier Scientific Publishing Co.: New York, 1981.
2. Brassel, S.; Eglinton G. In "Coastal Upwelling: Its Sediment
    Record", Part A; Suess, E.; Thiede, J., Eds.; NATO CONFERENCE
    SERIES, Plenum Press: New York, 1983; pp. 545–571.
3. Hunt, J. M. "Petroleum Geochemistry and Geology"; W. H. Free-
    man and Co.: San Francisco, CA, 1979.
4. Tissot, B. P.; Welte, D. H. "Petroleum Formation and Occur-
    rence"; Springer-Verlag: New York, 1978.
5. "Proceedings of a Workshop on Scientific Problems Relating to
    Ocean Pollution," National Oceanic and Atmospheric Administra-
    tion, 1978.
6. "Oil in the Sea; Inputs, Fates, and Effects," U.S. National
    Academy of Sciences, 1985.
7. Bowen, V. T.; Noshkin, V. E.; Livingston, H. D.; Volchok,
    H. L. Earth Planet. Sci. Lett. 1980, 49, 411–434.
8. Sholkovitz, E. R. Earth-Science Rev. 1983, 19, 95–161.
9. Mackay, D.; Mascarenhas, R.; Shiu, W. Y. Chemosph. 1980, 9,
    257–264.
10. Brownawell, B. J.; Farrington, J. W. Geochim. Cosmochim. Acta,
    1985, in press.
11. Rappaport, R. A.; Eisenreich, S. J. Environ. Sci. Technol.
    1984, 18, 163–170.
12. "Polychlorinated Biphenyls," National Academy of Sciences,
    1979.
13. Miller, S. Environ. Sci. Technol. 1982, 16, 98A–99A.
14. Kimbrough, R. D., Ed. Elsevier/North Holland Biomedical
    Press: New York, 1980; 406 pp.
15. Brown, M. P.; Werner, M. B.; Sloan, R. J.; Simpson, K. W.
    Environ. Sci. Technol. 1985, 19, 656–661.
16. Duinker, J. C.; Hillebrand, M.T.J. Environ. Sci. Technol.
    1983, 17, 449–456.
17. Ballschmiter, K.; Zell, M. Frezenius Z. Anal. Chem. 1980,
    302, 20–31.
18. Mullin, M. D.; Pochini, C. M.; McCrindle, S.; Romkes, M.; Safe
    S. H.; Safe, L. M. Environ. Sci. Technol. 1984, 18, 468–476.
19. Gschwend, P. M.; Wu, S. Environ. Sci. Technol. 1985, 19,
    90–96.
20. Duinker, J. C.; Hillebrand, M.T.J.; Boon, J. P. Netherlands
    J. Sea Res. 1983, 17, 19–38.
21. Mackay, D. Environ. Sci. Technol. 1982, 16, 274–278.
22. Chiou, C. T. Environ. Sci. Technol. 1985, 19, 57–62.
23. "Acushnet River PCBs: Data Management, Final Report U.S. EPA
    Contract 68–04–1009, U.S. EPA Region I," Metcalf and Eddy,
    Inc., 1983.
24. Weaver, G. Environ. Sci. Technol. 1984, 18, 22A–27A.
25. Farrington, J. W.; Tripp, B. W.; Davis, A. C.; Sulanowski, J.
    Proc. Internatl. Symp. Utiliz. Coast. Ecosyst.: Planning, Pol-
    lution, Productivity, 1982, in press.
26. Gagosian, R. B.; Dean, J. P.; Hamblin, R.; Zafiriou, O. C.
    Limnol. Oceanogr. 1979, 24, 583–588.
27. Grice, G. D.; Harvey, G. R.; Bowen, V. T.; Backus, R. H. Bull.
    Environ. Contam. Toxicol. 1972, 1, 125–132.

28. Farrington, J. W.; Risebrough, R. W.; Parker, P. L.; Davis, A. C.; deLappe, B.; Winters, J. K.; Boatwright, D.; Frew, N. M. Technical Report WHOI-82-42, Woods Hole Oceanographic Institution, Woods Hole, MA, 1982.
29. Galloway, W. B.; Lake, J. L.; Phelps, D. K.; Rogerson, P. F.; Bowen, V. T.; Farrington, J. W.; E. D. Goldberg, Laseter, J. L.; Lawler, G. C.; Martin, J. H.; Risebrough R. W. <u>Environ. Toxicol. Chem.</u> 1983, 2, 395-410.
30. Brownawell, B. J.; Farrington, J. W. In "Marine and Estuarine Geochemistry"; Sigleo, A. C.; Hattori, A., Eds.; Lewis Publishers, 1985, in press.
31. Bush, B.; Simpson, K. W.; Shane, L.; Koblintz, R. R. <u>Bull. Environ. Contam. Toxicol.</u> 1985, 34, 96-105.
32. Report of the Massachusetts Division of Marine Fisheries, 1984.
33. Stegeman, J. J. In "Polycyclic Hydrocarbons and Cancer"; Gelboin, H. V.; Ts'o, P.O.P., Eds.; Academic Press: 1981; Vol. 3, pp. 1-60.
34. Ballschmitter, K.; Buchest, H.; Bihler S.; Zell, M. <u>Frezenius Z. Anal. Chem.</u> 1981, 306, 323-329.
35. Shulte, E.; Acker, L. <u>Naturwiss.</u> 1974, 61, 79-81.
36. Shaw, G. R.; Connel, D. W. <u>Aust. J. Mar. Freshw. Res.</u> 1982, 33, 1057-1070.
37. "Nederlandse Staatscourant nr. 239 6 December 1984" available from State Institute for Quality Control of Agricultural Products (RIKILT), Wageningen, The Netherlands.

RECEIVED October 11, 1985

12

Polychlorinated Biphenyls and Hydrocarbons

Distributions among Bound and Unbound Lipid Fractions of Estuarine Sediments

H. R. Beller[1] and B. R. T. Simoneit

Environmental Geochemistry Group, College of Oceanography, Oregon State University, Corvallis, OR 97331

Selective extraction was used to operationally determine the quantitative and qualitative distributions of
PCB's and saturated hydrocarbons among free lipid
(FL), humic acid (HA), and humin (HU) fractions of
four contaminated estuarine sediments. In all samples, over 90% of the total sedimentary PCB's and
hydrocarbons were extracted with FL fractions. Bound
(HA and HU) and free assemblages of these compounds
may have derived from different sources. Two polar,
chlorinated pollutants also detected in this study,
hexachlorophene (HCP) and pentachlorophenol (PCP),
were proportionately more concentrated in bound fractions than the non-polar compounds; HCP was detected
only in HA fractions and was probably chemically bound
to refractory organic matter. Selective extraction
is a promising technique for investigating strongly
bound polar pollutants, such as HCP, which apparently
are not recovered by conventional solvent extraction.

Suspended and bottom sediments are widely regarded as a sink for
PCB's and other hydrophobic organic pollutants released into aquatic
systems. The mechanism commonly proposed to explain the affinity of
PCB's for sediments is equilibrium sorption or partitioning, which
is a function of the aqueous solubility of PCB isomers and the
attractiveness of the sedimentary matrix to PCB's (1-4). Field and
laboratory studies indicate that sedimentary organic matter plays a
fundamental role in PCB-sediment associations (5-12) and that humic
substances are important components of the sedimentary organic matrix for such associations (6,13-16). For example, Choi and Chen
(6) found that sedimentary PCB and DDT concentrations were linearly
related to the humic acid content of sediments from Los Angeles
Harbor. Pierce et al. (15) determined that humic acid could account

----

[1]Current address: Tetra Tech, Inc., 11820 Northup Way, Bellevue, WA 98005.

0097-6156/86/0305-0198$06.00/0
© 1986 American Chemical Society

for over half the sorption capacity of marine sediments in a laboratory experiment involving p,p'-DDT; similar results were obtained in
an experiment with hexachlorobutadiene (13). More recent studies
have demonstrated that PCB isomers and DDT can become associated
with or incorporated into dissolved humic substances (17-19) and
Lichtenstein et al. have shown that p,p'-DDT can become partially
"bound" (solvent inextractable) in wetted soils after brief incubation (20). While the incorporation of synthetic, hydrophobic pollutants into refractory organic matter has been implicated by these
studies, it has not been explicitly investigated in the field.

The objective of this study was to determine, by selective
extraction, whether or not non-polar, non-ionic pollutants (e.g.,
PCB's and petroleum hydrocarbons) can become "bound" in natural
sediments, and, if so, whether their qualitative distributions provide information about the incorporation process.

The incorporation of non-polar compounds into sedimentary organic matter is of interest in a pure geochemical sense, but also
has practical environmental implications. "Bound" pollutants would
be undetected in many environmental studies because such studies
typically employ simple solvent extraction; the quantitative significance of bound pollutants must be ascertained to produce an accurate assessment of contamination. Also, the biological availability
of sedimentary pollutants could depend on the nature of their association with organic matter. For example, "bound" PCB's entrapped
within humic matrices might be less accessible to benthic fauna than
"free" PCB's.

Methods

Figure 1 presents a flow diagram of the selective extraction procedure. This procedure was designed to effect exhaustive extractions
at each stage and to treat FL, HA, HU fractions as similarly as
possible. Dried sediment samples (ca. 90 to 140 g, dry wt) were
Soxhlet extracted with azeotropic toluene/methanol (1:3, v/v) for
over 95 hr (ca. 115 cycles) with a solvent change at 30 hr; all
solvents used were distilled-in-glass and of high purity (Burdick
and Jackson). Humic substances were extracted with five successive
one liter additions of solvent-cleaned 0.2 N KOH. The combined
aqueous extract was filtered through a Whatman GF/A glass fiber
filter (to remove suspended sediments from the extract), acidified
(to precipitate the humic acid), and re-filtered (to separate humic
and fulvic acids). The humic acid precipitate was rinsed through
the filter with 0.2N KOH, resulting in a basic humic extract without
fine particles. The filters and trapped particles were added to the
residual sediment. The purpose of saponifying humic fractions was
to disperse the polymers in the presence of organic solvents, thus
releasing entrapped or sorbed compounds. The FL extracts were
saponified to make them comparable to HA and HU fractions. The
strong base used to saponify FL, HA, and HU fractions converted DDT
(dichlorodiphenyltrichloroethane) to DDE (dichlorodiphenyldichloro-
ethylene) by dehydrohalogenation; thus, DDT is reported as DDE in
this study. Separatory funnel extractions of saponified extracts
were carried out at >pH 12 and <pH 2. The organic extracts from
liquid-liquid extraction were rotoevaporated to near dryness and

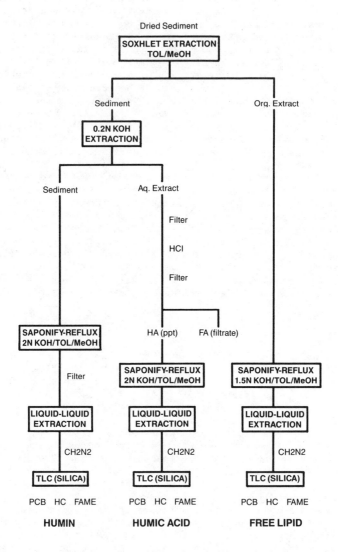

Figure 1.   Flow diagram of selective extraction procedure.

treated with diazomethane according to the procedure of Fales et al. (21). Extracts were methylated to facilitate the analysis of fatty acids (results not reported in this paper). Derivatized extracts were subjected to preparatory thin layer chromatography (TLC) on Analtech silica GHL plates; the elution solvent was 6.6% diethyl ether/hexane. Bands were visualized by exposing the plates to iodine vapor. Saturated hydrocarbons and chlorinated compounds were collected together in a broad band directly beneath the solvent front. These hydrocarbon/PCB fractions were dissolved in hexane and treated with metallic mercury to remove elemental sulfur prior to gas chromatographic analysis.

Examination of TLC fractions beneath hydrocarbon/PCB bands indicated that certain PCB isomers did not elute with the collected bands and were not analyzed. However, collection techniques were consistent and quantitative and qualitative intrasample comparisons are valid inasmuch as they are based on the same pool of isomers.

Chlorinated compounds were analyzed by gas chromatography–electron capture detection (GC–ECD) on a Vista 44 GC System equipped with a DB-1701 fused silica, capillary column (30 m, 0.25 mm i.d.; J & W Scientific). Helium and nitrogen were the carrier and make–up gases, respectively. The injector and detector temperatures were 260°C and 285°C. The column oven was temperature programmed from 65°C (held 1 min) to 230°C at 25°C/min, held at 230°C for 7.4 min, and then heated to 245°C at 10°C/min. The linear velocity at maximum temperature was ca. 50 cm/sec. All GC injections in this study were splitless with the inlet purged 1 minute after injection. The retention time of decachlorobiphenyl under these conditions was approximately 26 minutes.

Non-chlorinated hydrocarbons were analyzed by gas chromatography–flame ionization detection (GC–FID) with a Hewlett–Packard Model 5890 GC fitted with a 30 m, 0.25 mm i.d. DB-5 column (J & W Scientific). As for the ECD analyses, helium and nitrogen were the carrier and make–up gases. The column oven was programmed from 65°C (held 5 min) to 130°C at 10°C/min and then heated to 275°C at 5°C/min. The linear velocity at 275° was ca. 20 cm/sec. Injector and detector temperatures were 290°C and 325°C.

Gas chromatography–mass spectrometry (GC–MS) was carried out on selected samples either to confirm compounds identified by retention times or to investigate the identities of other peaks of interest. Samples were run on a Finnigan Model 9610 GC coupled to a Finnigan Model 4021 quadrupole MS. Mass spectrometric data were acquired and processed with a Finnigan–INCOS 2300 Data System. The MS was operated in electron impact mode with 70 eV electron energy. Gas chromatographic conditions were comparable to those used for GC–FID.

PCB quantification is difficult because the ECD has a variable response to the 209 possible isomers depending on their degree of chlorination, and secondarily, on their chlorine substitution patterns (22–26). While this study only required relative (intrasample) PCB quantification, detector response still had to be standardized so FL, HA, and HU assemblages with distribution maxima in different chlorine content ranges could be compared reliably. The co-elution of isomers and the cost of authentic isomer standards precluded isomer-specific quantification. Thus, PCB peaks were grouped according to their retention times (strongly related to

number of chlorine substituents) and the summed area of each group
was corrected with an appropriate response factor. Total PCB concentrations were estimated as the sum of all individual group concentrations. Retention time boundaries for each homolog group (di-
through decachlorobiphenyl) were constructed from the retention
times of peaks of known chlorine content in a 1:1:1 (wt/wt/wt)
standard mixture of Aroclor 1242:1254:1260 (standard used in 25).
This technique was possible because retention times were very reproducible and virtually all sample peaks corresponded to peaks in the
standard. Non-PCB peaks (e.g., methylated chlorinated phenols) were
excluded from this quantification. Relative response factors were
estimated for each homolog group using two types of data from recent
literature: the abundances of isomers in Aroclors 1242, 1254 and
1260 (27,28) and the response factors of each of these isomers (24).
Forty five isomers including representatives from all homolog groups
were chosen that account for over 70 mole percent of the chlorobiphenyls in each of the Aroclor mixtures. Average homolog group
response factors relative to decachlorobiphenyl, the co-injection
standard, were calculated from the 45 isomers. Selected isomer
standards were used to verify that literature values were applicable
to the GC-ECD system used in this study.

This quantification technique was tested on a commercial mixture of known composition, Aroclor 1242. The weight percentages of
each homolog class in the Aroclor were determined and compared to
approximate values provided by the manufacturer (Table I). Agreement between the determined and manufacturer's values was considered
sufficient for this study.

Table I.   Homolog Composition of Aroclor 1242 (wt. %)

	1Cl	2Cl	3Cl	4Cl	5Cl	6Cl	7Cl	8Cl	9Cl
Manufacturer	1.0	16.0	43.0	27.0	9.0	4.0	–	–	–
Determined	–	14.0	45.7	24.5	9.5	2.9	2.1	1.3	0.1

Saturated hydrocarbons were quantified with perdeuterated tetracosane as a co-injection standard. Total hydrocarbon concentrations were determined by electronically integrating the area above a
blank baseline; resolved hydrocarbon concentrations were determined
by integrating peak areas above an unresolved envelope. Individual
n-alkanes were identified by comparison of sample peak retention
times to those of an external standard mixture.

Sedimentary organic carbon contents were determined by wet
oxidation and dry combustion according to the procedure of Weliky et
al. (29).

Sample Descriptions

All samples consisted of fine-grained, organic matter-rich sediments
from areas of known PCB contamination.

Samples NB(0-3) and NB(29-31) were derived from the 0-3 cm and
29-31 cm horizons of a large volume box core taken at Station 67 of
Summerhayes et al. (30) in Buzzards Bay, near New Bedford, Mass.
The sedimentation rate at this site, which borders a dredged naviga-

tion channel, is unknown. The organic carbon contents of NB(0-3)
and NB(29-31) were 5.76% and 4.79%. The Acushnet River and adjacent
areas of Buzzards Bay are severely contaminated with PCB's due to
chronic releases from two capacitor manufacturing plants (31-33).
     Sample HR (3.70% organic carbon) was collected with a Shipek
grab sampler from a marginal cove in the Hudson River approximately
60 river miles north of the southern tip of Manhattan (1). The
sample consisted of the top 10 cm of sediment, which according to a
^{137}Cs profile of this core, encompassed 25 years of sedimentation
(Dr. R. Bopp, pers. comm.). PCB contamination in the Hudson River,
due predominantly to two capacitor manufacturing plants, is well
documented (1,34,35).
     Sample LA (3.71% organic carbon) was a composite of 0-2 cm
sections of sediment cores taken from near the Terminal Island
sewage outfall in Los Angeles-Long Beach Harbor (33°43'51" N ,
118°14'27" W).
     Sulfides, indicative of sulfate-reducing conditions, were prev-
alent in samples NB(0-3), NB(29-31), and LA.

Results and Discussion

Chlorinated hydrocarbons: PCB's and DDE. Table II displays the
distributions of PCB's and DDE among FL, HA, and HU fractions (as
percent of total sedimentary concentrations) and the total sedimentary concentrations on a dry wt. basis. The tabulated concentrations are intended only to denote relative trends in PCB distribution and are not considered absolute. Over 90% of the total PCB's
and DDE is associated with the FL fraction. This finding is not in
accord with sorption experiments (e.g., 15), which indicate that
over half of the PCB's or DDE should be associated with humic substances. Thus, sorption may not be a controlling factor in the
distributions of PCB's among organic matter fractions. While exhaustive Soxhlet extraction may have removed some humic acid-asso-
ciated PCB's along with free PCB's, a trial humic acid extraction of
a subsample that had not been solvent extracted indicated that such
procedural artifacts could not explain the discrepancy between the
results of this study and those of Pierce et al. (15).

Table II.   Relative   PCB and DDE Distributions[a,b]

Sample	FL	HA	HU	Total
NB(0-3)	91%	0.96%	8.1%	1.3 ppm PCB
NB(29-31)	99%	0.12%	1.2%	6.5 ppm PCB
HR	98%	0.87%	1.4%	1.4 ppm PCB & DDE
LA	99.9%	-	0.10%	100 ppb DDE
	ca. 100%			91 ppb PCB

[a] as percent of total sedimentary concentrations

[b] Total percentages may deviate from 100 due to rounding error.

     Qualitative PCB distributions of the New Bedford samples sug-
gest that free and bound PCB's may have derived from different
sources. Figure 2 presents capillary ECD chromatograms of the FL

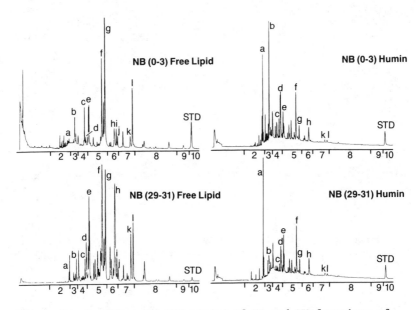

Figure 2. Capillary GC-ECD traces of FL and HU fractions of
samples NB(0-3) and NB(29-31). Peak a is pentachlorophenol (as
methyl ether derivative); the remaining peaks are PCB's. STD -
co-injection standard, decachlorobiphenyl. Approximate elution
ranges of di- through decachlorobiphenyl are delineated.

and HU fractions of samples NB(0-3) and NB(29-31). PCB's with five
or more chlorine substituents, which are highly resistant to microbial degradation (36), provide a good basis for qualitative comparison of sample fractions. The relative intensities of peaks f
through 1 in the two FL fractions (Figure 2) are markedly different
than in the corresponding HU fractions. Furthermore, the relative
intensities of several peaks differ considerably between the two FL
fractions (e.g., peak clusters g,h and k,l), yet these differences
are not apparent in the two HU fractions . In fact, the assemblages
of higher chlorinated isomers (peaks f through 1) in the HU fractions are very similar and may derive from the same source. If the
PCB sources for the FL and HU fractions of each sample were the
same, then a strong qualitative similarity between FL and HU fractions of a given sample would be expected, barring selective incorporation processes. Even selective incorporation or selective diagenetic processes could not explain the similarity of the HU fractions since the FL fractions reflect ambient assemblages of the same
PCB peaks at different relative concentrations.

   Given that PCB mixtures are subject to numerous types of environmental modification and that the agents of modification cannot be
specified with certainty for a given sample, it is not possible to
definitively assign sources to free and bound PCB assemblages based
on qualitative information. However, it seems likely that free
PCB's were derived from the untreated, PCB-contaminated effluents
that were discharged in large quantities by capacitor manufacturers
near the New Bedford site and upriver from the Hudson River site.
PCB's, often combined with oils for industrial use, could have been
components of particle surface coatings that were extractable with
organic solvents.

   Possible sources for bound PCB's are more problematic. However, a qualitative trend may yield useful information about bound
PCB sources: in samples NB(0-3), NB(29-31), and HR, the bound
fractions displayed narrower ranges of homolog groups and greater
relative concentrations of less chlorinated isomers than did FL
fractions (Table III). (HA fractions are not included because
procedural blanks revealed possible laboratory contamination.)

Table III. PCB Homolog Compositions of FL and HU Fractions (as wt. %)

Sample	Fraction	2Cl	3Cl	4Cl	5Cl	6Cl	7Cl	8Cl	9Cl	10Cl
NB(0-3)	FL	22	10	5.7	49	5.0	6.2	2.0	0.28	–
	HU	23	45	15	14	2.8	0.39	0.08	–	–
NB(29-31)	FL	6.7	6.7	8.0	49	21	6.9	2.2	0.09	–
	HU	30	12	15	34	6.8	1.3	0.29	–	–
HR	FL	9.4	31	19	22	7.1	2.7	4.4	1.4	3.4
	HU	18	46	19	13	1.8	0.75	2.0	0.61	–

   One possible explanation for this trend is that bound PCB's
derived from an atmospheric source and were associated with soot
particles. Several physicochemical properties of PCB's could favor
the generation of atmospheric assemblages enriched in less chlorinated isomers; laboratory studies of PCB's demonstrate that there is
a trend toward higher vapor pressure with decreasing chlorine
content (37) and that photodecomposition selectively destroys higher

chlorinated isomers (38). Furthermore, a possible source of atmospheric PCB's exists in the New Bedford area, as significant atmospheric loadings from incineration of PCB-contaminated sewage sludge
have been reported near the sample site (32,33). PCB's associated
with soot particles would be operationally defined as bound if they
were inextractable with organic solvents but were released by saponification.

Environmental transformation processes, such as microbial
degradation, may have affected qualitative PCB distributions and
may, in part, account for the greater abundances of less chlorinated
isomers in bound relative to free PCB assemblages. Oxidative
microbial activity tends to selectively degrade less chlorinated
PCB's (36,39-41) whereas reductive microbial dehalogenation is
thought to preferentially affect more highly chlorinated PCB's (42).
Thus, the effect of microbial activity would have depended on
sedimentary redox conditions and on the relative accessibility of
free and bound PCB's to microbes.

Saturated Hydrocarbons. Saturated hydrocarbons, like chlorinated
hydrocarbons, tended to distribute predominantly in the FL fractions
of sediments (Table IV). Similar quantitative results were reported
in studies of estuarine (43) and lacustrine (44) sediments.

Table IV.   Relative Saturated Hydrocarbon Distributions[a]

Sample	FL	HA	HU	Total
NB(0-3)	99.8% (10)[b]	0.11%	0.11% (45)	719 ppm
NB(29-31)	99.8% (10)	0.04%	0.17% (28)	637 ppm
HR	99.5% (18)	0.05%	0.48% (17)	204 ppm
LA	99.9% (10)	0.05%	0.09% (67)	64 ppm

[a] as percent of total sedimentary concentrations

[b] parenthetical values are resolved/total as percent

Several lines of evidence suggest that all of the samples were
from petroleum-contaminated areas: 1) FL fractions of all samples
contained a large unresolved complex mixture (UCM) of hydrocarbons
with a broad boiling point range, which is a strong indication of
petroleum contamination (45,46), 2) the levels of total hydrocarbons
were generally above levels consistent with biogenic production,
and 3) a series of (17α H, 21β H)-hopanes, considered to be good
markers of petroleum contamination (47), were the predominant triterpenoids in FL fractions from all study areas.

A general pattern of hydrocarbon distribution characterizes
samples NB(0-3), NB(29-31), and LA: the UCM is the predominant
component of FL fractions but is much less significant in HU fractions (see resolved/total values in Table IV; qualitative HA distributions will not be discussed as procedural blanks indicated possible laboratory contamination). This trend is most pronounced in

sample LA (Figure 3), the HU fraction of which contains n-alkanes characteristic of terrestrial plant waxes (a predominance of n-alkanes from $C_{23}-C_{31}$ with a high odd:even preference; 48-50) whereas the FL fraction has a large UCM with superimposed terrestrial alkanes. The NB samples display a similar trend but the HU fractions are more complicated (Figure 4). Peaks a and b, major components of the HU fractions, yielded mass spectra with molecular ions at m/z 362 and base peaks at m/z 81. These unidentified compounds may be cyclic olefins with elemental compositions of $C_{26}H_{50}$. Sample HR is an exception to this trend and its FL and HU hydrocarbon distributions are similar, as can be inferred from the resolved/unresolved values in Table IV.

One possible explanation for the general trend in hydrocarbon distribution is that detrital particles (e.g., vascular plant fragments) containing some solvent-inextractable hydrocarbons are major contributors to the HU hydrocarbon fractions, whereas the FL hydrocarbons (mostly petroleum hydrocarbons) exist as easily extractable components, perhaps as surface coatings on particles (43). A size fractionation study supports the idea that within a sedimentary matrix, hydrocarbons of different sources can be associated with distinct particle types or surfaces (51). If a portion of hydrocarbons in plant detritus were entrapped and unavailable to solvents during extraction, this material might be released during saponification.

Chlorinated Phenols: HCP and PCP. Two phenolic pollutants derivatized as methyl ethers, pentachlorophenol (PCP) and hexachlorophene (HCP or 2,2'-methylenebis(3,4,6-trichlorophenol)), eluted with PCB's during TLC separation. The identities of these compounds in samples were confirmed by comparison of retention times and mass spectra to those of methylated standards. Salient features of these mass spectra are described by Buhler et al. (52).

In the two samples in which its presence was confirmed, HR and NB(0-3), HCP occurred only in HA fractions and was a major chlorinated component of these fractions. Figure 5 depicts this distribution pattern in sample HR. Note that the chlorinated phenols, HCP and PCP, were essentially the only chlorinated compounds in the HA fraction. HCP may have occurred in the HA fractions of samples LA and NB(29-31) but was not confirmed due to low concentrations. HCP was not detected in any blanks.

The occurrence of HCP in HA fractions and its absence in FL fractions suggests that the pollutant was strongly associated with organic matter and was probably deposited in the sediments in bound form. HCP may have been covalently bound to organic matter and released hydrolytically during base treatment. Laboratory studies support these observations: Miller et al. (53) demonstrated that HCP covalently binds to rat tissue protein (in vitro) and Mathur and Morley (54) showed that a structurally similar compound, methoxychlor (2,2'-bis(p-methoxyphenyl)1,1,1-trichloroethane), strongly associated with a synthetic humic acid.

While some HCP may have been recovered with FL fractions if Soxhlet extractions had been performed at lower pH's, PCP distributions (Table V) indicate that the 3:1 methanol/toluene mixture was capable of efficiently extracting phenolic compounds and that inter-

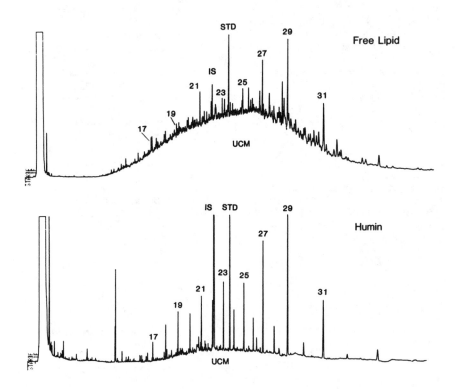

Figure 3. Capillary GC-FID traces of saturated hydrocarbons in
FL and HU fractions of sample LA. STD - co-injection standard,
perdeuterated tetracosane; IS - internal standard, p-terphenyl;
UCM - unresolved complex mixture. Chain lengths of odd-numbered
n-alkanes are denoted.

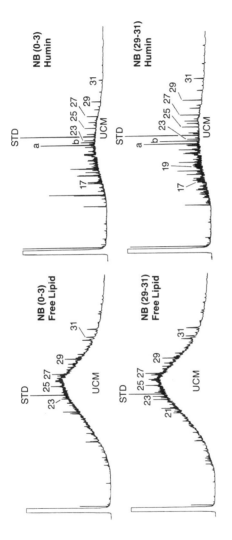

Figure 4.   Capillary GC-FID traces of saturated hydrocarbons in FL and HU fractions of samples NB(0-3) and NB(29-31). Legend as in Figure 3.   Peaks a and b are discussed in text.

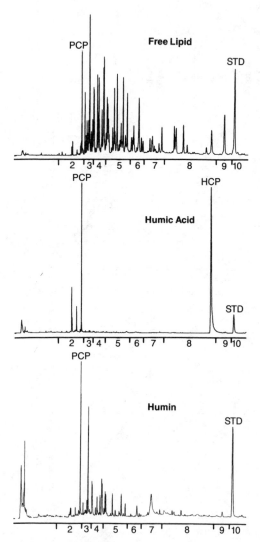

Figure 5. Capillary GC-ECD traces of FL, HA, and HU fractions of sample HR. PCP - pentachlorophenol (as methyl ether derivative); HCP - hexachlorophene (as dimethyl ether derivative); STD - decachlorobiphenyl. Approximate PCB homolog elution ranges are delineated.

sample differences in phenolic distributions were controlled by
factors other than the pH of Soxhlet extraction (which was the same
for all samples).
The detection of HCP in environmental samples is noteworthy not
only because of its unique distribution, but also because it is
seldom reported in environmental literature. Apparently, only two
papers document environmental HCP distributions and both were written in the early to mid 1970's. HCP concentrations in the ppb range
were reported for municipal sewage effluents (52) and for water and
sediments near sewage outfalls (55). The commercial use of HCP, a
powerful bacteriostatic agent and germicide, has been greatly reduced since a ban was instituted by the FDA in 1972. However, HCP
is very persistent: a laboratory study revealed that HCP was apparently not degraded in river water and had a "half life" of 290 days
in estuarine sediments at 22 °C (56). Furthermore, the environmental release of HCP after 1972 is plausible since some hospitals
still use the chemical and may discharge their effluents through
municipal sewage treatment plants.
　　PCP, a general biocide widely used as a wood preservative, may
also have been chemically bound to sediments, although the evidence
for this is not as strong as that for HCP. Table V, which presents
PCP concentrations for FL, HA, and HU fractions of all samples,
shows that PCP was a more significant component of bound fractions
than were the less polar PCB's. Similarly, Murthy et al. found that
PCP and a microbially methylated metabolite, pentachloroanisole,
bound to HA and HU fractions of aerobic and anaerobic soils over a
24 day laboratory incubation period (57).

Table V. Relative PCP Distributions (as % of total sed. conc.)

Sample	FL	HA	HU
NB(0-3)	38%	17%	45%
NB(29-31)	74%	3%	23%
HR	86%	9%	5%
LA	94%	2%	3%

Conclusions

1) Over 90% of total sedimentary PCB's and petroleum hydrocarbons
were recovered with FL fractions.

2) Qualitative PCB and hydrocarbon distributions among organic matter fractions suggest that free and bound assemblages may derive
from different sources in a given sediment.

3) Polar pollutants, such as chlorinated phenols, can occur largely
as chemically bound components of refractory organic matter.

4) Selective extraction could be a useful technique for investigating chemically bound pollutants, which are not detectable by
conventional solvent extraction.

Acknowledgments

Samples and background information were generously provided by Bruce
Brownawell and Dr. J. Farrington (New Bedford samples), Dr. R. Bopp
(Hudson River sample), and Tony Phillips (Los Angeles sample). We
thank Dr. F. I. Onuska for donating the 1:1:1 Aroclor standard and
Dr. F. Prahl, James Butler, and Orest Kawka for their helpful com-
ments on the investigation and manuscript.

Literature Cited

1. Bopp, R.F.; Simpson, H.J.; Olsen, C.R.; Kostyk, N. Env. Sci.
    & Tech. 1981, 15, 210-16.
2. Chiou, C.T.; Peters, L.J.; Freed, V.H. Science 1979, 206, 831-
    32.
3. Dexter, R.N.; Pavlou, S.P. Mar. Chem. 1978, 7, 67-84.
4. Pavlou, S.P.; Dexter, R.N. Env. Sci. & Tech. 1979, 13, 65-71.
5. Abdullah, M.I.; Kveseth, N.J.; Ringstad, O. Water, Air Soil
    Poll. 1982, 18, 485-97.
6. Choi, W.W.; Chen, K.Y. Env. Sci. & Tech. 1976, 10, 782-86.
7. Halcrow, W.; Mackay, O.W.; Bogan, J. Mar. Poll. Bull. 1974,
    5, 134-36.
8. Hiraizumi, Y.; Takahashi, M.; Nishimura, H. Env. Sci. &
    Tech. 1979, 13, 580-84.
9. Karickhoff, S.W.; Brown, D.S.; Scott, T.A. Water Research
    1979, 13, 241-48.
10. Schwarzenbach, R.P.; Westall, J. Env. Sci. & Tech. 1981, 15,
    1360-1367.
11. Steen, W.C.; Paris, D.F.; Baughman, G.L. Water Research 1978,
    12, 655-57.
12. Wildish, D.J.; Metcalfe, C.D.; Akagi, H.M.; McLeese, D.W. B.
    Env. Contam. Tox. 1980, 24, 20-26.
13. Diachenko, G.W. Ph.D. Thesis, University of Maryland, 1981.
14. Haque, R.; Schmedding, D. J. Env. Sci. Health 1976, B11, 129-
    137.
15. Pierce, R.H., Jr.; Olney, C.E.; Felbeck, G.T., Jr. Geochim.
    Cosmochim. Acta 1974, 38, 1061-1073.
16. Sawhney, B.L.; Frink, C.R.; Glowa, W. J. Env. Qual. 1981, 10,
    444-48.
17. Carter, C.W.; Suffet, I.H. Env. Sci. & Tech. 1982, 16, 735-40.
18. Hassett, J.P.; Anderson, M.A. Env. Sci. & Tech. 1979, 13,
    1526-1529.
19. Landrum, P.F.; Nihart, S.R.; Eadie, B.J.; Gardner, W.S. Env.
    Sci. & Tech. 1984, 18, 187-92.
20. Lichtenstein, E.P.; Katan, J.; Anderegg, B.N. J. Agric. Food
    Chem. 1977, 25, 43-47.
21. Fales, H.M.; Jaouni, T.M.; Babashak, J.F. Anal. Chem. 1973,
    45, 2302-2303.
22. Ballschmiter, K.; Zell, M. Fresenius Z. Anal. Chem. 1980, 302,
    20-31.
23. Duinker, J.C.; Hillebrand, M.T.J. Env. Sci. & Tech. 1983,
    17, 449-56.
24. Mullin, M.D.; Pochini, C.M.; McCrindle, S.; Romkes, M.; Safe,
    S.H.; Safe, L.M. Env. Sci. & Tech. 1984, 18, 468-76.

25. Onuska, F.I.; Kominar, R.J.; Terry, K.A. J. Chromatography
    1983, 279, 111-18.
26. Duinker, J.C.; Hillebrand, M.T.J.; Palmork, K.H.; Wilhelmsen, S.
    B. Env. Contam. Tox. 1980, 25, 956-64.
27. Albro, P.W.; Corbett, J.T.; Schroeder, J.L. J. Chromatography
    1981, 205, 103-11.
28. Albro, P.W.; Parker, C.E. J. Chromatography 1979, 169, 161-66.
29. Weliky, K.; Suess, E.; Ungerer, C.A.; Muller, P.J.; Fischer, K.
    Limnology and Oceanography 1983, 28, 1252-59.
30. Summerhayes, C.P.; Ellis, J.P.; Stoffers, P.; Briggs, S.R.;
    Fitzgerald, M.G. WHOI Tech. Report, WHOI-76-115, 1977.
31. Metcalf and Eddy, Inc. "Acushnet Estuary PCB's Data Management-
    Final Report"; for EPA Region I, 1983.
32. Weaver, G. Env. Sci. & Tech. 1984, 18, 22A-27A.
33. Weaver, G. "PCB Pollution in the New Bedford, Mass. Area: A
    Status Report", Mass. Coastal Zone Management, 1982.
34. Bopp, R.F.; Simpson, H.J.; Olsen, C.R.; Trier, R.M.; Kostyk, N.
    Env. Sci. & Tech. 1982, 16, 666-76.
35. Hetling, L.; Horn, E.; Tofflemire, J. "Summary of Hudson River
    PCB Study Results", N.Y. State D.E.C., Tech. Paper #51, 1978.
36. Furukawa, K.; Tonomura, K.; Kamibayashi, A. Appl. Env.
    Microbiol. 1978, 35, 223-27.
37. Bopp, R. J. Geophys. Research 1983, 88, 2521-2529.
38. Hutzinger, O.; Safe, S.; Zitko, V. "The Chemistry of PCB's";
    CRC Press: Cleveland, Ohio, 1974.
39. Furukawa, K.; Matsumura, F. J. Agr. Food Chem. 1976, 24, 251-
    56.
40. Clark, R.R.; Chian, E.S.K.; Griffin, R.A. Appl. Env.
    Microbiol. 1979, 37, 680-85.
41. Tucker, E.S.; Saeger, V.W.; Hicks, O. B. Env. Contam. Tox.
    1975, 14, 705-13.
42. Brown, J.F., Jr.; Wagner, R.E.; Bedard, D.L.; Brennan, M.J.;
    Carnahan, J.C.; May, R.J.; Tofflemire, T.J. Northeastern
    Environmental Science 1984, 3, 167-179.
43. Van Vleet, E.S.; Quinn, J.G. J. Fish. Res. Board Can. 1978,
    35, 536-43.
44. Cranwell, P.A. Org. Geochem. 1981, 3, 79-89.
45. Boehm, P.D.; Quinn, J.G. Est. Coast. Mar. Sci. 1978, 6, 741-94.
46. Farrington, J.W.; Quinn, J.G. Est. Coast. Mar. Sci. 1973, 1,
    71-79.
47. Simoneit, B.R.T.; Kaplan, I.R. Mar. Env. Research 1980, 3, 113-
    128.
48. Eglinton, G.; Hamilton, R.J. Science 1967, 156, 1322-1335.
49. Simoneit, B.R.T. In "Chemical Oceanography"; Riley, J.P.;
    Chester, R., Eds.; Academic Press: N.Y., 1978; Vol. 7, Chap. 39,
    pp. 233-311.
50. Smith, D.J.; Eglinton, G.; Morris, R.J. Gecochim. Cosmochim.
    Acta 1983, 47, 2225-2232.
51. Thompson, S.; Eglinton, G. Geochim. Cosmochim. Acta 1978, 42,
    199-207.
52. Buhler, D.R.; Rasmusson, M.E.; Nakaue, H.S. Env. Sci. &
    Tech. 1973, 7, 929-34.
53. Miller, A., III; Henderson, M.C.; Buhler, D.R. Molecular
    Pharmacol. 1978, 14, 323-36.

54. Mathur, S.P.; Morley, H.V. B. Env. Contam. Tox. 1978, 20, 268–
     74.
55. Sims, J.L.; Pfaender, F.K. B. Env. Contam. Tox. 1975, 14, 214–
     20.
56. Lee, R.F.; Ryan, C. Proc. Workshop: Microbial Degradation
     Pollut. Mar. Envi., 1979, pp. 443–50. EPA–600/9–79–012.
57. Murthy, N.B.K.; Kaufman, D.D.; Fries, G.F. J. Env. Sci. Health
     1979, B14, 1–14.

RECEIVED September 16, 1985

Fate of Carbonized Coal Hydrocarbons in a Highly Industrialized Estuary

Terry L. Wade[1], Mahlon C. Kennicutt II[1], and Elizabeth G. Merrill[2]

[1]Department of Oceanography, Texas A&M University, College Station, TX 77843
[2]Department of Environmental Regulations, Jacksonville, FL 32207

Fluorescence and GC/MS analyses show that carbonized coal hydrocarbons are widespread contaminants of sediments in the Elizabeth River, Norfolk, Va. The highest levels are found in the vicinity of suspected sources and generally decrease with increased distance from these sources. Parent aromatic compounds are the predominant hydrocarbon component of carbonized coal and can be uniquely detected even in the presence of petroleum hydrocarbons. Carbonized coal products are a chronic source of priority pollutant polynuclear aromatic hydrocarbons in the Elizabeth River.

Determination of the quantities and types of anthropogenic hydrocarbons in the marine environment is essential in order to understand their fates and long- and short-term effects. Of particular interest are the polynuclear aromatic hydrocarbons (PNA's), because some of these compounds are known carcinogens. One of the major sinks for hydrocarbons released in the coastal environment is their incorporation into bottom sediments (1). Sediments near large urban areas may contain high concentrations of anthropogenic hydrocarbons due to their proximity to source areas (2). Resuspension of contaminated sedimentary material by natural (tides, storms, etc.) or by artificial (dredging, shipping, etc.) means can disperse these pollutants to areas much larger than were originally effected.

The Elizabeth River is located in a highly industrialized urban area and empties into lower Chesapeake Bay. The river is a major shipping channel for the Norfolk area and is regularly dredged to deepen channels or build new docking facilities. The industries along the river that could be considered potential sources of hydrocarbons include marine shipping terminals, ship drydock facilities, sewage treatment plants, wood preserving facilities, a coal-fired electric power plant, dredging operations, and a dredge disposal site.

Due to the large number of creosoted docking facilities, the existence of wood preserving plants, and the occurrence of creosote

spills in the past, creosote may be a major contributor of PNA's to
Elizabeth River sediments.

The objectives of this study were to characterize the hydrocarbon signature in carbonized coal products (creosote and coal
tar), and then using this signature, determine if carbonized coal
products are a contaminant of Elizabeth River sediments, and
whether this contamination is transported into Chesapeake Bay. The
carbonized coal product signature was developed using capillary gas
chromatography (GC), gas chromatography/mass spectroscopy (GC/MS),
and total scanning fluorescence.

Methods and Materials

Sediment samples were collected using a grab sampler (which samples
approximately the top 10 cm) aboard Old Dominion University's
research vessels, ODU-1 and Linwood Holton. The sample locations
are shown in Figure 1. The samples were stored frozen in clean,
solvent-washed jars until analyzed. Creosoted wood samples were
collected from areas adjacent to the Elizabeth River (Figure 1).
Three samples of refined creosote and one sample of coal tar were
also analyzed. Creosote samples from Atlantic Wood Industries, the
remaining operative creosoting facility on the Elizabeth River,
were not available. The woodstove soot sample was obtained from a
domestic woodstove in which predominantly hardwoods were burned.
The diesel stack soot sample was taken from the Linwood Holton.
The No. 2 Fuel Oil sample is of the type used locally for home
heating. The Kuwait Crude Oil was kindly supplied by Robert Brown
of Mote Marine Lab, Sarasota, Florida.

Sediment samples were thawed and mixed to ensure homogeneity.
The dry weight and percent water content of the sediments were
determined by drying approximately 5 g of the sediment at 105-110°C
for several hours until a constant weight was reached. Grain-size
distribution was determined by wet sieving and pipette analysis.

For hydrocarbon GC and GC/MS analysis, sediment or other solid
sample (such as creosoted wood, soot, etc.) was placed in a 50-ml
centrifuge tube; internal standards, n-eicosane (n-C20) and 3-
methylfluoranthene, were added and the sample was saponified/
extracted (Metholic KOH) (3). Sediment hydrocarbon concentrations
determined using the above test tube extraction technique were
within the range of concentrations determined by Soxhlet and
reflux/saponification extraction of larger sediment samples (4).
Liquid samples (such as creosote, oil, etc.) were analyzed by
dissolving the sample in dichloromethane. The aliphatic (f1) and
aromatic (f2) hydrocarbons were separated from other organics by
thin-layer chromatography (TLC) (5,6).

The TLC sample fractions were injected on a Hewlett-Packard
Model 5830 gas chromatograph equipped with a 25-m SE-54 fused
silica capillary column, utilizing a flame ionization detector
(FID). The signal from the FID was recorded by a Hewlett-Packard
Model 18850A reporting integrator. The gas chromatograph was
programmed from 70°C to 300°C at 10°C per minute.

Quantitative determinations of hydrocarbon concentrations were
made by comparing integrator area counts of the internal standard
with integrator area counts of the peaks or planimetry of the

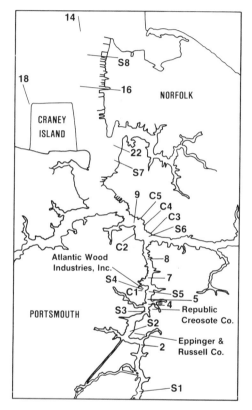

Figure 1. Sampling locations for creosoted wood (C) and sediments (S) and the location of the wood preserving facilities on the Elizabeth River, Norfolk, VA.

unresolved complex mixture. Qualitative determination of hydrocarbons in the samples were made by comparison of retention times
of known compounds, co-injection with known compounds, and by
GC/MS. A Finnigan OWA-Model 20 mass spectrometer coupled to a
Sigma 3B gas chromatograph was used for the GC/MS analysis.
Procedural blanks and standards were run systematically throughout
the analysis period to determine if contamination had occurred and
to ensure the proper functioning of the gas chromatograph. The
sample values reported are corrected for the presence of these
procedural blanks.

Samples for fluorescence analyses (15 g) were lyophilized,
ground to a uniform size with a mortar and pestle, and Soxhlet
extracted with hexane for 12 hours. All glassware and alundum
thimbles were pre-cleaned with Micro cleaning solution, washed with
nanograde solvents, and combusted at 500°C for at least 4 hours.
The extracts were concentrated to a volume of about 5 ml using a
Buchi Rotovapor R. Care was exercised at all times to ensure that
the extract was not brought to complete dryness to prevent
volatilization of lighter sample components. The volume of the
extract was brought up to 7 ml and stored at 4°C in the dark until
further analysis. This treatment minimizes photolytic losses and
the chemical interactions of the extracted compounds. A total
system blank was routinely run for every set of samples processed
and checked by both fluorescence and gas chromatography to ensure
acceptable blank levels.

The extract was scanned to acquire a three-dimensional
fluorescence spectra of emission wavelength, excitation wavelength,
and intensity using a Perkin-Elmer Model 650-40 UV-spectro-
fluorometer. This fluorometer is controlled by a Perkin-Elmer
Model 3500 or 3600 data station. The total fluorescence
excitation-emission wavelength array is filled for each sample by
sequential stepping of the excitation monochromator over the
wavelength range of interest and scanning with the emission
monochromator. Standard wavelength intervals are 200-500 nm for
both excitation and emission.

Results

GC and GC/MS. A detailed discussion of the results of the GC and
GC/MS analyses will be presented elsewhere (3). Gas chromatograms
of hydrocarbon distributions provide a visual comparison of the
similarities and differences between carbonized coal products, oil
products, and sediment samples. Chromatograms of the aliphatic
(f1) and aromatic (f2) fractions of Creosote A, the oil spill
sample, and sediments from Station S6 are shown in Figures 2, 3,
and 4, respectively. The f1 and f2 fractions may contain resolved
peaks and a UCM. The f1 fraction contains resolved peaks including
n-alkanes, pristane and phytane. The resolved f2 peaks were
divided into three groupings (Figure 2). The first group (PP)
consists of the 16 Environmental Protection Agency priority
pollutant PNA's (7). The second group consists of 25 other major
resolved peaks (MRP) that were found in most of the creosote
samples. These peaks were identified by comparison of peak
retention times, co-injection with standards, and/or confirmation

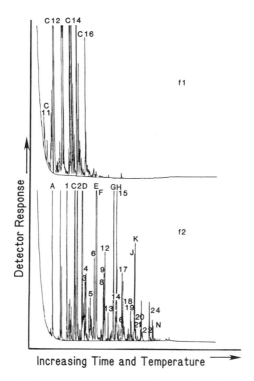

Figure 2. Gas chromatograms of the aliphatic (f1) and aromatic (f2) fractions of Creosote A. The f1 fraction contains identified n-alkanes in the range of n-C11 through n-C16.

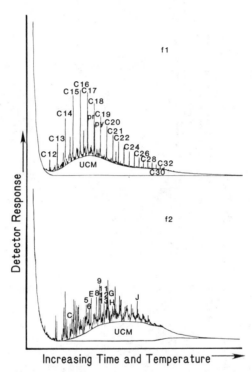

Figure 3. Gas chromatograms of the aliphatic (f1) and aromatic
(f2) fractions of the oil spill sample. The f1 fraction
contains identified n-alkanes in the range of n-C12 through n-
C32 and pristane (pr) and phytane (py). UCM-unresolved complex
mixture.

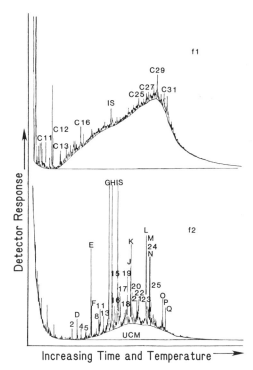

Figure 4.   Gas chromatograms of the aliphatic (f1) and aromatic
(f2) fractions of the sediment sample from Station S6.   The f1
fraction contains identified n-alkanes in the range of n-Cll
through n-C31.   UCM-unresolved complex mixture, IS-internal
standard.

by GC/MS. Specific peak identities and the level of identification
(8) are given elsewhere (3). The third group of peaks consists of
all of the remaining resolved peaks (RRP). All carbonized coal
products had chromatograms similar to that shown by Figure 2. Few
n-alkanes of higher molecular weight than n-C16 were detected and
n-C15, normally present in unweathered petroleum and biogenic
samples (9), was not usually detected. The carbonized coal
products contained 2% or less of their hydrocarbons in the form of
an unresolved complex mixture, with the majority (41-63%) of the
hydrocarbons present as resolved PP (Figure 2). Roof tar and coal
tar, which are by-products of the coal tar distillation process
(10), had distributions similar to that of creosote, but with a
greater relative percent of the higher molecular weight compounds.

In contrast, the fl fraction of the oil spill sample (Figure 3)
has a homologous series of n-alkanes from n-C12 through n-C31 with
the dominant peaks in the n-C15 through n-C18 range. There is also
a UCM in both the fl and f2 fractions. There are resolved peaks in
the f2 fraction including some PP, but they constitute less than 1%
of the hydrocarbons on a weight percent basis. The Kuwait Crude
Oil has no detectable aliphatic UCM which is unusual for weathered
petroleum products (11). Woodstove soot, No. 2 Fuel Oil, diesel
stack soot, the oil spill sample, and Kuwait Crude Oil all
contained a substantial percentage of their f2 hydrocarbon in the
form of a UCM. With the exception of woodstove soot, the f2
fractions of these samples contained the highest percentages of RRP
and the lowest percentages of PP. Woodstove soot contained 21% PP
and 14.2% RRP.

The chromatogram of the sediment sample extract (Figure 4) has
a hydrocarbon distribution that could be the result of weathered
petroleum products mixed with carbonized coal products. There is a
UCM present in both the fl and f2 fractions, similar to the oil
spill sample (Figure 3). The resolved fl fraction contained lower
molecular weight (n-C16 and below) n-alkanes, but n-C15 was not
detected, as with the creosote sample (Figure 2). The higher
molecular weight resolved n-alkane peaks, n-C25,27,29 and 31 are
indicative of a biogenic input (12). The resolved peaks in the f2
fraction are also similar to resolved f2 peaks found in the
creosote sample (Figure 2).

Hydrocarbon concentrations determined by GC for the sediment
samples ranged from 0.1 to 2.9 mg/g. The concentrations are within
the range of values previously reported for the Elizabeth River
(13, 14). Hydrocarbons have been shown to be preferentially
associated with fine-grained sediments (5, 15). The f2 fractions
normalized to fine-grain sediment content (16) have decreasing
aromatic hydrocarbon concentrations with increasing distance from
the area of the creosoting facility sites, indicating the
probability of an aromatic hydrocarbon source in this area. The fl
normalized fraction does not have a similar decreasing concentration pattern, but instead has varied concentrations along the
length of the Elizabeth River, indicating the presence of multiple
sources for the aliphatic hydrocarbons which do not add sufficient
quantities of aromatic hydrocarbons to influence the decreasing
concentration pattern of the f2 fraction (5).

Fluorescence Analysis. Aromatic compounds constitute a very high percentage of carbonized coal products (17), and therefore these compounds may provide a useful fingerprint for carbonized coal products. Fluorescence methods are particularly useful for the detection and measurement of complex organic compounds containing one or more aromatic functional groups. Petroleum contains significant amounts of aromatics with one to four (or more) aromatic rings and their alkylated analogues. The aromatic composition of an oil provides a distinctive fingerprint which can be used in conjunction with other analyses to provide additional typing information (18, 19). The distribution of aromatic compounds in carbonized coal products may provide a unique fluorescence fingerprint.

Fixed wavelength and synchronous scanning fluorescence suffer from non-selectivity and are generally ineffective in structural elucidation (particularly for mixtures). Despite the ability to select both the excitation and emission wavelengths, the conventional luminescence methods have limited applicability since most spectra of complex mixtures often cannot be satisfactorily resolved. The use of a computer-controlled total scanning fluorometer can overcome many of the limitations of previous methods.

The three-dimensional fluorescence spectra of a creosote sample, a sediment sample from the vicinity of the creosote facilities (S5), a sediment sample from the entrance to the Elizabeth River (S8), and No. 2 Fuel Oil are shown in Figure 5. The fluorescence spectra for the creosote (a) and the sediment extract from Station S5 (b) are very similar, while the extract from station S8 has no emission at lower wavelengths (Table I). This difference can be quantified based on a statistical point-to-point comparison of spectra (20). On a scale of 0 to 1, a similarity index (SI) of 1 indicates that the two spectra are identical and a SI of zero means they are completely dissimilar. This type of comparison uses more than 800 individual spectral intensities under the given instrumental conditions. Fluorescence SI between a creosote sample and sediment extracts near the spill range from 0.90 to 0.81. The SI decreases away from the spill to ~ 0.65 at the mouth of the Elizabeth River and to ~ 0.40 in the bay proper.   Different creosotes also showed significantly different fluorescence signatures, suggesting that multiple creosote sources may be discernible by fluorescence fingerprinting. Three-dimensional fluorescence spectra provide a unique fingerprint for creosote. The fluorescence spectra of a No. 2 Fuel Oil (d) is significantly different from the creosote (SI = 0.6) and sediment spectra (SI = 0.63 to 0.14) (Table I).

The maximum fluorescence intensities of emission spectra at selected wavelengths are given in Table I. There is a general pattern of higher intensities at stations near the creosoting facilities (S3 and S5), and lower intensities at stations both upstream and downstream (S1 and S8). Maximum fluorescence intensity for individual compounds is related to concentration, though it is not always linear when mixtures are analyzed. At six stations both fluorescence and total aromatic hydrocarbon concentration measured by GC were available (Table I). The maximum

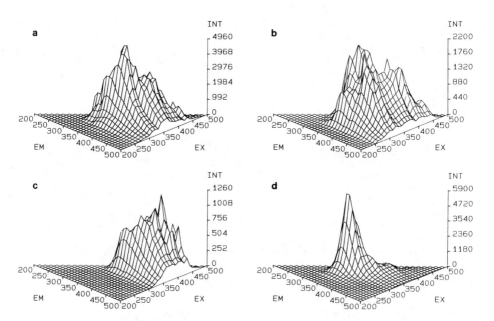

Figure 5.   Three-dimensional total scanning fluorescence of a)
creosote, b) S5, c) S8, d) No. 2 Fuel Oil.

TABLE I. Total Aromatic Hydrocarbon Concentrations, Fluorescence Maximum Intensities at Selected Emission Wavelengths, and Similarity Index*

Sta.	µg/g	[------------------Emission wavelength (nm)------------------]						SI*
		320	340	360	380	400	420	
S1	0.8	38	196	783	991	824	645	0.89
2		2,388	10,200	17,940	26,460	26,940	32,940	0.85
S3	49.0	15,960	109,200	190,200	187,200	152,400	175,200	0.81
4		15,080	41,200	87,600	109,600	88,000	69,200	0.91
5		7,920	18,060	43,400	58,400	57,400	50,400	0.87
S5	420.0	96,400	298,000	388,000	410,000	414,000	326,000	0.89
7		240	5,820	26,400	53,000	61,400	76,600	0.60
8		1,470	14,790	61,200	107,400	124,500	129,600	0.72
S6	66.0	21,480	35,400	50,000	59,200	50,400	43,200	0.86
9		26,800	38,800	62,000	72,000	63,600	78,400	0.82
S7	12.0	9	302	761	1,120	1,530	1,590	0.65
22		210	3.090	6,990	8,380	8,440	7,640	0.89
16		7	80	1,020	1,400	1,880	1,990	0.62
S8	4.4	1	35	585	700	744	872	0.63
18		0	7	145	324	490	482	0.38
14		0	1	98	556	556	689	0.36
δ**		0.9955	0.9611	0.9210	0.9345	0.9611	0.9030	

*SI = similarity index (20) based on creosote from Warner Graham (3).

**Correlation with total aromatic hydrocarbon concentration determined by GC analysis.

intensity of fluorescence emission was found to be linearly
related to total aromatic hydrocarbon concentration with
correlation coefficients that range from 0.9030 to 0.9955 (Table
I).
The fluorescence data support the conclusions from the GC data
that creosote is a contaminant of Elizabeth River sediments.
Because fluorescence is more sensitive and a less labor-intensive
analytical technique, it is possible to process more samples at a
lower cost than by GC techniques. Total scanning fluorescence,
therefore, would be a useful tool for prescreening large numbers of
samples to determine which ones should be further analyzed by GC.

Discussion and Conclusions

The characteristic PNA distribution in creosote can be altered by
environmental processes. PNA's introduced into the marine environment may experience biological uptake, microbial degradation,
volatilization, dissolution and dilution, photo-oxidation, and
sedimentation (5, 21, 22). Microbial degradation in the water
column and evaporation may be the primary removal processes for the
lower molecular weight aromatics, such as naphthalenes,
anthracenes, and phenanthrenes (23). For the higher molecular
weight aromatics, such as chrysenes, benzanthracenes and
benzpyrenes, removal processes are dominated by sedimentation and
photo-oxidation (23). Once deposited in the sediments, the PNA's
including benz(a)anthracene, chrysene, fluorene, and anthracene are
readily degraded at the sediment/water interface; whereas the
higher molecular weight PNA's including benz(a)pyrene and
dibenz(a,h)anthracene show only slight degradation at the
sediment/water interface (22). When PNA's are added to a natural
aqueous environment, the removal processes may be dominated by
processes such as dissolution and dilution with ultimate
sedimentation (22). Degradation and removal is less important once
the PNA's reach the sediment where they can remain unaltered for
years (21, 24). Preferential removal processes for the lower
molecular weight PNA's in creosote, before reaching the subsurface
sediments, could lead to the PNA, GC, and fluorescence distributions seen in the sediments from Stations S5, S6, S1, S8, and S7.
The fact that high concentrations of PNA's remain in these
sediments may be due to the large inputs, or to toxic materials
contained in creosote (such as phenols) which would surpress
microbial activity. Given the observed preservations, the
sediments may act as a chronic source of PNA's to the water column
by resuspension.
Differences in the physical properties of creosote and oils
may also affect PNA distributions. When creosote was mixed with
seawater in the laboratory, three phases were formed: one more
dense than seawater, one dissolved in seawater, and one less dense
than seawater. Analysis of the phase more dense than seawater
produced a gas chromatogram indistinguishable from that of intact
creosote. At a spill site, the phase with a density greater than
seawater may be rapidly removed to the sediments with only slight
alteration. The water-soluble compounds may then be slowly leached
into the water column and act as a chronic PNA source. Petroleum,

which is less dense than seawater, would be subject to more weathering and dispersal processes than creosote before reaching the sediments.

Since the sediment samples appeared to contain a mixture of carbonized coal products and weathered petroleum products, a plot of the percent PP (main component of carbonized coal products) was plotted against the percent fl UCM (main component of weathered petroleum products). This plot (Figure 6) shows a linear relationship (correlation of -0.955) which would be expected if the changes in the percent composition were due to simple mixing of these two end members. Sediments that are contaminated predominantly by weathered petroleum (25) are calculated to contain 89% on a weight basis of their hydrocarbons as a UCM and less than 0.4% as PP (5). These percentages are in good agreement with our X-intercept of 88% UCM. The Y-intercept (39%) should reflect the carbonized coal end member which ranges from 41 to 63% PP. The reason this percentage is low could be due to preferential weathering (such as dissolution) of the PP compounds in the sediments.

Carbonized coal products have a unique fingerprint by both GC and fluorescence analyses. Both these fingerprints confirm that sediments from the Elizabeth River are contaminated with carbonized coal products and allow for the detection of carbonized coal hydrocarbons, even in the presence of petroleum-derived hydrocarbons. Fluorescence allows for the rapid analysis of more samples and shows the contamination within the Elizabeth River to be widespread. Carbonized coal products in the sediments may constitute a chronic long-term source of PNA's to the water column.

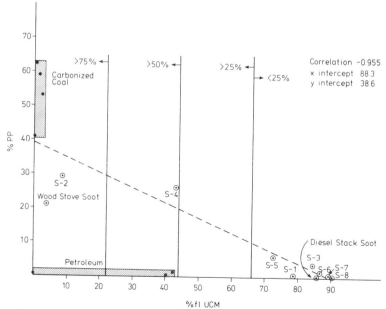

Figure 6. Graph showing the linear relationship between %PP and %fl UCM of the sediment samples. Source sample areas are identified. Solid lines indicate the percentage contribution of carbonized coal products in the mixing relationship.

Acknowledgments

Partial funding for this research was provided by National Science
Foundation Grant OCE 8301538.

Literature Cited

1. "Petroleum in Marine Environment"; National Academy of
   Sciences, Washington, D.C., May 21-25, 1973.
2. Wakeham, S.G.; Farrington, J.W. In: "Contaminants and
   Sediments"; Baker, R. A., Ed.; Ann Arbor Science Publishers;
   Ann Arbor, MI, 1980; Vol. 1, pp. 3-32.
3. Merrill, E.G.; Wade, T. L. Environ. Sci. Technol. 1985, 19,
   597.
4. Merrill, E. G. M.S. Thesis, Old Dominion University, Norfolk,
   VA, 1984.
5. Wade, T. L.; Quinn, J. G. Org. Geochem. 1979, 1, 157.
6. Moore, W. E.; Effland, M. J.; Roth, H. G. J. Chromatogr. 1968,
   38, 522.
7. Fed. Regist, 1979, 44, 69-494.
8. Environ. Sci. Technol. 1984, 18, 203A.
9. Blumer, M.; Guillard, R. R. L.; Chase, T. Mar. Biol. 1971, 8,
   183.
10. Berkowitz, N. "An Introduction to Coal Technology"; Academic
    Press; New York, 1979.
11. Farrington, J. W.; Meyers, P. A. In "Environmental Chemistry";
    Eglinton, G., Ed.; The Chemical Society: London, 1975; Vol. 1,
    Chap. 5.
12. Ehrhardt, G.; Blumer, M. Environ. Pollut. 1972, 3, 179.
13. Bieri, R. H.; Hein, C.; Huggett, R. J.; Shou, P.; Slone, H.;
    Smith, C.; Su, C. "Toxic Organic Compounds in Surface Sediments
    from the Elizabeth and Patapsco Rivers and Estuaries"; VA
    Institute of Marine Science: Glouchester Pt., VA, 1982.
14. Virginia State Water Pollution Control Board, "The Elizabeth
    River: An Environmental Perspective"; 1983, VA SWPCB Basic Data
    Bulletin 61.
15. Meyers, P. A.; Quinn, J. G. Nature (London) 1973, 244, 23.
16. Brown, R. B.; Wade, T. L. Water Res. 1984, 18, 621.
17. McNeil, D. Rec. Annu. Conv. Bd. Wood Preserv. Assoc. 1959,
    136-150.
18. Brooks, J. M.; Kennicutt, M. C. II; Barnard, L. A.; Denoux, G.
    J.; Carey, B. C. In "Proc. 15th Annual Offshore Technol.
    Conf.", OTC 8624, Offshore Technology Conference, 1983, pp.
    393-400.
19. Kennicutt, M.C. II; Brooks, J.M. Mar. Pollut. Bull. 1983, 14,
    335.
20. Kennicutt, M.C. II; Brooks, J.M.; Denoux, G.J. In "Analysis of
    North Slope Crude"; Magoon, L.B., and Claypool, G.E., Eds.;
    Custom Editorial Productions, Inc., Cincinnati, OH, 1985;
    Chap. 15.
21. Lee, R. F.; Ryan, D. Can. J. Fish. Aquat. Sci. 1983, 40, 86.
22. Farrington, J. W.; Quinn, J. G. Est. Coast. Mar. Sci. 1973,
    1, 71.
23. Lee, R. F.; Gardner, W. S.; Anderson, J. W.; Blaylock, J. W.;
    Barwell-Clarke, J. Environ. Sci. Technol. 1978, 12, 832.
24. Gschwend, P.M.; Hites, R. A. Geochim. Cosmochim. Acta 1981,
    45, 2359.
25. Zafiriou, O. C. Est. Coast. Mar. Sci. 1973, 1, 81.

RECEIVED January 27, 1986

Hydrocarbon Contamination from Coastal Development

Richard H. Pierce[1], Robert C. Brown[1], Edward S. Van Vleet[2], and Rosanne M. Joyce[2]

[1] Mote Marine Laboratory, Sarasota, FL 33577
[2] Department of Marine Science, University of South Florida, St. Petersburg, FL 33701

Hydrocarbon analyses were obtained for samples of
surface sediment, oysters and water from four areas in
Charlotte Harbor, Florida. Each area represented a
different type of coastal development activity
including: residential development canals; municipal
and industrial impact; commercial fishing and marine
industry facilities; and a nondeveloped control area.
Characterization of hydrocarbons was performed using
high resolution glass capillary GC-FID chromatograms of
the aliphatic (saturated) and aromatic/olefinic
(unsaturated) hydrocarbon fractions and by GC-MS
analysis of select aromatic components. Residential
canal systems contained petroleum contamination
resulting from marina and highway service station
activities, contamination indicative of crankcase oil
which diminished with distance from the source. Total
hydrocarbon content of canal sediment ranged from more
than 50 ug/g air-dry sediment at a marina to less than
5 ug/g at nonimpacted areas. Sediment collected near
municipal-industrial activities exhibited high concentrations of a broad range of heavy, residual fuel oils
(85 ug/g sediment). The highest contamination was
observed at commercial fishing port areas, contamination indicative of a low to mid-boiling range fuel
oil attaining a concentration of 142 ug/g sediment,
compared to less than 5 ug/g in unimpacted areas.
Oyster samples generally reflected the contamination
observed in sediment. Water samples exhibited different hydrocarbon patterns than oysters or sediment,
consisting primarily of terrigenous and marine biogenic
material. Biogenic hydrocarbons with chromatographic
patterns that mimic some petroleum characteristics were
observed in certain areas, showing the importance of
obtaining pre-oil spill data for accurate
interpretation of oil spill impact. Additional
information from time series sampling is needed to
ascertain rates of hydrocarbon input and degradation.

0097-6156/86/0305-0229$06.00/0
© 1986 American Chemical Society

This project was undertaken to identify petroleum hydrocarbon contamination resulting from coastal development activities in the Charlotte Harbor estuarine system located on the Southwest Florida coast (Figure 1). The overall goal was to assess present petroleum contamination and to provide baseline data to prove the extent and duration of additional impact from future coastal development and oil spills.

Environmental impact from large oil spills in the coastal marine environment is readily observable and easily documented. The extent and duration of contamination after the surface oil has been removed and weathered, however, are not so easily ascertained especially when pre-spill conditions are not known (1-6). Problems also arise when attempting to document chronic petroleum discharges from various coastal development activities. The detection of petroleum hydrocarbons is hindered by the presence of recently biosynthesized (biogenic) hydrocarbons and hydrocarbons resulting from combustion (pyrogenic). Source identification also is complicated by differential weathering through degradation, evaporation and solubilization of petroleum components (1-6).

The Charlotte Harbor estuarine system was chosen for this study because it is one of the largest and possibly one of the least contaminated estuaries in Florida (7). Recently, however, the area has become one of the most rapidly developing in the United States (8). A recent review of scientific information about Charlotte Harbor emphasized the potential for environmental impact from development and various sources of petroleum contamination (9).

A two-year investigation was undertaken to determine the composition and concentration of hydrocarbons in sediment, water and commercially important marine organisms throughout Charlotte Harbor. The entire project was detailed in an unpublished report to the Florida Department of Natural Resources (10). The first-year project revealed limited petroleum contamination associated with certain land-use activities (6,10). The second-year study, the primary subject of this presentation, investigated the problem areas in more detail to characterize the type of petroleum contamination and to identify probable sources.

Methods

Study Location. Charlotte Harbor is located on the Southwest Florida coast between 27°05'N and 26°27'N latitude (Figure 1). During the first year study, certain areas were identified as containing petroleum contamination (6,10). Three areas were chosen for more intensive study during the second year of the investigation including: Port Charlotte-Punta Gorda development canal systems (Area A, Figure 1); Caloosahatchee River downstream from Fort Myers (Area B, Figure 1); and the Matanzas Pass area behind Fort Myers Beach (Area C, Figure 1). A nondeveloped area in an aquatic preserve at Sanibel Island was chosen as the control area (Area D, Figure 1).

The study areas are described as follows:

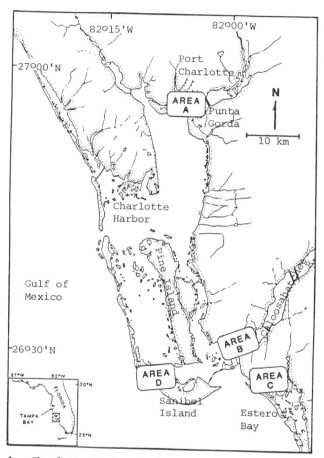

Figure 1. Charlotte Harbor hydrocarbon study, year-2 sampling areas location.

AREA A. This area represents impact from coastal residential development, extending into the mouth of the Peace River and upper Charlotte Harbor. Primary sources of contamination are expected to be stormwater runoff (automobile related petroleum), municipal sewage and small boat marina activities. Surface sediment was collected from fourteen sites, water from eight sites and oysters from three sites, representing conditions in the development canals at Port Charlotte across the mouth of the Peace River and into the canals of Punta Gorda.

AREA B. This area includes the Caloosahatchee River at Fort Myers and the mouth of the river at San Carlos Bay, downstream from Fort Myers. It is particularly susceptible to oil spills because of the Intracoastal Waterway (ICW) and the Caloosahatchee River, which would carry any spill downriver from the Fort Myers area. It was chosen to represent an area which receives chronic input from municipal and industrial activities and represents a high risk area for future oil spills. Surface sediment and oysters were collected from various sites.

AREA C. Matanzas Pass, behind Fort Myers Beach on Estero Island, represents impact from commercial fishing and heavy marine industry operations. The entire area from Carlos Bay through Matanzas Pass to Big Carlos Pass in Estero Bay was studied in an attempt to understand the distribution of contamination from the developed marine industrial area out into the nondeveloped Estero Bay. Sediment samples were collected from eight sites, oysters from five sites and water from four sites.

AREA D. This area includes sediment from three sites approximately one mile offshore from Sanibel Island and Captiva Island and oysters and sediment from mangrove fringes within the wildlife preserve on the estuarine side of the island, representing nonimpacted control areas.

Sampling Technique. Surface sediment (top 5 cm) was collected as a composite of at least three grabs with a Petite Ponar sediment sampler, to provide 500 g wet weight of relatively undisturbed sediment from each site. Each sediment grab was placed in a stainless steel tray and the top 5 cm of sediment retrieved with a stainless steel scoop and placed in a precleaned glass jar with aluminum lined caps. Jars were placed on ice for transport and stored at 4°C until analyzed. Sufficient sediment was collected for duplicate hydrocarbon analysis with extra sediment for additional analyses if necessary. Efficacy of storage at 4°C was verified by analysis of samples before and after storage for several months and by analysis of standard intercalibration sediment after storage for several months.

The oyster, Crassostrea virginica, was the species used to monitor uptake by sessile filter-feeding organisms. At least two dozen large oysters were collected from each organism sampling site. These were rinsed in ambient water, wrapped in aluminum foil and stored in plastic bags on ice for transport to the laboratory where they were then frozen in their shells until ready for analysis.

Near surface water samples were collected with a precleaned metal sampler. Samples of 10 to 15 liters were collected at each site and placed in 20 liter glass carboys for in-field extraction as described below. No attempt was made to differentiate dissolved from particulate fractions so that total water column hydrocarbons were observed.

Hydrocarbon Analysis. Sediment samples were thawed and mixed to provide a homogeneous sample. Approximately 150 g wet weight was placed in a Soxhlet extraction apparatus and saponified-extracted with benzene /0.5 $\underline{N}$ KOH-methanol (50/50) (ca. 250 ml total) for 24 hours, or until the extraction solution was clear ($\underline{6,11,12}$). An internal standard consisting of ca. 50 ug each of 5,α-androstane and o-terphenyl was added prior to extraction to provide assurance of extraction efficiency, separation of saturated and unsaturated fractions and to provide a standard reference for the gas chromatographic data system. Methylstearate (the methyl ester of stearic acid) was added to select samples to verify saponification efficiency.

The saponified solution was extracted with 3 x 50 ml hexane and the resulting hexane-benzene solution was washed with distilled water to remove residual KOH-MeOH. The benzene-hexane was then reduced to a volume of ca. 0.1 ml by rotary evaporation, followed by purging with N_2 gas in a warm bead bath (45°C) and repeated addition of hexane to replace benzene with hexane. A final sample volume of 1 ml hexane was added to a column of 2 g neutral alumina (80/20 mesh), 2 g silica gel (100/200 mesh) and 1 g sodium sulfate (granular) for clean-up and separation into aliphatic (saturated) and aromatic/olefinic (unsaturated) fractions. Alumina, silica gel and sodium sulfate were activated at 500°C for 4 hours and stored at 110°C until used. The saturated fraction was eluted with 3 bed volumes (bv) of hexane and the unsaturated fraction was eluted with 3 bv of hexane and benzene (50/50). Each fraction was then reduced to 0.5 ml volume under a stream of N_2 gas as described above in preparation for GC analysis.

Oysters were thawed, opened with a clean knife, the liquid drained off, and a composite of several whole oysters collected. The entire sample of each site was homogenized and divided into four subsamples. Two were subjected to the analytical scheme, the third archived and frozen for further analysis, if necessary, and the fourth dried at 103°C to obtain the dry weight.

For analysis, ca. 10-15 g wet-weight tissue was transferred to a tared, hexane washed 250 ml boiling flask. The internal standard mixture was added at this point in acetone solution. A 50 ml portion of 2.0 $\underline{N}$ aqueous KOH was added to the flask containing the tissue for saponification under reflux for 4 hours (h) or until the tissue was well digested. An equal volume of saturated NaCl solution was added to the mixture, and the solution extracted with three 50 ml portions of hexane. The extracts were combined, reduced in volume with a vacuum rotary evaporator and transferred to a cleaned vial. The remaining solvent was reduced to 1 ml volume in preparation for separation into saturated and unsaturated fractions and GC analysis as described above for sediments.

Water samples were processed in the field using a portable water extractor consisting of a stirring blade operated by a battery powered reversible drill which fits into the 20 liter glass carboy ($\underline{10}$). The extraction procedure consisted of placing 10 to 15 liters

of a water sample in the carboy. The internal standard mixture was
added (5,α-androstane and o-terphenyl in acetone solution), followed
by 500 to 750 ml of CH_2Cl_2. The mixture was stirred vigorously for
5 min and allowed to settle for one hour (until phases separated).
Water was siphoned off the top and the CH_2Cl_2 recovered for
hydrocarbon analyses as described above for sediments. Prior to
field use, the extractor was evaluated for recovery and precision by
spiking distilled water with 1 mg l^{-1} of Kuwait crude oil and
comparing recovery with that obtained from separatory funnel,
liquid-liquid extraction (10).

Gas chromatographic analysis was performed with each fraction
using a Varian Vista 6000 gas chromatography (GC) system coupled
with a Vista 401 data system (Varian Instruments, Sunnyvale, CA). A
flame ionization detector (FID) was used and the sample was
separated on a glass capillary column (30 m x 0.25 mm with WCOT
SE-30), temperature programmed from 100°C to 280°C at 8°/min, and
held at 280°C for 10 min, which allowed observation of the n-alkane
homologous series from n-C_{13} through n-C_{32}. The injector was
operated in the splitless mode with septum purge after 30 sec. The
carrier and make up gas, was N_2. Injection volume was 2 ul; with an
attenuation of 4 x 10^{-11}.

Approximately 10% of the sediment and organism samples were
analyzed by combined high resolution gas chromatography-mass
spectrometry. Samples were analyzed on a Hewlett-Packard Model
5992B computerized GC-MS system (Hewlett-Packard, Avondale, PA)
equipped with a 30 m DB-5 fused silica capillary column. Running
conditions were as follows: carrier gas = helium; column flow
rate = 1.7 ml min^{-1}; injection port temperature = 240°C; splitless
injection mode; column oven temperature programmed from 90-250°C at
4°C min^{-1}; electron multiplier voltage = 1200-2000 eV; GC-MS run in
selected ion monitoring mode; dwell time = 100 msec for each ion.
Samples were first analyzed by electron impact to obtain the total
ion spectra to check for interfering ions. Samples were then
analyzed by selected ion monitoring to identify and quantitate
specific polynuclear aromatic hydrocarbons (PNA) homologues.

<u>Quality Assurance</u>. The quality assurance program consisted of a
sample chain of custody verification, precautions to guard against
and to detect sample contamination, verification of precision and
accuracy and intercalibration with two other laboratories to verify
results from our extraction and analysis procedures. This program
is detailed in the unpublished Final Report of the overall study to
the FL Department of Natural Resources (10). Interlaboratory
calibration consisted of three phases: 1) instrument intercalibration by comparing results of standard hydrocarbon solutions;
2) methods intercalibration by comparing results of analyses of a
standard sediment sample (Duwamish-I), supplied by Dr. William
MacLeod of the NOAA, Northwest and Alaska Fisheries Center, Seattle,
WA (13); and 3) sample intercalibration, consisting of the exchange
among the three participating laboratories of three sediment samples
from the study areas.

<u>Results and Discussion</u>

Petroleum hydrocarbon identification was based primarily on the
GC-FID chromatogram patterns of the saturated and unsaturated

hydrocarbon fractions. Additional information regarding petrogenic vs. pyrogenic or biogenic hydrocarbons was provided by GC-MS analysis of specific PNA series from the unsaturated fraction with select samples (3,5,14,15).

For comparison, GC-FID chromatograms of representative petroleum contaminants are shown in Figure 2 and chromatograms of hydrocarbons from various marine plants (biogenic) are given in Figure 3. Key parameters used for hydrocarbon characterization are listed in Table I.

Area A. Samples from Area A included sediment, water and oysters from Port Charlotte canals (Sites A-1 through A-4), the Peace River (Sites A-5 through A-9), and Punta Gorda residential canals (Sites A-10 through A-14). Sediment hydrocarbons from both canal systems showed a heavy residual crankcase oil-like petroleum contamination from marina and highway service station activities (Site A-1, and A-12, Figure 4). This contamination diminished in residential canals away from pollution sources and did not appear to be exported to nearby bay and river sediment which exhibited a totally different hydrocarbon pattern somewhat representative of hydrocarbons in red mangrove leaves (Site A-6, Figure 4, Table I).

Oysters collected within the canal systems near the marinas exhibited the crankcase oil-like patterns, whereas those away from the marina exhibited biogenic material patterns.

Water samples in the Port Charlotte canal system yielded mangrove-hydrocarbon patterns similar to that found in sediment and oysters outside the canals along the Peace River and Alligator Bay, whereas water collected from the Peace River exhibited very little discernable hydrocarbon pattern. Water in Punta Gorda canals more closely reflected the hydrocarbon patterns found in sediment and oysters, i.e. highly contaminated at the marina diminishing to very low levels within about a mile into or out of the canals.

Localized contamination of water and sediment was observed near a docking area for commercial fishing boats, at Fisherman's Wharf along the Peace River. This pattern was unique to the area showing a bimodal unresolved complex mixture (UCM) and a total sediment hydrocarbon content of 90 ug/l (Site 10, Table I).

Analysis of select PNA to verify petroleum origin showed that all sediment contained mixed petrogenic and pyrogenic material. The relative abundance of PNA in marina sediment, however, was much greater (180 ng/g benzanthracenes) than that found in sediment from non-marina areas (<0.1 ng/g benzanthracenes).

Oyster samples reflected specific PNA patterns similar to the adjacent sediment. Because of the low concentration of PNA's in water, results from water PNA analyses were inconclusive.

Area B. Area B samples included sediment collected from the Caloosahatchee River at Fort Myers, sediment collected downstream, and sediment and oysters collected at the mouth of the river (Figure 3, Table I). Sediment collected near the municipal-area exhibited a very large, bimodal UCM, indicative of heavy industrial heating oil transported by tanker or oil barge (Site B-1, Figure 5). The sample contained primarily highly weathered oil as evidenced by the lack of the large unresolved complex. The resolved component

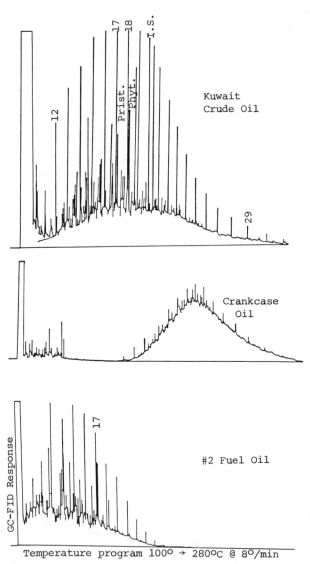

Figure 2. GC-FID chromatograms of reference petroleum products. Numbers indicate n-alkane carbon number represented by the adjacent chromatographic peak. I.S. indicates the position of the internal standard (5,α-androstane) chromatographic peak.

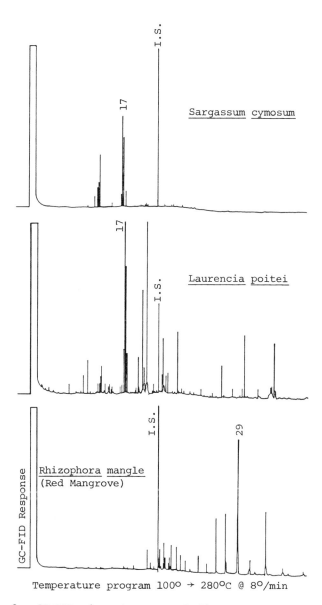

Temperature program 100° → 280°C @ 8°/min

Figure 3. GC-FID chromatograms of the saturated hydrocarbon fraction from marine and coastal flora. Numbers indicate n-alkane carbon number represented by the adjacent chromatographic peak. I.S. indicates the position of the internal standard (5,α-androstane) chromatographic peak.

Table I. Hydrocarbon characterization parameters for Kuwait crude oil, Duwamish River sediment intercalibration standard, and select surface sediment samples from study areas A, B, C and D.

SAMPLE	TOTAL (µg/g)		RATIOS				KEY HYDROCARBONS (ng/g)				N-ALKANES	
	Sat.	Unsat.	Resol/ Unres.	Prist/ Phyt.	C17/ Prist.	C18/ Phyt.	1500	1700	2085	2900	Homol. Ser.	CPI
Kuwait	2,800	--[1]	0.4	0.6	5.6	3.0	58,000	47,000	--	--	C_{12-31}	1.1
Duwamish	29	7	0.1	1.3	0.8	1.2	23	50	4	122	C_{15-30}	1.4
A-1	55.6	4.5	0.10	1.0	4.8	0.9	93.2	217.8	29.8	851.2	C_{16-21}	2.6
A-6	4.5	0.5	0.2	--	--	1.6	--	--	--	73.1	C_{18-25}	2.0
A-10	61.8	19.0	0.28	1.1	1.7	1.4	--	338.3	61.7	2266.9	C_{16-30}	3.3
A-12	48.2	4.4	0.2	1.2	1.2	1.5	--	107.6	--	1625.8	C_{23-30}	3.0
B-1	52.3	35.0	0.5	40	0.7	--	1.2	0.8	4.5	15.0	C_{13-30}	1.5
B-3	7.0	1.7	0.2	0.2	2.0	2.0	19	75	--	45.0	C_{13-30}	3.6
B-6	4.4	2.9	1.2	3.8	1.9	0.8	13	100	114	72	C_{14-30}	4.7
C-2	8.9	0.2	0.4	--	--	1.2	19.0	43.1	10.3	302.6	C_{15-30}	4.2
C-4	142.2	16.5	0.1	--	--	0.3	317.0	--	--	--	--	--
C-7	3.5	1.6	0.4	--	--	1.3	10.3	24.0	13.7	73.2	C_{15-27}	3.3
D-2	1.3	0.1	--	--	--	--	8.7	10.4	41.6	34.8	--	--
D-5	1.1	0.1	--	--	--	1.0	15.7	55.6	13.2	--	--	--

[1]-- Below detection limit.

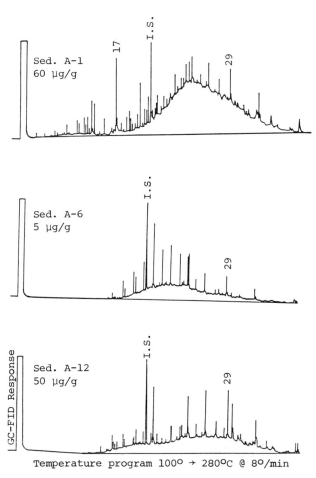

Temperature program 100° → 280°C @ 8°/min

Figure 4. GC-FID chromatograms of the saturated hydrocarbon fraction of sediments collected from sampling Area A. Numbers indicate n-alkane carbon number represented by the adjacent chromatographic peak. I.S. indicates the position of the internal standard (5,α-androstane) chromatographic peak.

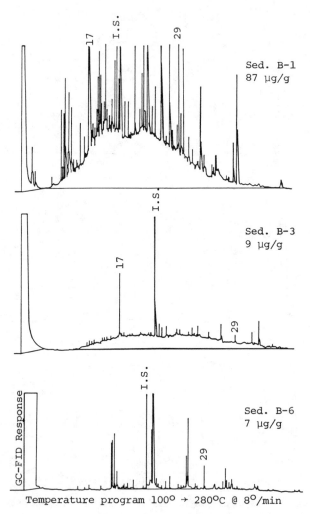

Figure 5. GC-FID chromatograms of the saturated hydrocarbon fraction of sediments collected from sampling Area B. Numbers indicate n-alkane carbon number represented by the adjacent chromatographic peak. I.S. indicates the position of the internal standard (5,α-androstane) chromatographic peak.

contained a considerable amount of biogenic hydrocarbons indicated by the predominance of specific odd-numbered carbon compounds. Sediment collected approximately 10 km downstream showed diminished concentrations of weathered oil (Site B-3, Figure 5, Table I).

Sediment and oyster samples collected approximately 25 km downstream at the mouth of the river, produced chromatograms indicating primarily biogenic hydrocarbons (Site B-6, Figure 5), providing excellent background data for comparison in the event of a future oil spill.

Area C. Sediment, water and oyster samples were collected from the marine industry port areas in Matanzas Pass and from nonimpacted sites extending to   both ends of the Pass to observe the distribution of petroleum pollutants. Samples collected adjacent to the fishing fleet docks exhibited a preponderance of No. 2 fuel oil-like hydrocarbons with some crankcase oil-like GC-FID patterns (Site 4, Figure 6, Table I). Sediment and oysters provided the best evidence for petroleum contamination, whereas hydrocarbons in water samples were dominated by biogenic compounds. Sediment and oyster contamination dropped to background biogenic hydrocarbon levels within one mile of the marine industrial activities (Site 2 and Site 7, Figure 6). The petrogenic nature of contaminated samples was verified with GC-MS analysis of select PNA compounds.

Area D. Samples from Area D were analyzed to provide background hydrocarbon data from low organic content Gulf of Mexico samples (Sites D-1, D-2 and D-3) and from high biogenic input mangrove fringe areas (Sites 4 and 5). These samples provided excellent insight about non-petroleum contaminated biogenic hydrocarbons representative of nearshore Gulf of Mexico and barrier island mangrove fringe areas (Site D-2 and D-5, Table I).

A most important observation regarding biogenic hydrocarbons at Area D as well as other areas was that those sediments receiving high biomass input, such as in mangrove forests, exhibited chromatographic patterns of the saturated hydrocarbon fraction containing a UCM as well as a homologous series of compounds (not necessarily n-alkanes) in the C-18 to C-25 region (Figure 3, Red Mangrove; Figure 4, Sed. A-6; Figure 6, Sed. C-7). Sediments receiving low biogenic input, such as Gulf of Mexico samples (Table I, Sed. D-2) and areas of limited deposition (Figure 5, Sed. B-6), show characteristic marine biogenic chromatographic patterns. Such characterization of endemic biogenic hydrocarbons is needed to establish the natural hydrocarbon content in specific areas of critical concern. These data are essential for accurate interpretation of petroleum impact from coastal development and oil spills. Additional information should be obtained over time to establish the dynamic relationships regulating hydrocarbon input, accumulation and degradation within the various coastal marine environments.

Data Interpretation. Interpretation of these results is based on specific distinctions among biogenic, pyrogenic and petrogenic hydrocarbons. Biogenic hydrocarbons exhibit discrete sets of n-alkanes and alkenes, whereas petroleum contains the homologous series of n-alkanes, branched and cyclic alkanes and substituted polynuclear aromatic hydrocarbons. A predominance of specific

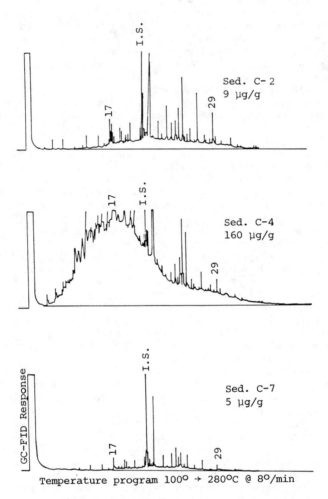

Figure 6. GC-FID chromatograms of the saturated hydrocarbon fraction of sediments collected from sampling Area C. Numbers indicate n-alkane carbon number represented by the adjacent chromatographic peak. I.S. indicates the position of the internal standard (5,α-androstane) chromatographic peak.

compounds such as pristane (2,6,10,14-tetramethylpentadecane), pentadecane (n-C_{15}), and heptadecane (n-C_{17}) are indicative of marine biogenic sources, while pristane in the presence of the isoprenoid, phytane, and a homologous series of n-alkanes, indicates petroleum hydrocarbons (16-18). Hydrocarbons of terrigenous flora exhibit a strong odd/even carbon predominance index (CPI) in the n-C_{23} through n-C_{31} n-alkane region (18-20). Petrogenic hydrocarbons are characterized by a UCM of hydrocarbons in the aliphatic fraction, which is comprised primarily of branched and cyclic saturated hydrocarbons too numerous to be resolved by gas chromatographic techniques (21-23). Various petrogenic hydrocarbon sources include tanker washings, storage transfer spills, outboard motors, sewage and land runoff. Classification may include refined or crude oil, recent or weathered oil. Source identification can be accomplished through specific diagnostic compounds in the aliphatic and aromatic/olefinic fractions, aided by GC-MS analyses to verify compound identification (24-26).

Petroleum that has been recently introduced into the marine environment is indicated by a smooth alkane distribution (CPI = 1) over a UCM (14,26,27). Crude oil GC patterns are characteristic of geographical origin, but generally exhibit a wide boiling range of n-alkanes and UCM, sometimes with a bimodal UCM distribution indicative of tanker spillage (26). Refined petroleum distillates favor low-boiling components, whereas residual oils show a predominance of the higher boiling compounds (28-30).

Interpretation of the aromatic/olefinic hydrocarbon distribution has recently been developed to help characterize hydrocarbon sources. Olefinic compounds are not abundant in petroleum; thus their predominance would indicate biogenic input (4,11,20,31).

Although much information can be obtained from GC-FID, one of the most diagnostic characteristics of a petroleum sample is the molecular nature of the aromatic hydrocarbon fraction which can only be determined by GC-MS (6,18,24,32). The reason for this is that in addition to the many PNA parent compounds, there also exist C_1-C_3 alkylated homology series for each parent PNA. In addition, each of these alkylated homologs can be present as a number of different isomers for each parent structure.

Polynuclear aromatic hydrocarbons have routinely been used in order to differentiate suites of compounds in petroleum from those originating from combustion processes (24,25,33). This differentiation is based upon the relationship between the parent aromatic compounds (such as phenanthrene) and the alkylated homologs such as methyl phenanthrene, dimethyl phenanthrene and trimethyl phenanthrene, designated C_1P, C_2P, and C_3P, respectively. Under conditions of relatively low temperatures and slow heating, such as in petroleum formation, the alkylated homologs are formed in the greatest abundance. Under conditions of higher temperature and more rapid heating, such as during combustion, the alkylated homologs undergo thermal cracking, resulting in a greater abundance of the nonalkylated parent compounds. Sporstol et al. (34) have shown that the naphthalene homologs show a similar homolog distribution but the dibenzothiophenes do not. These authors suggest that the benzothiophene alkyl homolog series is not a sufficiently sensitive tool for distinguishing petroleum from combustion PNA sources.

Overton et al. (35) have shown that the ratios of alkyl phenanthrenes to alkyldibenzothiophenes are useful indicators for tracing petroleum hydrocarbon sources in environmentally impacted samples. The ratios of C_1, C_2 and C_3 alkyl homologs for these PAH's have been shown to be uniquely characteristic of specific oils and are not rapidly altered by weathering processes. In an attempt to identify weathered petroleum between biogenic and pyrogenic hydrocarbons, we determined the concentrations of individual PNA compounds and reported the phenanthrene to dibenzothiophene homolog ratio as suggested by Overton et al. (35). In addition, the total amount of C_0-C_3 naphthalenes, phenanthrenes, dibenzothiophenes and pyrenes as determined by GC-MS was calculated for each sample and used to determine the ratios of alkylated homolog to parent PHA compounds. These homolog distributions were also plotted and then used to differentiate petrogenic from pyrogenic hydrocarbons.

In summary, hydrocarbon analyses of water, sediment, and oysters collected from the four study areas described above have provided information regarding hydrocarbon contamination from specific coastal development activities as well as background data from unimpacted areas. Although these data establish areas of existing petroleum pollution with identified source material, the baseline data represent samples collected at one point in time. Baseline information needed to assess the full impact of an oil spill should include samples collected over time as well as space. These additional data are needed to establish rates of hydrocarbon input and degradation during different seasonal and environmental conditions.

Conclusions

1) Petroleum contamination from specific land use activities was identified through GC-FID hydrocarbon characterization supported by GC-MS analysis of specific PNA compounds.

2) Commercial marina and highway service station activities were found to provide a source for crankcase oil-like contamination to residential canal systems.

3) Residential canal systems not exposed to major highways or marinas did not exhibit the oil contamination.

4) Hydrocarbon contamination from commercial fishing port facilities was primarily from No. 2 fuel oil. Oysters within the intertidal region were contaminated up to one mile away.

5) Samples collected near municipal-industrial activities at Fort Myers exhibited petroleum contamination representative of heavy residual fuel oil used for electric power plant generators.

6) Samples collected several miles downstream from the municipal-industrial area, however, did not reflect significant petroleum hydrocarbon transport.

7) Samples from nonimpacted areas exhibited biogenic hydrocarbon chromatographic patterns which in high biomass areas, such as mangroves, exhibited chromatograms similar to certain petroleum components. These data provide information essential to avoid erroneous conclusions regarding impact following an oil spill.

8) Characterization of petroleum contamination from certain land use activities and from biogenic sources has provided information for assessing impact from continued coastal development activities and to assess the duration and extent of impact in the event of a major oil spill in the areas studied. However, time series samples are needed to establish rates of hydrocarbon input and degradation.

Acknowledgments

This project was funded by the Florida Department of Natural Resources. The authors are grateful to Alan Huff, DNR Project Coordinator. Special thanks to Stephanie Boggess, Shawn Murphy, Jim Mullin and Kathy Peck for assisting with sample collection and analysis.

Literature Cited

1. "Oil in the Sea: Input, Fates and Effects", National Academy of Sciences and National Research Council, National Academy Press, Washington, D.C., 1985.
2. Atlas, R. M. *Environ. Int.* 1981, 5, 33-38.
3. Boehm, P. D. IXTOC Oil Spill Assessment, Final Report. BLM-YM-PT-82-005-3331, U.S. Department of Interior, Bureau of Land Management: Washington, D.C., 1982.
4. Farrington, J. W. In "Petroleum in the Marine Environment"; Petraleis, L.; and Weiss, F. T., Eds.; ADVANCES IN CHEMISTRY SERIES No. 185, American Chemical Society: Washington, D.C., 1980; pp. 1-22.
5. Pierce, R. H.; Anne, D. C.; Saksa, F. I.; and Weichert, B. A. In "Wastes in the Ocean, Volume 4: Energy Wastes in the Ocean"; Duedall, I. W.; Kester, D. R.; and Park, P. K., Eds.; John Wiley and Sons, Inc.: New York, 1985, Chap. 7.
6. Van Vleet, E. S.; Pierce, R. H.; Brown, R. C.; and Reinhardt, S. B. *Org. Geochem.* 1984, 6, 249-257.
7. Wang, J. C. S.; and Raney, E. C. Distribution and fluctuations in the fish fauna of the Charlotte Harbor Estuary, Florida. Mote Marine Laboratory Collected Papers No. 112, Sarasota, FL, 1971.
8. Department of Community Affairs. Final recommendations of the Governor's Nomination Review Committee, Charlotte Harbor Estuarine Ecosystem Complex, Tallahassee, FL, 1981.
9. Estevez, E. D. A review of scientific information: Charlotte Harbor (Florida) estuarine ecosystem complex, Final Report to Southwest Florida Regional Planning Council, Fort Myers, FL. Mote Marine Laboratory Review Series No. 3, 1981.
10. Pierce, R. H.; Brown, R. C.; and Van Vleet, E. S.; Charlotte Harbor Hydrocarbon Study, Year-2, Final Report. Florida Department of Natural Resources, St. Petersburg, FL, 1983.
11. Farrington, J. W.; and Tripp, B. W. In "Marine Chemistry in the Coastal Environment"; Church, T., Ed.; ACS SYMPOSIUM SERIES No. 18, American Chemical Society: Washington, D.C., 1975; pp. 267-284.
12. Gearing, P.; Gearing, J. N.; Lytle, T. F.; and Lytle, J. S. *Geochim. Cosmochim. Acta.* 1976, 40, 1005-1017.
13. MacLeod, W. D.; Prohaska, P. G.; Gennero, D. D.; and Brown, D. W. *Analyt. Chem.* 1982, 306-392.

14. Farrington, J. W.; and Meyers, P. A.; "Environmental Chemistry"; Eglinton, G., Ed.; The Chemical Society: United Kingdom, 1975; pp. 109-136.
15. Boehm, P. D.   Interpretation of sediment hydrocarbon data, Volume II, Chap. 10, Final Report for BLM-MAFLA Study 1977/1978 project period, 1978.
16. Blumer, M.; Guillard, R. R. L.; and Chase, T. Mar. Biol. 1971, 8(30), 183-189.
17. Ehrhardt, M.; and Blumer, M. Environ. Pollut. 1972; 3, 179-194.
18. Farrington, J. W.; and Tripp, B. W. Geochim. Cosmochim. Acta. 1977, 41, 1627-1641.
19. Bieri, R. H.; Cueman, M. K.; Smith, D. L.; and Su, C-W. International J. of Environ. Analyt. Chem. 1978, 5(4), 293-310.
20. Boehm, P. D.; and Quinn, J. G. Estuarine and Coast. Mar. Sci. 1978, 6, 471-494.
21. Boehm, P. D.; and Fiest, D. L. Proc. Conf. Prelim. Sci. Results Pierce/Researcher Cruise to the IXTOC-I Blowout, NOAA Office of Marine Pollution Assessment, Rockville, MD, 1980.
22. Wakeham, S. G.; and Farrington, J. W. In "Contaminants and Sediments, Volume 1: Fate and Transport, Case Studies, Modeling, Toxicity"; Baker, R. A., Ed.; Ann Arbor Science, Ann Arbor, MI, 1980, Chap. 1.
23. Brown, R. C.; Pierce, R. H.; and Rice, S. A. Mar. Pollut. Bull. 1985, 16(6), 236-240.
24. Youngblood, W. W.; and Blumer, M. Geochim. Cosmochim. Acta. 1975, 39, 1301-1314.
25. Hites, R. A.; LaFlamme, R. E.; Windsor, J. G.; Farrington, J. W.; and Deuser, W. G. Geochim. Cosmochim. Acta. 1980, 44, 873-878.
26. Boehm, P. D.; Barak, J. E.; Fiest, D. L.; and Ekskus, A. A. Mar. Environ. Res. 1982, 6, 157-158.
27. Pierce, R. H., Jr.; Cundell, A. M.; and Traxler, R. W. Appl. Micro. 1975, 29(6), 646-652.
28. Butler, J. N.; Morris, B. F.; and Sass, J.   Pelagic tar from Bermuda and the Sargasso Sea.   Bermuda Biological Station Special Publication No. 10, Bermuda. 1973.
29. Thompson, S.; and Eglinton, G. Mar. Pollut. Bull. 1978, 9, 133-136.
30. Traxler, R. W.; and Pierce, R. H.   Standard and intercomparison criteria: tar balls and particulate matter, NBS special publication No. 409, Marine Pollution Monitoring (Petroleum) Symposium, 1974, pp. 161-162.
31. Keizer, P. D.; Dale, J.; and Garden, D. C. Geochim. Cosmochim. Acta. 1978, 42, 105-172.
32. Giger, W.; and Schaffner, C. Analyt. Chem. 1978, 50, 243-249.
33. LaFlamme, R. E.; and Hites, R. A. Geochim. Cosmochim. Acta. 1978, 42, 289-303.
34. Sporstol, S.; Gjos, N.; Lichtenthaler, R. G.; Gustavsen, J. O.; Urdal, K.; and Oreld, F. Environ. Sci. Technol., 1983, 17(5), 282-286.
35. Overton, E. B.; McFall, J.; Mascarella, S. W.; Steel, C. F.; Antonine, S. A.; Politzer, I. R.; and Laseter, J. L., American Petroleum Institute: Washington, D.C., 1981, 541-546.

RECEIVED September 16, 1985

15

Distribution of Trace Organics, Heavy Metals, and Conventional Pollutants in Lake Pontchartrain, Louisiana

Edward B. Overton[1], Michael H. Schurtz[2], Kerry M. St. Pé[2], and Christian Byrne[3]

[1]Institute for Environmental Studies, Louisiana State University, Baton Rouge, LA 70803
[2]Louisiana Department of Environmental Quality, Baton Rouge, LA 70804
[3]Center for BioOrganic Studies, University of New Orleans, New Orleans, LA 70148

Polynuclear aromatic hydrocarbons from urban runoff
were found at elevated levels in nearshore sediment
samples from Lake Pontchartrain. Concentrations decreased with distance from the New Orleans shoreline
and approached background levels three to six miles
offshore. Quantitative profiles for individual PAH
isomers differed significantly between nearshore
and offshore sediments. Similar trends were observed
with chlorocarbons and lead, but concentrations of other
heavy metals did not decrease with distance from the
shoreline. Salt water intrusion causes stratification
over the southeastern portion of the Lake in the
summer. This stratification contributes to and
exacerbates bottom anoxic conditions during warm
weather months.

The objectives of this year long study were twofold and included:
(A) determination of the occurrence and distribution of chemicals
in Lake Pontchartrain, Louisiana (particularly in the southern
portion of the Lake near New Orleans) that resulted from
anthropogenic activity; and (B) an overall water quality
assessment of the Lake to ascertain any phenomena, in particular
urban runoff related water pollution, that may be adversely
affecting the ecological balance of the Lake. The strategies
developed to address these objectives included: (1) Collection of
biota and sediment samples and their analyses for a broad
spectrum of chemical substances which included, but were not
limited to, those designated as priority pollutants by the U.S.
EPA. (2) Examination of the temporal and spatial variation of
important estuarine water quality parameters, as well as
conventional pollutants, that contribute to nutrient enrichment
and other impairments such as increased turbidity and dissolved
oxygen depression.

0097–6156/86/0305–0247$07.00/0
© 1986 American Chemical Society

Field and laboratory work was begun in March of 1982. Field work was undertaken by biologists with the Louisiana Department of Environmental Quality. Analyses of biota and sediment samples for anthropogenic substances were performed at the Center for Bio-organic Studies at the University of New Orleans. Water Quality and conventional pollutant analyses were provided by the laboratory staff of the Department of Environmental Quality.

Lake Pontchartrain is a shallow, open water embayment of a major estuarine system in southeastern Louisiana. It has a surface area of approximately 630 square miles (1.60×10^9 m^2). Depths in offshore areas (greater than one mile from shore) range typically from 12-18 feet (4 to 6 m) with the bottom exhibiting a gradual relief. Localized depressions proximate to three tidal passes, which connect the lake to the Gulf of Mexico, slope to depths up to 40 feet (12m). The passes themselves have depths approaching 100 feet (30m).

Lake Pontchartrain receives drainage from an area of approximately 5,000 square miles located mostly to its north. Several rivers drain the Louisiana coastal plain terraces while two other rivers drain the former Mississippi River floodplain southwest and west of the lake. From the south, Lake Pontchartrain receives extensive wetland (marsh and swamp) drainage. The greater New Orleans area, which includes Jefferson and Orleans Parishes, is now leveed off from the lake. However, this metropolitan area, home for over a million people, is drained by pumping from an extensive network of man-made canals into Lake Pontchartrain.

In addition to man-made stormwater drainage canals along the southern shoreline, Lake Pontchartrain is influenced by a deep draft (12m) channel known as the Mississippi River Gulf Outlet (MRGO). Highly saline waters are introduced directly into Lake Pontchartrain via this canal.

Because of the close proximity of the Greater New Orleans Metropolitan Area to Lake Pontchartrain, with no intervening buffer zone, the southern portion of the lake should logically be the most affected by anthropogenic impacts. As such it should serve as a "worst case" barometer of the lake's environmental health. Therefore the major focus of this study was directed toward the southern area of Lake Pontchartrain.

Sampling Sites

Primary sampling stations for the study were established at 16 fixed localities in the southern region of Lake Pontchartrain. These stations were selected from an approximate 100 square mile (2.6×10^9 m^2) grid adjacent to the southern shoreline extending 20 statute miles along Jefferson and Orleans Parishes and extending to six miles offshore at its widest point (see Figure 1). Eight of these stations were located in the immediate nearshore area adjacent to the mouths of major drainage canals. Each was established approximately .23 miles (370m) north of the respective canal's mouth. LORAN C coordinates for each were recorded during the first sampling cruise and then used during subsequent cruises to re-occupy the same sampling locality. Ten

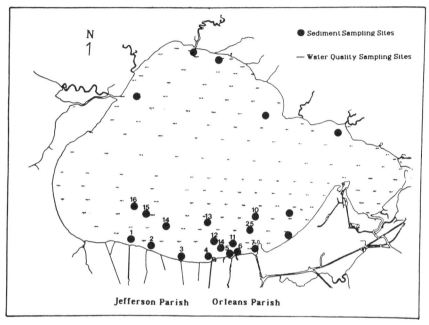

Figure 1. Map of Lake Pontchartrain showing locations of the 148 water quality sampling stations and the locations which were also sampled for sediments.

additional stations were established offshore within the 100
square mile study area. On a monthly basis, March 1982 to March
1983, field and laboratory water quality analyses were taken at
16 of the primary stations. Field WQ readings were taken during
the summer months over the entire area of Lake Pontchartrain at
148 sampling stations (see Figure 1). Field water quality
measurements were made at one meter (m) intervals from one meter
below the surface to 0.5m above the lake bottom. Readings were
taken with Martek Instruments, Inc. (Irvine, CA) multiprobe
analyzers.

On a quarterly basis between March 1982 and March 1983 lake
sediment and biota samples were collected from ten of the sixteen
primary stations and analyzed for anthropogenic organic chemicals
and trace metals. Stations were selected alternately so that
each was sampled at least once during the study period. Four
stations designated LP07, LP10, LP11, and LP12, were sampled
during each quarter. Sediment samples consisted of three
separate dredge grabs with Petite Ponar samplers (Wildlife Supply
Company). Sediment grabs were composited in stainless steel
buckets (3 gallon) that were washed and then rinsed with
nanograde hexane between each sampling station. Aliquots of
composited sediments were placed in wide-mouth quart jars, pre-
rinsed with nanograde hexane. Nearshore sediments were well
sorted fine silts with between 1 and 2 percent organic carbon
content. Offshore sediments were fine silts or clays and
generally contained between 2 and 3 percent organic carbon
content.

Analytical Methods

Analytical methods were designed to screen a relatively large
number of samples for a broad spectrum of organic compounds and
selected heavy metals. In the field, all samples were placed on
ice immediately following collection. Upon delivery to the
laboratory, samples were frozen and maintained at -10°C. Just
prior to analyses, the samples were thawed and thoroughly
homogenized.

The analytical procedure for trace organics included
extraction of all samples, fractionation of the extracts by
liquid-solid chromatography, and instrumental analyses of the
fractions by fused silica capillary gas chromatography using
flame ionization (general) detection, and, for some samples,
electron capture (halogen specific) detection. Selected
fractions were then chosen for detailed instrumental analyses
using computerized high resolution gas chromatography-mass
spectrometry (GCMS) (1).

All GCMS data files were examined with a general search
procedure developed for scanning GCMS data for anthropogenic
chemicals at trace levels. Hard copies of the mass spectral data
were examined manually to verify computer matches and identify
compounds not selected by the computer program. Identified
compounds were then quantitated by multiplying their peak area
with appropriate response factors obtained from analyses of
quantitative standards under identical instrumental tunes and

conditions. Standard compounds were not available for all the
substances identified in samples. Quantitative response factors
were estimated for those compounds not contained in analytical
standards. It is important to note that the use of estimated
response factors gives data that can be compared with equivalent
data within a given data base. Comparison of the data with other
results outside of the data base should be done within the
context of the limitations imposed by quantitative GCMS analyses.
In general, these limitations in the quantitative results are not
significantly large when compared with variabilities observed
between samples collected from marine environments. Experience
from analyses of many replicate samples and participation in
round-robin type interlaboratory calibration programs (2) has
produced data which indicates the analytical variabilities are
compound dependent and range from 5% to 30%. This analytical
variability is generally small when compared to variabilities
found between samples collected in the marine environment.

EPA approved atomic adsorption methods were used in all
trace metal analyses (3).

Standard quality control procedures were followed. These
included: a) careful washing of all glassware in strong
oxidizing solutions and with Type I water; b) frequent analyses
of glassware and reagent blanks; c) analyses of procedural
blanks with each batch of samples; and d) calibration of
instruments before each set of analyses by analyzing standard
solutions. Analytical proficiency in the analysis of biota and
sediment samples have been demonstrated by participation in
round-robin type interlaboratory calibration exercises (2).

Discussion of Results

Salinity. Concentration profiles for top and bottom salinities
and dissolved oxygen concentrations at four selected stations are
shown in Figure 2. Similarities in surface salinity are
apparent, but the surface dissolved oxygen concentrations varied
appreciably. The maximum surface salinity at all four stations
occurred in the early Fall (September and October, 1982). The
minimum surface salinity at all stations was measured in
February, 1983. Surface salinity readings ranged from 1.3 ppt
(parts per thousand) at stations LP10 to 9.0 ppt at station LP07.

Seasonal similarities in bottom salinities between all
stations were not apparent. Bottom salinities differed
appreciably in both magnitude and the time at which increases
began. Significant increases in bottom salinities at the station
nearest to the MRGO began between April and May, 1982. This was
followed by a marked salinity increase at stations 1.5 and 4.2
miles offshore between June and July, 1982. The station 4.1
miles offshore in Jefferson Parish (the most remote from the
MRGO) did not exhibit the abrupt bottom salinity increases that
were noted at the other stations. Bottom salinities of 7 to 9
ppt and up generally resulted in stratification with significant
decreases in bottom dissolved oxygen concentrations.

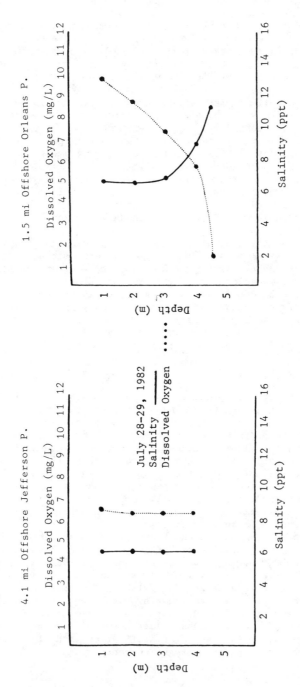

Figure 2. Top and bottom salinity and dissolved oxygen profiles at two stations in Lake Pontchartrain over a 12 month annual cycle.

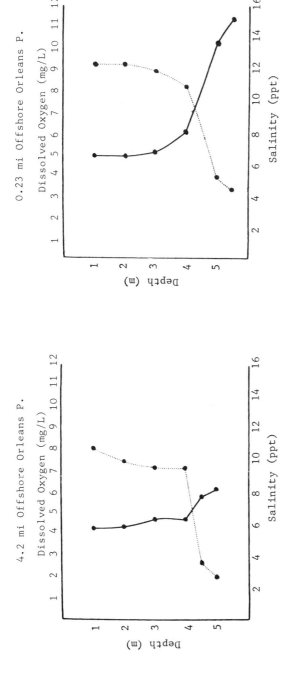

Figure 2. Top and bottom salinity and dissolved oxygen profiles at two stations in Lake Pontchartrain over a 12 month annual cycle.

Data collected at 0.5 m above the Lake's bottom, from the
148 lakewide stations in July and August 1982, are presented as
isopleths in Figure 3. Highest bottom salinities in Lake
Pontchartrain, during the two month sampling period, occurred at
and near the MRGO-Lake connection. Much lower salinities were
recorded in the northern sections of the Lake due to freshwater
inputs from northshore rivers and passes. Bottom dissolved
oxygen concentrations taken in July were lowest (1.0 ppm) in the
south central area of the lake and 3.0 ppm along the south shore
and in mid-lake regions. The lowest dissolved oxygen readings
recorded in August were found to be centralized in mid-lake
areas. In general, low DO readings closely follow regions of
significant stratification during the summer months when mixing
of the shallow lake from weather fronts was minimized.

Examination of the isohalines shown in Figure 3 clearly
indicates that the major source of highly saline waters was the
MRGO. Salinities at the mouths of two natural passes, which are
also contributors of saltwater into Lake Pontchartrain, are not
as high as at the mouth of the MRGO. The MRGO complex which
provides a direct man-made connection for highly saline waters
from the Gulf of Mexico to enter Lake Pontchartrain, contributes
significantly to stratification. On the other hand, the two
natural passes allow inflow of waters characterized by salinities
that are closer in concentration to those in Lake Pontchartrain.

Trace Organics. High resolution gas chromatograms from analyses
of the saturated hydrocarbon fractions of four sediment samples
collected from the lake in January 1983 are shown in Figure 4.
The samples collected off Jefferson Parish contained saturated
hydrocarbon profiles typical of terrestrial biogenic hydrocarbon
inputs (4). These inputs are characterized by odd carbon number
hydrocarbons, in the range of n-C24 to n-C30, and have greater
abundances than their even carbon numbered homologs. The samples
collected off Orleans Parish contained large complex unresolved
mixtures (CUM) which are indicative of weathered petrogenic
hydrocarbons. The source of the extremely high levels of
weathered petrogenic hydrocarbons in the nearshore Jefferson-
Orleans Parish sample is unclear. However, the hydrocarbon loads
in samples from this station were not typical of those found at
any of the other sampling stations in Lake Pontchartrain.
Consequently, these levels cannot be attributed to the normal
types and loads of hydrocarbons found in urban runoff and
probably resulted from the spillage of weathered petrogenic
hydrocarbons into the drainage canal situated near this sampling
station. Because the levels of petrogenic hydrocarbon were so
elevated at this station, we have chosen to consider all data
from this station as anomalous. Consequently, data from this
station have been excluded from further discussions of the
distributions and sources of organic contaminants in Lake
Pontchartrain.

Figures 5 through 7 show concentrations of several important
classes of organic compounds, as well as specific organics as a
function of distance offshore and date of sample collection. The
types of compounds, which are represented as sub-classes of the

Isohalines (0.5 m off bottom)   D.O. Isopleths (0.5 m off bottom)

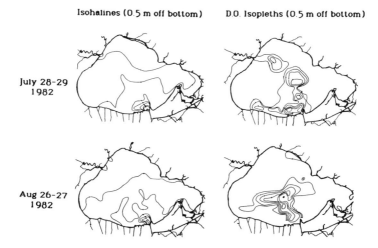

July 28-29
1982

Aug 26-27
1982

Figure 3. Isohalines and dissolved oxygen isopleths in Lake Pontchartrain on consecutive months in the summer of 1982.

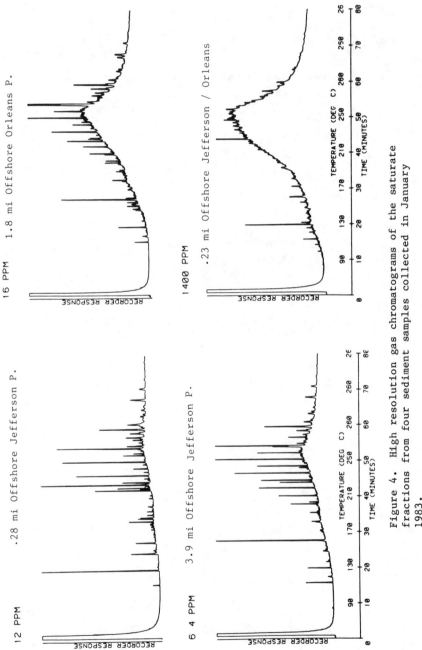

Figure 4. High resolution gas chromatograms of the saturate fractions from four sediment samples collected in January 1983.

total organics found in mass spectra data, are shown in Table I
and include: TIC (total identifiable compounds; PPAH (parent
polynuclear aromatic hydrocarbons); TAAH (total alkyl aromatic
hydrocarbons); C/N (chlordane and nonachlor); PCB
(polychlorinated biphenyls); and DDT (sum of the DDT
environmental degradation products). Total hydrocarbon
concentrations (THC) were derived from GC data. The quantities
of these various substances generally decrease, in a general
exponential fashion, with distance from the southern shoreline,
reaching background levels at distances of 3 to 6 miles offshore.
This is as expected since the majority of organic substances
found in these samples have extremely low water solubility (5).
Consequently, they are generally transported in the marine
environment as particulate matter (6, 7). There are several
exceptions to this general observation. For example, perylene
concentrations showed no downward trend with distance from the
shore. This compound is produced by both high temperature
combustion and natural diagenetic processes (8). Consequently,
it is not surprising that no significant trends were observed
relating perylene concentrations with distance from shore. The
most abundant class of organics found in Lake Pontchartrain
sediments was the unsubstituted polynuclear aromatic hydrocarbon
(PPAH). Figure 8 shows the quantities and relative distributions
of nine different PAH compounds at three different sampling
stations. These stations were chosen to show the PAH
distributions with distance from the southern shoreline. As a
general rule, similar distributions were noted at other sampling
sites as a function of distance offshore. Offshore PAH levels
were characterized by elevated concentrations of perylene while
near-shore samples contained mostly four ring PAH compounds.
      Table II shows the concentrations of selected trace organics
in biota samples examined during this study. In general, highest
concentrations were observed in the spring. Chlorocarbon
concentrations were comparable to other areas along the Gulf
Coast and generally lower than commonly encountered along the
eastern seaboard (9).

Heavy Metals. The concentrations of Barium, Lead and Cadmium,
with distance from the shoreline, are shown in Figure 9. Lead
concentrations generally decreased with distance from the
shoreline but there was considerable variability in this trend.
Lead is transported as particulate matter in the marine
environment (10, 11). Consequently, lower but highly variable
concentrations offshore are to be expected. Other heavy metals
studied did not exhibit any detectable trend with distance from
the shoreline. Even the highly insoluble heavy metal, Ba, did
not show significantly lower concentrations in offshore areas.

Conclusions

Factors which are currently affecting environmental quality in
Lake Pontchartrain are generally those related to urban
development and urban pollution, altered land use patterns, and
hydrologic modification within the lake's watershed. Paramount

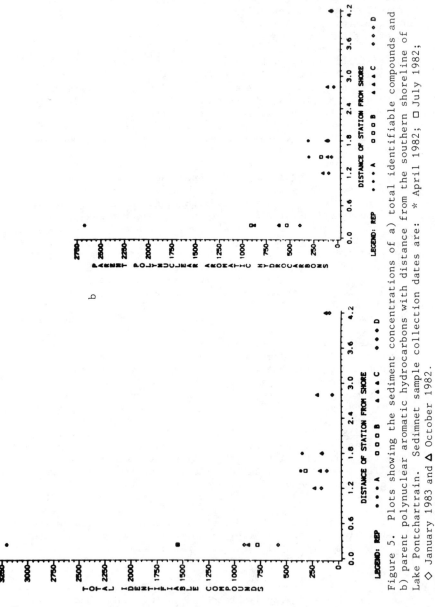

Figure 5. Plots showing the sediment concentrations of a) total identifiable compounds and b) parent polynuclear aromatic hydrocarbons with distance from the southern shoreline of Lake Pontchartrain. Sedimnet sample collection dates are: * April 1982; □ July 1982; ◇ January 1983 and △ October 1982.

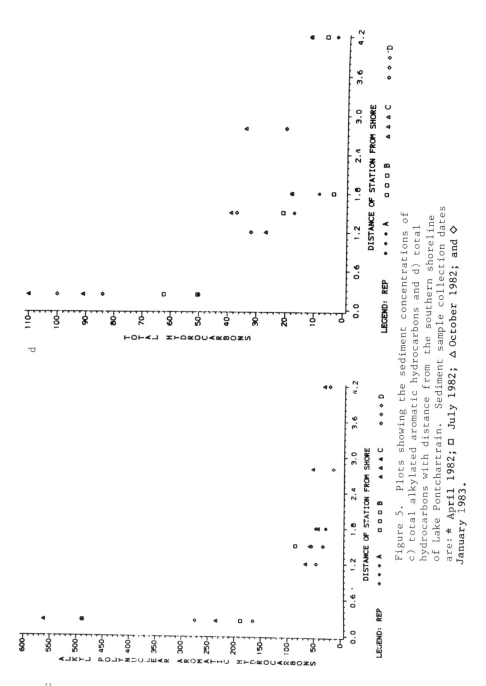

Figure 5. Plots showing the sediment concentrations of c) total alkylated aromatic hydrocarbons and d) total hydrocarbons with distance from the southern shoreline of Lake Pontchartrain. Sediment sample collection dates are: * April 1982; □ July 1982; ▲ October 1982; and ◇ January 1983.

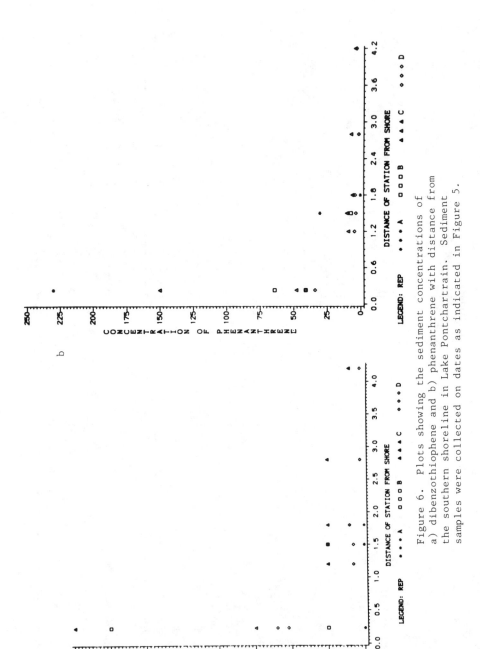

Figure 6. Plots showing the sediment concentrations of
a) dibenzothiophene and b) phenanthrene with distance from
the southern shoreline in Lake Pontchartrain. Sediment
samples were collected on dates as indicated in Figure 5.

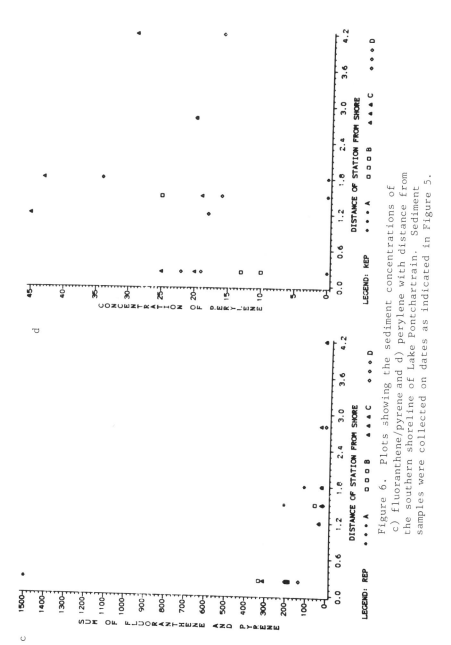

Figure 6. Plots showing the sediment concentrations of c) fluoranthene/pyrene and d) perylene with distance from the southern shoreline of Lake Pontchartrain. Sediment samples were collected on dates as indicated in Figure 5.

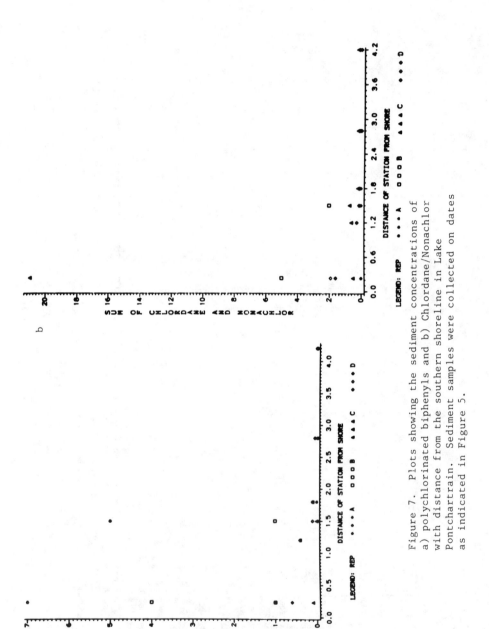

Figure 7. Plots showing the sediment concentrations of
a) polychlorinated biphenyls and b) Chlordane/Nonachlor
with distance from the southern shoreline in Lake
Pontchartrain. Sediment samples were collected on dates
as indicated in Figure 5.

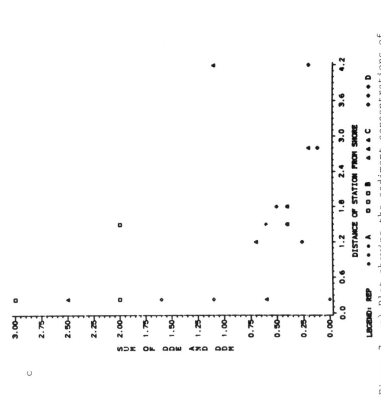

Figure 7. c) Plot showing the sediment concentrations of total DDT environmental degradation products with distance from the southern shoreline in Lake Pontchartrain. Sediment samples were collected on dates as indicated in Figure 5.

Table I. Specific Compounds Found in Lake Pontchartrain Sediment Samples
as well as the Identities of Compounds that Comprise the Various
Sub-classes of Pollutants whose Concentration Trends are Highlighted
in this Report.

Compound name	TIC	PPAH	TAAH	FP	Phen	DBT	DDT	PCB	C/N
o naphthalene	X	X							
alkyl naphthalenes	X		X						
o acenaphthene	X	X							
o acenaphthylene	X	X							
biphenyl	X	X							
alkyl biphenyls	X		X						
o fluorene	X	X							
alkyl fluorenes	X		X						
o phenanthrene	X	X			X				
o anthracene	X	X							
alkyl phenanthrenes	X		X						
dibenzothiophene	X	X				X			
alkyl dibenzothiophenes	X		X						
o fluoranthene	X	X		X					
o pyrene	X	X		X					
alkyl pyrenes	X		X						
o benz(a)anthracene	X	X							
o chrysene	X	X							
alkyl chrysenes	X		X						
naphthylbenzothiophene	X	X							
alkyl naphthylbenzo- thiophenes	X		X						
o benzo Fluoranthenes	X	X							
benzo(e)pyrene	X	X							
o benzo(a)pyrene	X	X							
perylene	X	X							
o indo(1,2,3-od)pyrene	X	X							
o dibenzo(a,h)anthracene	X	X							
o benzo(ghi)perylene	X	X							
o diethyl phthalate	X								
o bis(ethylhexyl)phthalate	X								
o other phthalates	X								
o DDE	X						X		
DDM	X						X		
o chlordane	X								X
o nonachlor	X								X
o PCB	X							X	
Unknown MW = 252 BP = 237	X								

o USEPA Priority Pollutants

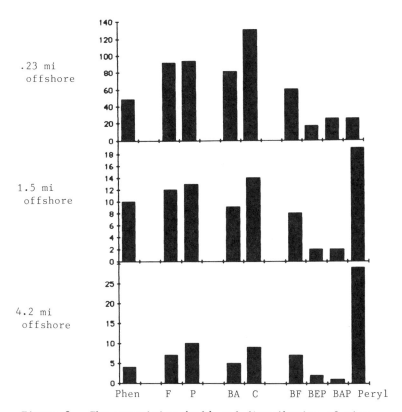

Figure 8. The quantities (ppb) and distribution of nine polynuclear aromatic hydrocarbons, at three sampling stations, in Lake Pontchartrain sediments.
Phen = phenanthrene, F = fluoranthene, P = pyrene, Bz = benzanthracene, C = chrysene, BF = benzofluoranthenes, BEP = benzo-e-pyrene, BAP =benzo-a-pyrene, Peryl = perylene.

Table II. Concentration of: Total identifiable Compounds/Parent
Polynuclear Aromatic Hydrocarbons (Top); PCBs/DDTs (Bottom) in the
Selected Biota Samples Expressed as PPB wet Weight.

Sample Location	April '82	July '82	Oct. '82	Jan '83
LP01		Shrimp *800/15 3/2	Crab 66/17 1/2	
LP02	Clam 360/114 6/6			
LP03			Crab 178/48 12/8	
LP05		Clam 570/190 130/13		
LP07	Croaker 2000/85 180/70		Oyster 140/63 4/1	Crab 44/5 ND/5
	Crab 1600/27 100/27			
LP10	Spot Fish 2000/280 200/130	Spot Fish 130/34 17/14		Croaker 33/4 1/4
LP15		Crab 66/4 2/4		Catfish *370/10 ND/1

*   Contaminated with diethylhexyl phthalate.

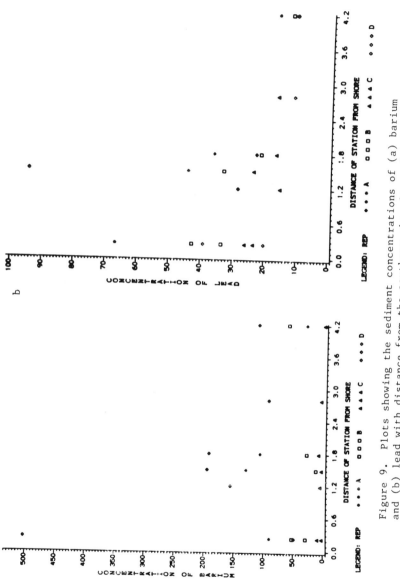

Figure 9. Plots showing the sediment concentrations of (a) barium and (b) lead with distance from the southern shoreline of Lake Pontchartrain. Sediment samples were collected on dates as indicated in Figure 5. Continued on next page.

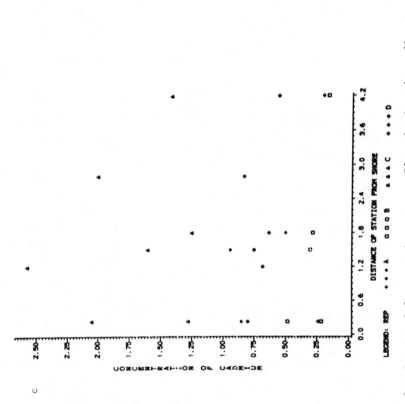

Figure 9. Continued from previous page. Plot showing the sediment concentrations of cadmium with distance from the southern shoreline of Lake Pontchartrain. Sediment samples were collected on dates as indicated in Figure 5.

among these is the tremendous rate of urbanization that has spread in recent decades from the metropolitan New Orleans area which occupies a major portion of the southern lakeshore. Because of its immediate proximity to the lake and the fact that much of this area is at or below sea level, a network of canals on the east side of the Mississippi River drains the area into the lake. These canals are catch basins for the street drainage, stormwater runoff, and point source discharges of treated, partially treated and untreated sewage. In addition, many discharges from small commercial facilities are made either to the sewage systems or directly to the canals. No major industrial facilities (e.g. petrochemical or organic chemical manufacturing) are permitted to discharge into Lake Pontchartrain. Therefore, the major chemical inputs of pollutants into the lake may be categorized as coming from three major source categories. These are (1) urban stormwater drainage, (2) discharges of domestic sewage that is generally treated to less than acceptable levels, and (3) discharges and spills from marine related facilities and marine vessels.

Examination of the data suggest that urban runoff is responsible for the greater variety and the higher levels of anthropogenic chemicals introduced into the lake. Sewerage discharges introduce pollutants that contribute to oxygen depletion and eutrophication.

The contribution of oil and other organic substances from marine associated facilities and vessels is more related to spill events, either accidental or intentional, rather than regular, long term input. Data reviewed to date indicate that specific areas are impacted by oils and other petroleum hydrocarbons.

Another factor of significant environmental influence is the alteration of salinity patterns in the southeastern and east-central region of the lake off the mouth of the MRGO. This hydrologic modification facilitates the intrusion of highly saline marine waters from the Gulf of Mexico into Lake Pontchartrain. This saltwater intrusion is most pronounced during the warm water months (May to October) of years with normal and below normal rainfall. Its effects reach their peak during the late summer (August/September) when prevailing southerly and southeasterly winds greatly facilitate the movement of marine waters up the MRGO and into Lake Pontchartrain.

Acknowledgments

The authors would like to express thanks and appreciation to: D. Sabins, W. Tucker, K. Cormier, C. Melchior, L. Racca, J. Dixon, L. Wellman and J. Akielaszek for collection of samples; J. McFall and S. Antoine for analyses of samples; P. Benoit, S. Stewart and D. Meyers for help in preparation of this manuscript; and the Louisiana Department of Environmental Quality for their financial support.

Literature Cited

1. Brown, P.W.; Ramos, L.S.; Uyeda, M.Y.; Friedman, A.J.;
    MacLeod, W.D. In "Petroleum in the Marine Environment";
    Petrakis L.; Weiss, F.T., Eds.; ADVANCES IN CHEMISTRY SERIES
    No. 185, American Chemical Society: Washington, D.C., 1980;
    p. 313.
2. MacLeod, W.D.; Prokaska, P.G.; Gennero, D.D.; Brown, D.W.
    Anal. Chem., 1982, 54, 386.
3. "Interim Methods for the Sampling and Analyses of Priority
    Pollutants in Sediments and Fish Tissue," U.S. Environmental
    Protection Agency, 1981.
4. Wakeham, A.J.; Farrington, J.W.; In "Contaminants and
    Sediments"; Baker, R.A., Eds.; Science Publishers Inc.: Ann
    Arbor, 1980; p. 3.
5. Lopez-Avila, V.; Hites, R.A. Environ. Sci. Technol. 1980, 14,
    1382.
6. Hamilton, A.E.; Bates, T.A.; Cline, J.D. Environ Sci.
    Technol. 1984, 18, 72.
7. Eganhouse, R.P.; Blumfield, D.L.; Kaplan, I.R. Environ. Sci.
    Technol. 1983, 17, 523.
8. Laflamme, R.E.; Hites, R.A. Geochim Cosmochem Acta. 1978,
    42, 289.
9. "Review of PCB levels in the environment" US Environmental
    Protection Agency, 1976.
10. Whipple, W.; Hunter, J.V. I. Water Pollution Control Fed.
    1977, 49, 15.
11. Boyden, C.R.; Afton, S.R.; Thorton, I. Estuarine and Coastal
    Marine Sci. 1977, 9, 303.

RECEIVED November 8, 1985

VOLATILE ORGANICS

16

Seasonal Cycles of Dissolved Methane in the Southeastern Bering Sea

Joel D. Cline[1], Charles N. Katz[2], and Kimberly Kelly–Hansen[3]

[1] Mobil Research and Development Corporation, Dallas, TX 75234
[2] InterOcean Systems, Inc., San Diego, CA 92123
[3] Pacific Marine Environmental Laboratory, National Oceanic and Atmospheric Administration, Seattle, WA 98115

Seasonal measurements of dissolved methane were made in the southeastern Bering Sea shelf waters between 1975-1981. This region is a large, coastal-shelf environment that is characterized by large variations in biological activity and climate. For these reasons, and others, this region is well suited to examine the seasonal distributions of methane and other biologically-generated gasses.

Maximum concentrations of methane were observed in the fall of 1981, followed by a minimum in spring. Surface concentrations, on the average, were about 400 nL/L(STP) in fall, dipping to concentrations near 100 nL/L(STP) in spring. Near bottom concentrations were generally higher throughout the region, and in particular over St. George Basin, where concentrations in excess of 2500 nL/L(STP) were observed.

The sea to air flux of methane ranged from 0.021 g CH_4 m^{-2} yr^{-1} in May 1981 to 0.34 g CH_4 m^{-2} yr^{-1} in October 1980; the mean flux was about 0.14 g CH_4 m^{-2} yr^{-1}. Based on a total area of 31.9 x 10^{10} m^2 and correcting for average ice coverage in winter, the annual transport of methane to the atmosphere for 1980-81 was 4.6 x 10^{10} g.

Seasonal and interannual variations in the concentration methane are large in the southeastern Bering Sea shelf. The driving force for these variations is the timing and flux of organic carbon as well as the severity (i.e., amount of ice cover) of the winters. These observations point to the need for seasonal observations in near shore coastal waters if meaningful budgets of methane, and presumably other trace gasses, are to be constructed.

Methane is an ubiquitous trace gas found in all marine and freshwaters. Its concentration in the surface waters of the open ocean is near saturation (1-3), however in some near shore areas, in anoxic basins, and in marine sediments the concentrations are significantly higher (4, 7) because of increased production rates. The highest production rates of methane are usually restricted to anoxic environments, but significant rates of production also have been observed in oxic marine water columns (8, 9).

The most complete set of published observations on the sources, sinks, and distributions of methane in coastal waters is that by researchers at Texas A&M University. Without detailing an exhaustive list of the observations, many of which deal with the thermogenic as well as biogenic sources of methane, the following papers are most relevant to our own discussion of coastal distributions of methane (7, 10-14).

Our intent here is not to summarize the extensive observations of methane in the marine environment, but rather to set the stage for a discussion of seasonal distributions in a high latitude shelf environment. There are few studies of the seasonal variations of trace gases in a marine environment, and to our knowledge, none in a high latitude, shallow sea, where biological influences are apt to be large. In this report, we will specifically examine the seasonal distributions of dissolved methane in the southeastern Bering Sea and qualitatively discuss the physical and biological forcing that results in the observed distributions.

Observations discussed in this report were made as a part of an environmental study of Alaskan coastal waters.

Methods

Sampling. Discrete water column samples were obtained with 5-L Niskin bottles attached to a General Oceanics Rosette fitted with Plessey Environmental Systems model 9040 CTD. A 1-L aliquot of seawater was transferred from the Niskin samplers into glass-stoppered bottles in such a way that air bubbles were not trapped. The bottles were then stored in the dark at approximately 5°C until analyzed; usually within one hour of sampling.

Analysis Concentration. The analysis of methane was accomplished routinely in the field using a purge and trap technique (15). The method involves removal of the dissolved gases from a 0.2L volume of seawater by helium purging. The gasses removed from solution are passed through Ascarite®, Drierite® and Tenax G.C.® traps to remove carbon dioxide, water vapor and heavy hydrocarbons respectively, before being concentrated on an activated alumina trap held at -196°C. After quantitative removal of the gasses from solution (approx. 6 minutes with a purge rate of 100 mL min^{-1}), the activated alumina trap was warmed to 100°C and the gasses backflushed into a gas chromatograph.

Gas Chromatography. Detection and quantitation of methane was carried out on a Hewlett-Packard 5710A gas chromatograph equipped with a flame ionization detector (FID). The column packing used

was activated alumina, 60-80 mesh (1.8m x 0.48 cm o.d.). The
column temperature was held isothermally at 100°C.
    Quantitation of the methane was obtained by comparing the
detector response of a methane gas standard with that of the
sample. The FID response was found to be linear with
concentration throughout the entire range of methane
concentrations encountered in the study. Standard gasses
prepared by Matheson Gas Products were intercalibrated with a
standard gas analyzed by the National Bureau of Standards (NBS).
Analytical precision was generally less than 1% while accuracy,
based on the NBS intercalibration, was 5%. The detection limit
of the method, based on a signal-to-noise ratio of 2 is
approximately 5 nL CH_4 L^{-1} seawater (STP).

Salinity-Temperature-Pressure (Depth). Conductivity, temperature
and pressure data were collected using a Plessey Systems CTD
with model 8400 data logger. These sensors were interrogated
five times per second for values of temperature, conductivity
(salinity), and pressure (depth). Data were recorded during the
down-cast using a lowering rate of 30 m min^{-1}. Niskin bottle
samples were taken on every other cast to provide temperature
and salinity calibration. Nominal precision of the salinity,
temperature and depth measurments was ± 0.02 g kg^{-1}, ± 0.02 °C
and ± 0.2 m, respectively.

Physical Setting

The survey area, which includes Bristol Bay, is a broad shelf
region located in the southeastern Bering Sea (Figure 1). It is
bounded on the south by the Alaska Peninsula, on the west by the
shelf break, and on the north by the coast of Alaska (16). The
area out to the 200 m isobath is about 32 x $10^{10}m^2$; the
area-weighted mean depth is about 70 m. The principal sources of
fresh water are the Kuskokwim and Kvichak rivers, which are
located at the northern and eastern boundaries (Figure 1).
Estimated discharge of these rivers is about 47 km^3 annually
(17), but there are also numerous small, diffuse sources of fresh
water along the entire coastline. Notable among them is the Port
Moller estuary located along the north side of the Alaska
Peninsula. Winter cooling results in a maximum of 60% of the
southeastern Bering Sea being covered with ice between the months
of December and April (16). This ice is largely wind driven from
the north, although during severe winters there is local ice
formation.
    Hydrographically, the region is divided into three domains:
coastal waters (z < 50m), middle shelf (50 m < z ≤ 100m), and the
outer shelf (z > 100m) (18). The greatest seasonal variations in
temperature and salinity are found in the coastal zone because of
its shallow depth and close proximity to sources of fresh water.
In contrast to the coastal regime, which is unstratified most of
the year because of wind and tidal mixing, the middle shelf is
thermally stratified much of the year. The outer shelf, principally St. George Basin (Figure 1), is characterized by salinites

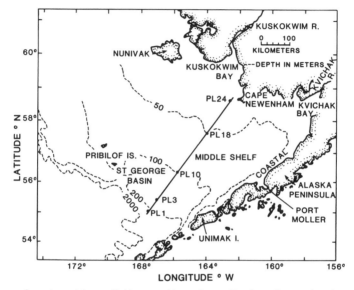

Figure 1.   Location of the southeastern Bering Sea, showing hydrographic regimes and bathymetry. Also shown in the position of a vertical section across the region (see Figure 4). Individual station locations, which varied from cruise to cruise, are shown on the areal distribution maps.

near 33°/∘∘ and a temperature range of 3-4°C. Surface temperatures and salinities are somewhat greater in summer. Currents over the shelf are generally weak (19, 20). Along the outer shelf in summer, current trajectories trend northwest at speeds of approximately 5 cm/sec. Across the inner shelf, currents are weak (~ 1 cm/sec) and counterclockwise.

Seasonal Distributions of Dissolved Methane

The vertical and horizontal distributions of dissolved methane are not only controlled by biological processes, but also by the local hydrography, which has a marked seasonal signature. Stratification strongly influences the vertical transport of methane and other gasses, and is largest where the buoyancy input is the greatest. Warming of the surface layers and the addition of fresh water from land drainage leads to a reduction in the density of the surface waters and subsequent stratification. Because stratification also affects in a significant way the vertical transport of methane, a brief description of the seasonal changes in salinity and temperature is instructive.

The coastal regime experiences large seasonal changes in water properties, because of its nearness to land and its shallow depth. Figure (2a) shows the temperature and salinity fields for the three seasons, superimposed on the lines of constant density $\{\sigma_t = (\rho - 1) \times 1000\}$. A large change in salinity is seen in August and is the result of fresh water runoff along the Alaska Peninsula. The lowest salinities are found near the Kvichak River, and the highest salinities are found near Unimak Pass. Relatively saline water entering Unimak Pass is driven cyclonically around the basin and gradually becomes less saline due to freshwater dilution. Winter and spring are characterized by isohaline conditions (30-32 °/∘∘), however the temperature rises from near freezing in winter to 4-8 °C in May.

The middle shelf is not so strongly influenced by seasonal changes in salinity because it is farther removed from land. However, temperatures there range from near freezing in winter to about 10 °C in August (Fig. 2b). The salinity range is 30-32 °/∘∘, about the same as observed in the coastal zone away from major sources of freshwater.

Methane: August, 1980. The surface distribution of dissolved methane (nL/L, STP) in August 1980 is shown in Figure 3a. The highest surface concentrations were found near the entrance to Port Moller and near Unimak Pass (see Figure 1). At the entrance to Port Moller, concentrations of dissolved methane were greater than 2500 nL/L (about 35 times the equilibrium value) and decreased along the coast toward the northeast. The direction of the methane plume marks the mean drift of the coastal current.

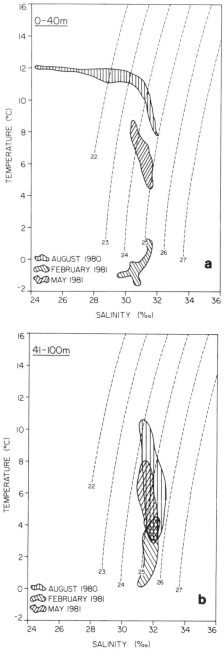

Figure 2.   Temperature and salinity fields for the coastal (a) and middle shelf (b) regions. Surface measurements at each station were averaged and plotted for the three observational periods.

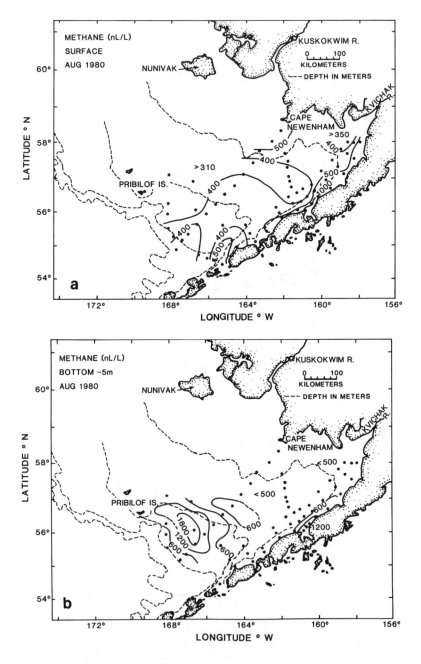

Figure 3. The surface (a) and near-bottom (b) concentrations of methane (nL/L; S.T.P.) observed in August, 1980. The near-bottom measurements were routinely taken 5 m above the bottom.

Over the remainder of the region, concentrations of methane
ranged from 300-500 nL/L. The small surface maximum observed near
Unimak Pass arises from vertical turbulence in Unimak Pass.
Near-bottom waters south of the Alaska Peninsula are enriched in
methane, which becomes entrained in the northerly flow through the
pass. Again, the surface concentrations of methane indicate the
mean surface current drift as water moves into the Bering Sea.

Distributions of dissolved methane near the bottom (i.e.,
5m above the bottom) are shown in Figure 3b. The largest accumulations were found over St. George Basin, a topographic depression
located on the outer shelf. These sediments are fine-grained and
relatively rich in organic carbon (TOC_{max} = 1%) (21). Contours of
methane concentration generally follow the basin bathymetry,
indicating weak circulation and a localized source of methane in
the basin. The maximum concentration of methane observed in
August 1980 was 2500 nL/L, decreasing to background concentrations
of 500 nL/L over the middle shelf.

In contrast, concentrations of methane over the middle shelf
and the coastal region were similar to the concentrations seen
in the surface waters of St. George Basin. Over most of the
middle shelf, near-bottom concentrations of methane ranged from
400-600 nL/L.

An example of cross-shelf variations in water properties
and their influence on the distribution of methane is shown in
Figure 4 (see Figure 1 for the location of the vertical
section). The estuarine character of the embayment is clearly
shown by the distribution of salinity, where low salinity water
at the surface moving seaward is being replaced by higher
salinity water at depth. High concentrations of methane are
evident in the bottom waters of St. George Basin (PL4-PL8) and
are the result of increased production and stratification.
There, vertical stratification (Figure 4c) restricts the
upward transport of methane. In the coastal zone (z < 50m),
the concentration of methane is homogeneous with depth,
indicating more intense mixing.

<u>Methane: January, 1981.</u> In winter, surface concentrations of
methane had fallen to 200-400 nL/L, or about 50% of that seen
the previous summer (Figure 5a). Once again, the highest
surface concentrations were found near the entrance to Port
Moller and northeast of Unimak Pass. Ice covered much of the
middle shelf and northern coastal regions, which restricted
observations to the southern reaches of the area. The maximum
surface concentration of methane observed was 2500 nL/L at the
entrance to Port Moller, once again demonstrating that this
estuary is a significant source of methane to the coastal
regime, even in winter.

High concentrations were again evident in the near-bottom
waters of St. George Basin (Figure 5b). In contrast to concentrations approaching 2500 nL/L seen the previous August (Fig. 3b),
maximum concentrations had decreased to about 1100 nL/L. Bottom
concentrations over the middle shelf were about 200-300 nL/L, the
same as the surface waters.

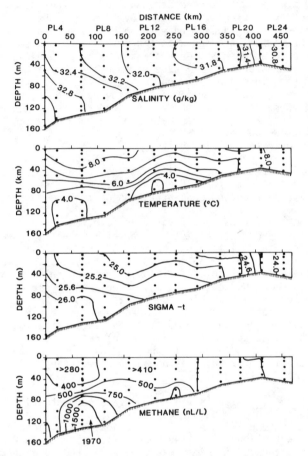

Figure 4.   The vertical distributions of salinity, temperature, sigma-t (i.e., density) and methane observed along PL1-PL24 in August, 1980. See Figure 1 for the location of this section. Sigma-t is equal to $(\rho-1) \times 1000$, where ρ is the specific-gravity seawater.

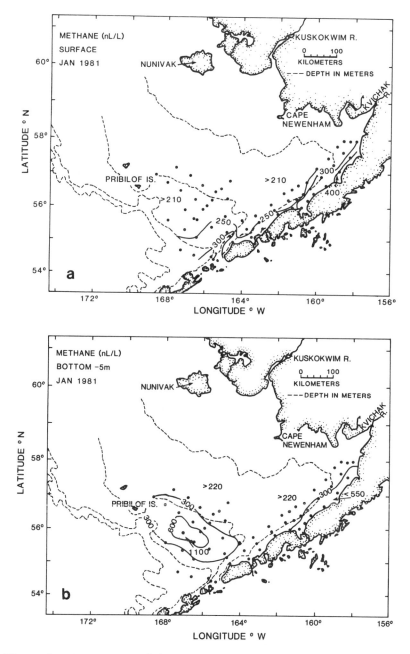

Figure 5. The surface (a) and near-bottom (b) concentrations of methane (nL/L) observed in January, 1981.

Methane: May, 1981. By late spring, surface concentrations had decreased to about 70 nL/L over much of the middle shelf, but remained high in the North Aleutian coastal waters (Figure 6a). The maximum concentration near Port Moller had decreased to about 600 nL/L.

Maximum concentrations of methane near the bottom of St. George Basin approached 1200 nL/L, which was about the same concentration seen in January (Figure 6b). Near-bottom concentrations over the middle shelf were only slightly greater than the concentrations seen in the surface waters.

Methane: September, 1975 and July, 1976. Concentrations of dissolved methane also were measured in the southeastern Bering Sea in the fall of 1975 and again during the following summer. These observations are shown for both surface and near-bottom waters in Figures (7a,b, and 8a,b).

As noted earlier, the highest surface concentrations of methane were near Port Moller, where the concentration exceeded 1500 nL/L in July 1976. Measurements were not obtained near the mouth of the estuary in September 1975, so comparable values were not available, but the coastal plume appears weaker than that seen in July 1976 (compare Figures 7a and 8a). Over the remainder of the embayment, concentrations were generally less than 100 nL/L, except as previously noted in the coastal zone.

Maximum concentrations of methane were again observed in the near-bottom waters of St. George Basin, however, the relatively large amounts observed in 1980-1981 were significantly lower in July 1976 (Figure 8b). The near bottom plume was more pronounced in October 1975, but maximum concentrations of 600 nL/L were below the concentrations seen in 1980-1981.

Processes Controlling the Distribution of Methane

Internal Sources and Atmospheric Exchange of Methane. Methane is produced by specialized groups of obligate anaerobic bacteria (22, 23). The formation of methane as a metabolic product results either from the microbial reduction of CO_2 with molecular H_2, or via the fermentation of acetic acid. More structurally complex substrates may also serve as electron acceptors/donors, but the end result of methanogenesis is to produce methane and CO_2 as end products (23).

Large production rates of methane are apparently only realized in the absence of dissolved oxygen and sulfate ions (6, 24), although methanogenesis and sulfate reduction should not be thought of as mutually exclusive processes (24, 25). While it may be macroscopically unrealistic to consider methane production in the presence of oxygen and/or sulfate ions, significant amounts of methane can be produced in oxic (9) and sulfate-bearing waters (24, 26), presumably within organic-rich microenvironments. These microenvironments might be organic particles in the water column, organic flocs at the sediment-water interface, or anaerobic metabolism in the gut of planktonic animals. For these reasons, it is reasonable to

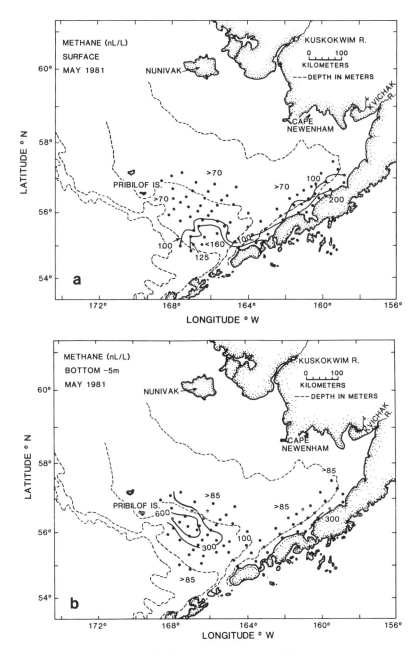

Figure 6. The surface (a) and near-bottom (b) concentrations of methane (nL/L) observed in May, 1981.

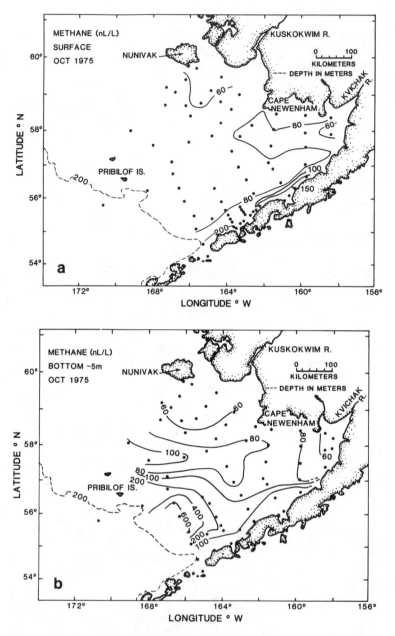

Figure 7.   The surface (a) and near-bottom (b) concentrations of methane (nL/L) observed in October, 1975.

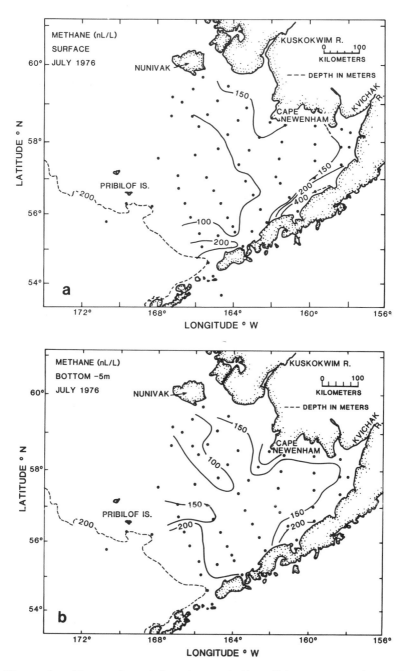

Figure 8. The surface (a) and near-bottom (b) concentrations of methane (nL/L) observed in July, 1976.

assume that the bulk of the methane production occurs in marine
sediments at depths where both oxygen and sulfate are absent.

However, much of the methane produced in bottom sediments
never reaches the atmosphere because it is oxidized to CO_2 by
microorganisms living in the surficial layers of the sediments
and in the oxic, overlying waters. The oxidation of methane
by sulfate reducers (or other organisms in the community) also
has been examined and it is the principal removal mechanism of
methane from shallow marine sediments (24, 25). Methane is
also oxidized by certain chemoautotrophic bacteria in the
presence of dissolved oxygen, although at much lower rates
compared to those observed in sediments (27).

Methane is also lost from surface waters by air-sea
exchange. If the surface concentration of methane exceeds its
equilibrium concentration, there will be a net flux to the
atmosphere. The empirical relationship commonly used to
quantify the transfer flux (F) is the stagnant-film boundary
layer model (28, 29).

$$F = v \{[CH_4] - [CH_4]'\} \qquad (1)$$

where $[CH_4]$ is the observed mixed layer concentration of
methane, $[CH_4]'$ is the equilibrium concentration, and v is the
exchange velocity, a quantity proportional to the thickness of
the hypothetical diffusive film at the sea surface. Later in the
discussion we will adopt this model to compute the flux from the
surface waters of the southeastern Bering Sea.

In the southeastern Bering Sea, the seasonal variations in
the concentration of methane are large, as shown in Table 1.
The seasonal range is about a factor of four; the highest
concentrations occur in late summer and the lowest in spring.
There did not appear to be much difference in the mean concentrations for each hydrographic area during a given season, which
was unexpected because the production rates and bottom sediment
organic carbon concentrations are rather different in that each
of the three hydrodynamic domains examined (21, 30). The
observed surface concentrations of methane represent a balance
between in-situ production and oxidation rates, vertical
transport rates, and air-sea exchange. Since the air-sea
exchange and in-situ oxidation rates are both proportional to the
seawater concentrations (27, 29), which are seasonally invariant
over the entire study area, uniform surface concentrations are
largely maintained by differences in local production rates and
vertical transport rates. Primary production rates control the
former, while the later is largely a function of vertical stratification and mixing.

The coastal zone is somewhat anomalous. There, the depth
of water is shallow (z < 50m) and the bottom sediments are
coarse-grained and low in organic carbon. Because of these
factors, the concentration of methane in the water column should
be near its atmospheric equilibrium solubility, however the
coastal waters receive large amounts of freshwater, which are
rich in dissolved methane, thus raising the concentration above
that expected from in situ sources along.

There is undoubtedly some water column production of methane, but its contribution relative to freshwater runoff is probably small.

Table I. Mean values for salinity, temperature, and methane in the three hydrographic regions of the southeastern Bering Sea. The number of values (n) used to compute the mean and standard deviation is also given. These observations were taken from the data report by Katz et al., (39).

Region	Season	n	Salinity °/°°	Temperature °C	CH$_4$ nL/L
Coastal	Aug. 1980	104	30.6 ± 1.8	10.0 ± 1.0	418 ± 78
(z ≤ 50 m)	Feb. 1981	182	31.2 ± 0.4	0.3 ± 1.2	288 ± 72
	May 1981	45	31.3 ± 0.2	6.3 ± 0.9	120 ± 43
Middle Shelf	Aug. 1980	92	31.6 ± 0.2	5.9 ± 0.8	400 ± 55
(50m <z ≤100m)	Feb. 1981	188	31.8 ± 0.2	2.2 ± 0.8	248 ± 52
	May 1981	48	31.8 ± 0.2	5.9 ± 1.0	88 ± 13
St. George	Aug. 1980	22	32.4 ± 0.3	8.0 ± 0.4	404 ± 68
(z > 100 m)	Feb. 1981	77	31.8 ± 0.2	3.4 ± 0.3	275 ± 65
	May 1981	38	31.9 ± 0.2	5.8 ± 0.6	103 ± 17

The surface waters over St. George basin are the most productive waters in the southeastern Bering Sea (30). Some of this carbon finds its way to the sea floor and is converted into methane, which is turbulently mixed upward into the pycnocline (see Figure 4). Stratification and higher net production over St. George Basin results in higher concentrations observed in the bottom waters. Because of this stratification (Figure 4), surface concentrations are balanced by vertical transport through the pycnocline and air-sea exchange, modulated to some degree by in-situ water column production and oxidation. In-situ oxidation of methane may be an important removal process in the pycnocline because the concentrations are relatively high and the mixing rate is low. Unfortunately we have no information on in-situ production and oxidation rates or the vertical transport parameters.

Surface waters over the middle shelf are characterized by methane concentrations similar to that seen over St. George basin (Table I). Here the water column and sediment production rates are lower than those in St. George Basin, principally because the primary production rates are lower. Also, vertical stratification is reduced, resulting in near bottom and surface concentrations being more nearly the same.

The sea to air seasonal flux of methane was computed for each hydrographic domain according to equation (1). Mean monthly averages were obtained by linear interpolation of the

seasonal data collected over the period 1980-81. These
calculations are summarized in Table II. The mean monthly flux
was computed for the entire survey region by summing the
contributions from each of the three hydrographic domains
(area-weighted).

In order to estimate the sea to air flux, the exchange
velocity must be known or estimated. We assume that the
exchange velocity is proportional to the quotient of the
molecular diffusivity of methane, D, and the hypothetical
thickness of the surface diffusive film, h* (29). The exchange
velocity was computed from estimates of the film thickness (31)
as a function of wind velocities (32) and temperature dependent
molecular diffusivities (33). The equilibrium concentration of
methane, $[CH_4]'$, was computed from solubility data (34) and the
temperatures given in Table III. The atmospheric partial
pressure of methane at the time the measurements were made was
about 1.66 ppm(v) (35) and when multiplied by the solubility
coefficient predicts an equilibrium solubility concentration
range of 61-70 nL/L, depending on temperature. These
calculations, together with the mean monthly methane flux, also
are summarized Table III.

The minimum flux of methane occurs in late spring when the
wind velocity and surface concentrations of methane are near
both relatively low. The maximum flux is observed in late fall
at a time when surface concentration and wind velocities are near
their maxima. The average flux, based on the observations made
in 1980-1981, is 5.1 ± 3.9 ng CH_4 m^{-2} s^{-1}, while the seasonal
range is about 12 ng CH_4 m^{-2} s^{-1}, with a factor of 20 difference
between the spring minimum and fall maximum.

The largest uncertainty in the calculation is the
dependence of the flux on the film thickness, or the wind
velocity. Intuitively one would expect the flux of methane to
be related to the magnitude of surface wind stress, however,
the dependence may not be the factor of five indicated in Table
III. Even if the wind had no effect on the methane transfer
flux, the seasonal variation in the concentration alone would
predict a factor of 3.5 between the spring minimum and fall
maximum.

The mean monthly transport of methane (Mt $month^{-1}$) is
shown in the last column of Table III. Summing the estimated
monthly contributions after correction for ice cover, the total
annual transport for the southeastern Bering Sea is 46 x 10^3, Mt,
based on a total area of 31.9 x 10^{10} m^2. Correction for ice
coverage was based on probability curves (36). We assumed that
the flux of methane was zero for probabilities ≥ 50%, which may
turn out to be too severe a restriction. On a statistical
basis, it is seen that the maximum ice cover occurs in March
(Table III), which may reduce the effective area for gas
transfer in that month to about 50 % of the total. Whether or
not the probability curves are representative of conditions in
1980-1981 is not known, however, the flux of methane under
partial to total ice cover is certainly not zero, although the
flux will undoubtedly decrease due to a reduction in both
surface area and the effective wind stress on the surface.

Table II.   Mean monthly values of methane and temperature computed
by linear interpolation of the mean values given in Table
I. The areas of the coastal zone, middle shelf, and St.
George Basin are 1.4×10^{11} m^2, 1.3×10^{11} m^2, and
4.6×10^{10} m^2, respectively.

Month	CH$_4$ nL/L	T °C	CH$_4$ nL/L	T °C	CH$_4$ nL/L	T °C	CH$_4$ nL/L	T °C
	Coastal		Middle Shelf		St. George		Area-Weighted Avg.	
Jan.	350	2.8	300	3.6	330	4.3	320	3.8
Feb.	290	1.1	250	2.4	270	3.5	264	2.7
Mar.	230	2.6	200	3.3	220	4.1	213	3.6
Apr.	180	4.0	160	4.4	170	4.8	167	4.5
May	120	5.5	100	5.4	120	5.5	112	5.4
Jun.	200	7.0	150	6.4	150	6.2	157	6.4
Jul.	300	8.2	250	7.5	250	6.9	257	7.3
Aug.	380	9.5	350	8.5	350	7.6	354	8.2
Sept.	400	9.3	380	8.5	400	7.5	391	8.2
Oct.	390	7.7	370	7.3	370	6.7	372	7.1
Nov.	380	6.0	350	6.1	360	5.9	358	6.0
Dec.	360	4.4	330	4.8	350	3.1	343	4.0

Table III. Compilation of the mean monthly atmospheric flux, F, and transport, Q, of methane from the surface waters of the Bering Sea.

Month	$S^{(a)}$ m s⁻¹	$h^{*(b)}$ cm	T °C	$D^{(c)}$ cm² s⁻¹	$v^{(d)}$ cm s⁻¹	$[CH_4]'^{(e)}$ nL L⁻¹	F ng m⁻² s⁻¹	$A^{(f)}$ m²	Q Mt CH₄ month⁻¹
Jan.	9.1	25x10⁻⁴	3.8	0.84x10⁻⁵	3.4x10⁻³	68.1	6.1	24.3x10¹⁰	3.9x10³
Feb.	9.6	20	2.7	0.80	4.0	69.7	5.5	19.4	2.8
Mar.	8.1	35	3.6	0.84	2.4	68.1	2.5	18.0	1.2
Apr.	7.8	35	4.5	0.87	2.5	66.4	1.8	19.4	0.92
May	6.8	45	5.4	0.90	2.0	65.1	0.67	28.5	0.50
Jun.	5.8	70	6.4	0.95	1.4	63.4	0.94	31.9	0.79
Jul.	5.8	70	7.3	0.98	1.4	62.6	1.9	31.9	1.6
Aug.	7.2	50	8.2	1.02	2.0	61.1	4.2	31.9	3.5
Sept.	8.1	35	8.2	1.02	2.9	61.1	6.8	31.9	5.7
Oct.	9.8	20	7.1	0.98	4.9	62.6	10.8	31.9	9.0
Nov.	10.1	15	6.0	0.93	6.2	64.2	13.0	31.9	10.9
Dec.	9.0	25x10⁻⁴	4.0	0.85x10⁻⁵	3.4x10⁻³	67.7	6.7	30.7x10¹⁰	5.4x10³

Average Flux: 5.1 ± 3.9 ng m⁻² s⁻¹
Average Transport: 4.6 x 10¹⁰ g yr⁻¹

(a) Scaler mean wind velocity.
(b) Stagnant film thickness, (31).
(c) Molecular diffusion coefficient of methane, (32).
(d) Piston velocity, D/h*
(e) Equilibrium concentration of methane based on the solubility coefficients (34)
     and a CH₄ mixing ratio of 1.66 ppm(v), (35).
(f) Effective Area corrected for ice cover (see text).

The mean annual flux of methane from the shelf waters of the southeastern Bering Sea is about 0.14 g CH_4 m^{-2} yr^{-1}, or a factor of three greater than observed in the pristine shelf waters of the Gulf of Mexico (Table IV). However, air-sea exchange from the shelf waters of Texas and Louisiana are much larger because of thermogenic gas seepage and underwater venting from offshore petroleum operations (13). By way of comparison, the flux of methane from the Bering Sea shelf is considerably larger than the open ocean estimate by Ehhalt (37). His estimate, based on few measurements, was about 0.012 g CH_4 m^{-2} yr^{-1} and that value is probably excessive. The surface concentration of methane is probably no more than 10 % supersaturated on the average (Weiss, R., Scripps Institute of Oceanography, personal communication), which translates into a mean oceanic flux of < 0.003 g CH_4 m^{-2} yr^{-1}, again assuming a mean piston velocity of 2.1 m day^{-1}. If these numbers are reasonably correct, then we might expect other high latitude shelf environments to be equally important contributors to the marine methane flux, at least seasonally. On a global basis, ocean waters probably contribute less than 1% of the atmospheric transport of methane (35, 37), but in terms of marine sources of methane to the atmosphere, the coastal and shelf environments are the most important. Shelf waters are generally more productive than the open ocean and are much shallower, hence the time scale for cycling methane to the atmosphere is much shorter in shelf waters.

Table IV. A comparison of sea-air exchange fluxes for several different marine environments. The area of the open ocean is 3.34 x 10^{14} m^2 (42).

Region	Flux g CH_4 m^{-2} yr^{-1}	Investigators
S.E. Bering Sea	0.14	This study
Gulf of Mexico[1]	0.04	(13)
Texas/Louisiana[2]	0.85	(13)
Open Ocean	0.012	(37)
Open Ocean[3]	< 0.003	This study

[1] Does not include the Texas-Louisiana Shelf.
[2] Recalculated, assuming a piston velocity of 2.1 m d^{-1}.
[3] Assumes that the mean supersaturation of surface waters is $< 10\%$ (Weiss, R., Scripps Institute of Oceanography, personal communication).

Inter-Annual Variations of Dissolved Methane

Methane concentrations also were measured in Fall of 1975 and Summer of 1976. The average surface concentrations for the three regions are summarized in Figure 9. It is readily obvious

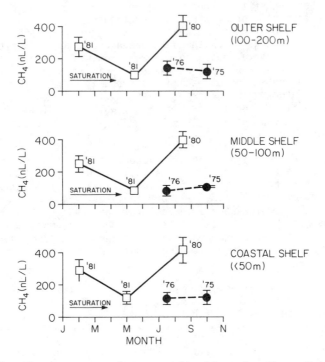

Figure 9.   The average surface concentration of methane for each
of the three hydrographic domains (see Table III).   The equilibrium concentration of methane ranged from about 62-68 nL/L,
assuming an atmospheric mixing ratio of 1.66 ppm(v).

that the concentrations of methane were suppressed in 1975-76 compared to 1980-81. Surface concentrations in October 1975 were only 100-150 nL/L, roughly the same as those seen in May 1981. The cause of the lower concentrations in 1975-76 is not precisely known, but one clue comes from the near-bottom concentrations of methane in St. George Basin. The maximum concentrations in October 1975 and July 1976 were only 600 nL/L and 400 nL/L, respectively (38), compared to concentrations in excess of 1800 nL/L in 1980-81 (Figures 3-5). While the lower concentrations were most evident in St. George Basin, near-bottom concentrations were also lower throughout the shelf waters of the southeastern Bering Sea in 1975-76.

The lower concentrations of methane correlate with lower sea surface temperatures (SST) and near-bottom temperatures over the middle shelf. Bottom temperatures over the middle shelf were < -1°C in October 1975 and July 1976, compared to temperatures of 3.5-4.5 °C observed in May 1981 (39). Water temperatures this low indicate local ice formation the previous winter. On the other hand, bottom temperatures in St. George Basin in May 1981 were about the same as those observed in October 1975 and July 1976 (17) indicating that these waters were less influenced by the abnormally cold winters than the middle shelf region. This is expected because the north-setting coastal current brings relatively warm Gulf of Alaska water across the outer shelf (19).

Nevertheless, the lower than normal SST and near-bottom temperatures signaled a significant short term climatic event over the eastern Bering Sea (40). The winters of 1974-75 and 1975-76 caused extensive ice formation over the region (41). For example, the southern limit of ice in April 1979 (a normal year), was near Nunivak Island (Figure 1), whereas it was southwest of the Pribilof Islands in April 1976 (36). It is hypothesized that the extensive ice coverage and its persistence into the late spring of 1975 and 1976 significantly affected the supply of carbon to the benthos, thus reducing the seasonal production rate of methane. This seems to be the case for St. George Basin and the middle shelf. Another important factor is the near-bottom water temperature over the middle shelf. Temperatures near and below freezing may have inhibited methanogenous in the near-bottom waters long after the ice had melted. We have independent evidence from Port Moller that methanogeneous is severely retarded in organic-rich sediments at temperatures near the freezing point of seawater (unpublished data), hence the low concentrations of methane over the middle shelf may be due to the combined effects of a reduced flux of carbon and cold bottom temperatures.

The apparent importance of the seasonal production of carbon to the water column inventory of methane suggests that the surface layers of the bottom sediments or the water column itself are important loci for methanogenesis. Deeper horizons within the sediment column are not going to be influenced by seasonal productivity and will generate methane at some integrated rate determined by the long term depositional rate of carbon.

Summarizing, we observed that the low surface concentrations of methane seen in October 1975 and July 1976 were correlated

with low water temperatures and severe ice conditions in the
winters of 1974-75 and 1975-76. Extensive ice coverage had
presumably reduced the flux of carbon below the euphotic zone,
thereby decreasing the amount of carbon available for
methanogenesis at the sediment-water interface and in the water
column. These observations indicate that high latitude shelf
environments experience large inter-annual variations in the
concentration of dissolved methane, and presumably other biogenic
gases as well.

Summary

Distributions of dissolved methane in the southeastern Bering Sea
were made in 1975-76 and again 1980-81. These observations show
large seasonal and inter-annual variations that appear to be
related to the magnitude and timing of primary production.
However, the relationship is not simple because minimum concentrations of methane correlate with the season of maximum productivity.
Seasonal and inter-annual production rates of methane suggest that
significant quantities are produced in the water column or from
surficial layers of bottom sediments. The relative proportions of
of methane produced in the water column and bottom sediments could
not be determined from these observations, however.
    The highest concentrations of dissolved methane are found in
the near-bottom waters of St. George Basin, which overlie organic-
rich sediments and are thermally stratified much of the year. Weak
circulation and mixing allows concentrations to reach 2,500 nL/L
(STP). High concentrations of methane also were observed near Port
Moller, but this source of methane is related to freshwater sources
within the estuary.
    Season distributions of methane also show that methane is a
qualitative descriptor of mean current flows in Unimak Pass, along
the North Aleutian shelf, and in the near-bottom waters of St.
Geoge Basin.

Acknowledgments

We are sincerely indebted to the Captains and crews of the
NOAA ships, DISCOVERER and SURVEYOR, without whose dedication
to purpose these measurements could not have been made. We
are especially grateful to Mr. Lee Ohler, Mr. Anthony Young,
and Ms. Susan Hamilton, who ensured that the measurements were
the best that could be obtained. Lastly, we would like to
thank the Mobil Corporation who graciously agreed to prepare
the manuscript and figures for publication.

This study was supported by the Bureau of Land Management,
through an interagency agreement with the National Oceanic and
Atmospheric Administration, while one of us (JDC) was an
oceanographer with NOAA.

Literature Cited

1. Lamontagne, R. A.; Swinnerton, J. W.; Linnenbom, V. J. Tellus, 1974 24, 71.
2. Lamontagne, R. S.; Smith, W. D.; Swinnerton, J. W. In "Analytical Methods in Oceanography"; Gibbs, Jr., T. R. P., Ed.; ADVANCES IN CHEMISTRY SERIES No. 147, American Chemical Society; Washington D. C., 1975; p. 163.
3. Swinnerton, J. W.; Lamontagne, R. A. Environ. Sci. Technol. 1974, 8, 657.
4. Atkinson, L. P.; Richards, F. A. Deep-Sea Res. 1967, 14, 675.
5. Lamontagne, R. A.; Swinnerton, J. W.; Linnenbom, V. J.; Smith, W. D. J. Geophys. Res. 1973, 78, 5318.
6. Reeburgh, W. S.; Heggie, D. T. Limnol. Oceanogr. 1977, 22, 1.
7. Bernard, B. B.; Brooks, J. M.; Sackett, W. M. J. Geophys. Res. 1978, 83, 4053.
8. Scranton, M. I.; Farrington, J. W. J. Geophys. Res. 1977, 82. 4947.
9. Scranton, M. I.; Brewer, P. G. Deep-Sea Res. 1977, 24, 127.
10. Frank, D. J.; Sackett, W. M.; Hall, R.; Fredericks, A. D. Amer. Assoc. Petrol. Geol. Bull. 1970, 54, 933.
11. Brooks, J. M.; Sackett, W. M.; Sackett, W. M.. J. Geophys. Res. 1973, 78, 5124.
12. Bernard, B. B.; Brooks, J. M.; Sackett, W. M. Earth Plant. Sci. Lett. 1976, 31, 48.
13. Brooks, J. M.; Sackett, W. M. In "Advances in Organic Geochemistry"; Compos, R.; Coni, J., Eds.; Revista Espanola de Micropaleontologia, Madrid, Spain; 1977; p. 455.
14. Brooks, J. M.; Gormly, J. R.; Sackett, W. M. Geophys. Res. Lett. 1974, 1, 213.
15. Swinnerton, J. W.; Linnenbom, V. J. J. Chromatogr. Sci. 1967, 5, 570.
16. Schumacher, J. D.; Kinder, T. H.; Pashinski, D. H.; Charnell, R. L. J. Phys. Oceanogr. 1979, 9, 79.
17. Kinder, T. H.; Schumacher, J. D. In "The Eastern Bering Sea Shelf: Oceanography and Resources, Vol. I"; Hood, D. W.; Calder, J. A., Eds.; Univ. Washington Press, Seattle, WA., 1981; p. 31.
18. Coachman, L. K.; Charnell, R. L. Deep-Sea Res. 1977, 24, 869.
19. Kinder, T. H.; Coachman, L. K. J. Geophys. Res. 1978, 83, 4551.
20. Coachman, L. K.; Charnell, R. L. J. Phys. Oceanogr. 1979, 9, 278.
21. Sharma, G. D. In "Oceanography of the Bering Sea; Hood, D. W.; Kelley, E. J., Eds.; Inst. Mar. Sci., Univ. Alaska, Fairbanks, AK, 1974; p. 517.
22. Wolfe, R. S. In "Advances in Microbial Physiology"; Hose, A. H.; Wilkinson, J. F., Eds.; Academic Press, New York, 1971; p. 107.

23. McCarty, P. L. In "Principles and Applications in
    Aquatic Microbiology"; Huekelekian, H.; Dondero, N. C.,
    Eds.; John Wiley, New York, 1964; p. 324.
24. Barnes, R. O.; Goldberg, E. D. Geology 1976, May, 297.
25. Oremland, R. S.; Taylor, B. F. Geochim. Cosmochim. Acta
    1978, 42, 209.
26. Martins, C. S.; Berner, R. A. Science 1974, 185, 1167.
27. Griffiths, R. P.; Caldwell, B. A.; Cline, J. D.; Broich,
    W. A.; Morita, R. Y. Appl. Environ. Microbiol. 1982,
    Aug., 435.
28. Liss, P. S. Deep-Sea Res. 1973, 20, 221.
29. Broecker, W. S.; Peng, T. H. Tellus 1974, 26, 21.
30. Goering, J. J.; Iverson, R. L. In "The Eastern Bering
    Shelf: Oceanography and Resources, Vol. II"; Hood, D.
    W.; Calder, J. A., Eds.; Univ. Washington Press,
    Seattle, 1981; p. 933.
31. Emerson, S. Limnol. Oceanogr. 1975, 20, 754.
32. Brower, Jr., W. A.; Searby, H. W.; Wise, J. L.; Diaz, H.
    F.; Prechtel, A. S. "Climatic Atlas of the Outer
    Continental Shelf Waters and Coastal Regions of Alaska,
    Vol. II, Bering Sea"; Univ. Alaska, Anchorage, AK,
    1977.
33. Bonoli, L; Witherspoon, P. A. In "Advances in Organic
    Chemistry"; Schench, P. A.; Havenaar, I., Eds.; Pergamon
    Press, New York, 1968; p. 373.
34. Yamamoto, S.; Alcaushas, J. B.; Grozier, T. E. J. Chem/
    Eng. Data 1976, 21, 78.
35. Khalil, M. A. K.; Rasmussen, R. A. J. Geophys. Res.
    1983, 88, 5131.
36. Webster, B. D. "A Climatology of the Ice Extent in the
    Bering Sea"; NOAA Tech. Memo., NWS AR-33, DOC/NOAA/EDS,
    Washington, D. C., 1981.
37. Ehhalt, D. H. Tellus 1974, 26, 58.
38. Cline, J. D. In "The Eastern Bering Sea Shelf:
    Oceanography and Resources, Vol. I"; Hood, D. W.;
    Calder, J. A., Eds.; Univ. Washington Press, Seattle,
    WA, 1981; p. 425.
39. Katz, C. N.; Cline, J. D.; Kelly-Hansen, K. "Dissolved
    Methane Concentrations in the Southeastern Bering Sea,
    1980-1981"; NOAA Data Rept. ERL/PMEL-6; DOC/NOAA/ERL,
    Boulder, CO., 1982.
40. Niebauer, H. J. In "The Eastern Bering Sea Shelf:
    Oeanography and Resources, Vol. I"; Hood, D. W.;
    Calder, J. A., Eds.; Univ. Washington Press, Seattle,
    WA, 1981; p. 23.
41. Niebauer, H. J. In "The Eastern Bering Sea Shelf:
    Oceanography and Resources, Vol. I"; Hood, D. W.;
    Calder, J. A., Eds.; Univ. Washington Press, Seattle,
    Washington, 1981; p. 133.
42. Sverdrup, H. U.; Johnson, M. W.; Fleming, R. H. "The
    Oceans"; Prentice Hall, Inc. Englewood Cliffs, N.J., 1942.

RECEIVED November 11, 1985

Stable Hydrogen and Carbon Isotopic Compositions of Biogenic Methanes from Several Shallow Aquatic Environments

Roger A. Burke, Jr., and William M. Sackett

Department of Marine Science, University of South Florida, St. Petersburg, FL 33701

Stable hydrogen (δD) and carbon (δ^{13}C) isotopic
compositions of methane gas bubbles formed in the
sediments of several shallow aquatic environments
were measured and found to range from -346°/oo to
-263°/oo and from -75.0^{o}/oo to -51.5^{o}/oo,
respectively. Evaluation of the δD data with a
previously published model implies that acetate
dissimilation accounts for about 50% to 80% of the
total methane production. δD-CH_4 and δ^{13}C-CH_4 are
generally inversely correlated; this indicates that
the observed isotopic variation is not solely due to
differential methane oxidation. δ^{13}C-CH_4 values
reported in this paper imply that methane produced
in these sediments is generally substantially more
^{13}C-depleted than the estimated average atmospheric
methane source. Methane with a δD near the
estimated atmospheric source average is produced in
some of these sediments; this apparent agreement may
be fortuitous as few relevant data are available.

The process of biological methane formation is of considerable
current interest because biogenic methane can accumulate in
commercially significant quantities in certain geological
situations, and methane emitted to the atmosphere may affect the
earth's climate. It has been estimated that more than 20% of the
world's proven gas reserves are of biogenic origin (1). Recent
studies (2,3) of polar ice cores indicated an apparent doubling of
atmospheric methane concentrations during the past few hundred
years. Because methane strongly absorbs infrared radiation within
the atmospheric window (700 to 1400 cm^{-1}) that transmits most of
the thermal radiation from the earth's surface to outer space (4),
the atmospheric methane increase is a potential contributor to
global warming. Various investigators (2,4,5) estimated the
contribution of methane to increasing global temperatures to be
about 20-40% of that attributed to the carbon dioxide increases.
  The microbial decomposition of complex organic matter under

anaerobic conditions to methane and carbon dioxide is a multi-step
process requiring the participation of at least three (6,7)
different trophic groups of bacteria. The final step in this
process, methanogenesis, is thought to occur mainly via two
pathways, CO_2 reduction and acetate dissimilation (8). Carbon-14
labelling studies have suggested that acetate dissimilation
normally accounts for about 60-70% of the total methane production
in sewage sludge digestors (9,10), paddy soils (11,12), and some
freshwater lake sediments (13-15) with CO_2 reduction responsible
for the remainder. In other freshwater lake sediments, however,
the CO_2 reduction pathway is apparently predominant, accounting
for ≳90% of the total methane production in Russian Lake
Kuznechikha (16) and about 75% of the methane generated in the
sediments of Blelham Tarn in the English Lake District (17).
Incubation of sediment samples from Cape Lookout Bight, North
Carolina (CLB), a small, nearshore marine basin with rapidly
accumulating, organic rich sediments, with $(1,2-^{14}C)$ sodium
acetate indicated that more than 50% of the CLB summertime methane
flux could be accounted for by acetate dissimilation (18). In
certain situations in which sediments receive a fairly specific
type of organic matter input or when methanogens are competing
with sulfate reducing bacteria for H_2 and acetate, a significant
fraction of the total methane production can result from
alternate substrates such as methanol and trimethylamine (19,20).

All of the aforementioned studies used C-14 labelled
substrates as tracers to estimate the relative contributions of
the methanogenic pathways to total methane production. An
alternate method of obtaining an estimate of relative pathway
contribution, based on measurements of the stable hydrogen
isotopic compositions of biogenic methane and the associated
water, has been proposed (21,22). Estimates obtained using this
method for methane generated from sewage sludge (21) and
freshwater lake sediments (22) are in good agreement with some of
the estimates of relative pathway contribution (9,10, 13-15)
obtained from C-14 labelling studies.

We report here measurements of the stable hydrogen isotopic
composition of methane and water, the stable carbon isotopic
composition of methane and carbon dioxide, and ancillary
parameters from several freshwater environments and from a few
locations within the Tampa Bay estuary. The stable isotopic
compositions determined in this study are reported as a
parts-per-mil ($^o/oo$) deviation (δ) from a standard with a known
stable isotopic ratio. The definition of the δ value is:

$$\delta = (R(Sample)/R(Standard) - 1) \times 1000 \qquad (1)$$

where R is D/H and $^{13}C/^{12}C$ and the standards are SMOW and PDB, for
hydrogen and carbon isotopic compositions, respectively. Using
these measurements and the model proposed by Woltemate et al. (22)
we can estimate the relative importance of the two primary
methanogenic pathways, acetate dissimilation and CO_2 reduction, to
methane production in these systems.

Methods

Gas samples were obtained by agitating the sediment with a rod and funneling the released gases into 125 ml serum bottles (Wheaton Scientific, Millville, N.J.) that were initially filled with lake water. The bottles were filled as completely as possible with gas and stoppered with a specially manufactured black rubber stopper (no. 2048-11800, Bellco Glass, Inc., Vineland, N.J.) that was then crimped with an aluminum seal (Wheaton). The samples were stored on ice during transit back to the laboratory where they were kept in a freezer until analysis. Water samples for δD determinations were collected in screw cap vials, the mouths of which were covered with parafilm to retard evaporation. Sediment samples were obtained by scraping surficial material into Whirlpak plastic bags and were frozen until analysis.

The methane gas samples were prepared for isotopic analyses with the system illustrated in Figure 1. Preparation involves gas chromatographic (GC) separation of methane from other gases followed by combustion to CO_2 and water in a vacuum line. Methane is combusted in a cupric oxide furnace ($\sim$850°C), to which 5-10 torr of oxygen has been added, as it is swept through by the GC carrier gas. Addition of oxygen gas to the system is necessary to assure complete combustion of the methane. We have found that the isotopic values of incompletely combusted methane are significantly more negative than the actual values; δD-CH$_4$ and δ^{13}C-CH$_4$ deviations of up to 60°/oo and 3.5°/oo, respectively, were observed. The CO_2 and water resulting from the combustion of methane are condensed by liquid nitrogen ($\sim$-196°C) in the trap immediately downstream from the combustion oven, the excess oxygen and helium are pumped away, and the CO_2 and water are cryogenically ($\sim$-90°C) separated. The δ^{13}C of the CO_2 is then measured using a Finnigan-Varian MAT 250 isotope ratio mass spectrometer (IRMS). The water from the combustion is reduced to hydrogen gas using the zinc metal method (23) and analyzed with the IRMS. The GC used is a Hewlett-Packard 5710 A equipped with a thermal conductivity detector and two 6 mm OD stainless steel columns (3 m grade 12 silica gel and 2 m molecular sieve 5A) connected in series through a Valco ten port switching valve configured for column sequence reversal and backflush of the silica gel column to the detector. This GC configuration allows baseline separation of H_2, O_2, N_2, CH_4 and CO_2. Helium carrier flow is 50 ml/min and the oven temperature is held constant at 90°C. The system described here is a flow-through system in which methane is combusted as it is swept through the furnace by the GC carrier gas. There is no Toepler pump or bellows to cycle the gas repeatedly through the furnace. As a result, it is reasonable to expect that the time the methane spends in the oven, which is determined by the GC carrier gas flow rate, could be an important variable. To test this, a methane working standard was analyzed several times using GC carrier gas flow rates of 25 ml/min and 50 ml/min. Isotopic values obtained at 50 ml/min (mean δD-CH$_4$ = -164°/oo, std.dev. = 1.0°/oo, n = 12; mean δ^{13}C-CH$_4$ = -44.2°/oo, std.dev. = 0.14°/oo, n = 12) were not significantly different from those obtained at a GC flow of 25 ml/min (mean δD-CH$_4$ = -164°/oo,

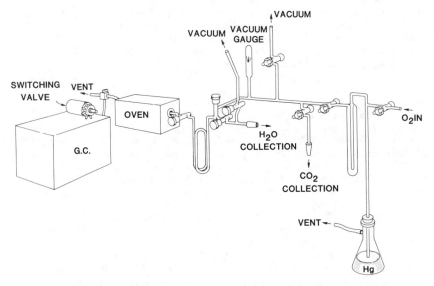

Figure 1. System for preparing methane for isotopic analyses.

std.dev. = 1.7^{o}/oo, n = 6; mean $\delta^{13}C-CH_4$ = -44.1^{o}/oo, std.dev. = 0.17^{o}/oo, n = 10). Repeated analyses of a second working standard, analyzed daily during the course of this work, yielded standard deviations (mean $\delta D-CH_4$ = -189.3^{o}/oo, std.dev. = 1.7^{o}/oo, n = 36; mean $\delta^{13}C-CH_4$ = -39.6^{o}/oo, std.dev. = 0.21^{o}/oo, n = 32) that are similar to those obtained above for the GC flow rate test and are comparable to those reported by Schoell (21).

The combustion efficiency of the GC-combustion system (Figure 1) was tested by comparing the methane concentration of the gas atmosphere in the vacuum line that resulted from the injection of ~ 2 ml of methane under normal operating conditions (three separate trials) to the methane concentration of the vacuum-line gas atmosphere after injection of ~ 2 ml of methane with the combustion furnace turned off (three separate trials). The trap immediately downstream from the combustion oven was immersed in a Dewar flask containing an isopropyl alcohol slush ($\sim -90^{\circ}$C). The resulting methane concentrations, which were obtained with a GC equipped with dual flame ionization detectors, indicated a combustion efficiency > 99%.

Water samples were prepared for δD determination by the zinc metal method as described by Coleman et al. (23). Repeated analyses of the water reference samples V-SMOW (n = 10, std. dev. = 1.4^{o}/oo) and NBS-1 (n = 10, std. dev. = 1.3^{o}/oo), performed during the course of this work, indicated that the uncertainty (one standard deviation) involved with the $\delta D-H_2O$ analyses was about 1.5^{o}/oo. Preparation of the carbon dioxide gas for analysis by IRMS involved cryogenic separation of the CO_2 from the other gases. This was accomplished by injecting an aliquot of the gas through a rubber septum into a vacuum line with a trap cooled by liquid nitrogen and then simply pumping away the liquid nitrogen non-condensable gases (O_2,N_2,CH_4). Water vapor was then removed cryogenically ($\sim -90^{\circ}$C) from the CO_2. The methane and carbon dioxide concentrations (mol%) of the gas samples were calculated from the GC-integration units (HP-3390A Integrator) following application of appropriate T.C. Weight Factors as given by McNair and Bonelli (24). Sedimentary organic matter was prepared for $\delta^{13}C$ analysis by IRMS in a Craig type combustion apparatus (25).

Model of Methane Formation

Figure 2 is a plot of $\delta D-H_2O$ versus $\delta D-CH_4$ for samples obtained from littoral zone sediments of several freshwater lakes and from several shallow (1 m or less water depth) areas of the Tampa Bay estuary. Also shown on Figure 2 are lines that describe predicted $\delta D-H_2O/\delta D-CH_4$ isotopic pairs resulting from varying the relative contributions to methane production of the acetate dissimilation and CO_2 reduction pathways. This model was originally proposed by Woltemate et al. (22) and used in that study to estimate that methyl group transfer (from acetate or other methyl group donors such as methanol) was responsible for about 76% of total methane production in the sediments of Wurmsee, a shallow lake near Hannover, FRG.

In constructing this model, Woltemate et al. (22) assumed that for the CO_2 reduction pathway all four methane hydrogens are

supplied by the environmental water (21,26), and for the acetate
dissimilation pathway only one hydrogen comes from water, with the
remaining three coming from the methyl group of acetate (27). As
a result, the slopes of the 0:100 (acetate dissimilation: CO_2
reduction, not shown here) and the 100:0 (not shown here) lines
are 1 and 0.25, respectively. The slopes of the intermediate
lines are obtained by multiplying the relative amount of each
pathway by the appropriate slope (1 or 0.25) and summing. For
example, the slope of the 80:20 line would be: 0.80(0.25) +
0.20(1) = 0.4. The y-intercept of the 80:20 line was obtained
from the sewage sludge incubation experiment of Schoell (21). In
that experiment, aliquots of sewage sludge were incubated in
plastic bottles spiked with different amounts of D_2O. The results
indicated a linear correlation between δD-H_2O and δD-CH_4 that was
described (21) by:

$$\delta D\text{-}CH_4 = 0.4\ \delta DH_2O - 323^{o}/oo \qquad (2)$$

The y-intercept of the 0:100 line resulted from the
observation that δD-H_2O/δD-CH_4 pairs obtained from the
measurement of natural samples, in which methane was presumably
formed via CO_2 reduction (21), fit the relationship (21, 28):

$$\delta D\text{-}CH_4 = \delta D\text{-}H_2O - 160(\pm10)^{o}/oo \qquad (3)$$

The y-intercepts of the remaining lines (22) are obtained by
assuming that the y-intercepts are spaced equally ((323-160)/4 =
$41^{o}/oo$) apart. Equations (2) and (3) are based on analyses of
methane formed under widely different environmental conditions.
Variations in the magnitude of isotopic fractionation, due to
variation in parameters such as the rate of reaction and time
(degree of isotopic equilibration), may introduce error into
predictions of relative methanogenic pathway selection obtained
with the model (Figure 2). We believe, however, that the model
yields results that are generally valid, and that application of
the model to our data yields useful interpretation.

Importance of Alternate Methanogenic Substrates

Lovley and Klug (29) reported that methanol and methylamines were
the precursors for less than 5 and 1%, respectively, of total
methane production in the sediments of eutrophic Lake Wintergreen,
Michigan. The likely explanation for this is the low abundance of
methanol and methylamine precursors relative to H_2 and acetate
precursors in the organic matter input to the sediments (29). Low
rates of methane production from methanol and methylamines in
near-surface, sulfate-rich marine sediments have been reported
recently (20,30). Although methane production from these
"noncompetitive" substrates (30) is of interest for several
reasons, we would argue that it is likely to generate large
quantities of methane (defined here as a significant fraction of
that required to achieve saturation) only under relatively unusual
conditions. Many marine organisms use methylamines or methylamine
precursors (31) in osmoregulation (32). In response to salt

stress, the halophyte Spartina alterniflora has been shown to concentrate up to 30% of its total leaf nitrogen in glycine betaine (33), which can be metabolized to acetate and trimethylamine in marine sediments (31). Oremland et al. (19) reported that metabolism of methanol and trimethylamine could account for the bulk of methane produced in salt marsh sediments containing 8 ml of methane per liter of wet sediment. The experiments that yielded these results were conducted in flasks containing 60 ml of San Francisco Bay water, 40 ml of sediment, and 10 ml of homogenized Spartina foliosa materials (19). While known (macroalgae (34)) and potential (mangroves, seagrasses) contributors of methylamine precursors (unknown concentrations) account for some organic input, sediments as rich in organic matter as those used above (19), are rarely encountered (35) in the Tampa Bay estuary. From all of this, we conclude that methane production from methanol and methylamines is quantitatively unimportant in the samples we have analyzed during this study; therefore, in applying the model (22) to the data presented here, we consider acetate to be by far the major source of methyl groups to the methanogens active in these sediments.

Relative Contributions of the Methanogenic Pathways

Our measurements of $\delta D-H_2O/\delta D-CH_4$ pairs from various freshwater and estuarine sediments (Figure 2) imply that about 50% to 80% of the total methane production in these sediments can be attributed to acetate dissimilation. This agrees with some earlier studies (13-15,22), but is in conflict with others that indicated that CO_2 reduction dominates methane production in other freshwater lake sediments (16,17). This variation in pathway importance between individual lakes and even between different locations within the same lake (Figure 2) may be attributable to natural qualitative and quantitative variations in the organic matter and bacteria occuring in these sediments as suggested by Belyaev et al. (16). The predominance of the acetate dissimilation pathway under estuarine-marine (Salinity (S) = 6–34°/oo) conditions inferred from our data has been found in other nearshore-marine circumstances (18,36). In contrast, the biogenic methane found in deep-sea sediments is thought to originate almost exclusively from CO_2 reduction (21, 37) and is isotopically distinguishable (δD = -170 to -190°/oo (38)) from the marine methane measured here (Figure 2). The $\delta D-CH_4$ values reported here support the idea that with respect to methane production, highly productive brackish and marine sediments can bear a greater resemblance to freshwater than deep-sea sediments if sulfate is depleted near the sediment – water interface due to high rates of respiration (30,39).

Methane Oxidation

An important area of study in organic geochemistry concerns the origin of natural gas; two of the more useful parameters in addressing this problem are $\delta D-CH_4$ and $\delta^{13}C-CH_4$. As discussed above, $\delta D-CH_4$ is useful in estimating relative pathway contribution in biological methane formation. Combined with information concerning the molecular composition of a natural gas, $\delta^{13}C$-CH_4 can be useful in determining whether the gas was formed biologically or from the thermocatalytic breakdown of complex organic matter (40). Consequently, processes that alter the isotopic compositions and confuse interpretation are of great interest. One process that can cause extensive isotopic fractionation is biological methane oxidation (41). Based on results obtained from a laboratory study by Coleman et al. (41), it can be calculated that aerobic oxidation of 40% of an aliquot of methane would leave the remaining 60% about 12.5°/oo and 125°/oo more ^{13}C- and D-enriched, respectively. If the percentage of methane oxidation and the degree of isotopic fractionation varied depending upon location within a given freshwater lake, for example, isotopic variation (Figure 2) could be created. The generally negative correlation between $\delta^{13}C$-CH_4 and δD-CH_4 indicated in Figure 3 indicates that the isotopic variation seen in our samples is not solely due to this process. The negative correlation does not mean that our samples were totally unaffected by in situ methane oxidation, but it does imply that some additional process (i.e., variation in pathway) contributed to the isotopic variation. All currently known examples of anaerobic methane oxidation, other than minor (2% of the rate of CH_4 production) oxidation by methanogenic bacteria (42) have taken place in the presence of dissolved sulfate, which is apparently required as an electron acceptor (39). Thus, it is unlikely that extensive anaerobic methane oxidation would occur in low sulfate freshwater environments. Methanogenic bacteria are strict anaerobes and produce methane only under highly reducing conditions (6). Methane produced under these conditions could only be oxidized by the aerobic process if it either migrated (as a bubble or in solution) to a less reducing location, or if the depth of oxygen penetration into the sediment increased. If a bubble moved upward from the depth at which it was formed, it would probably exit the sediments entirely because the distance between the methane production zone and the sediment is short (<50cm), and would not be sampled by the methods employed here. Likewise, methane that dissolved would not be sampled, and because the process of partial dissolution itself would not be expected to induce much isotopic fractionation (43), any methane remaining in bubble form should maintain its isotopic signature. Deeper penetration of oxygen into the sediments, because of a temperature induced decrease in oxygen consumption, might allow partial oxidation of trapped bubbles. This could be responsible for the slight increase (Table I) in the isotopic compositions of methane sampled from Crescent Lake (#2) on 1/23/85 compared to that sampled at the same location on 12/21/84. This slight difference has very little effect on any of our interpretations, and when considered along with the other arguments advanced above, argues against extensive isotopic fractionation of samples reported here (Figures 2,3) by aerobic methane oxidation.

Ancillary Data

All of the isotopic and ancillary data considered in this study

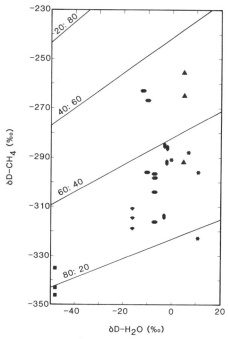

Figure 2. Plot of $\delta D-CH_4$ vs. $\delta D-H_2O$. Lines represent the percentage mix of the acetate dissimilation and CO_2 reduction methanogenic pathways. Symbols:●-Crescent Lake, FL; ◗-Mirror Lake, FL; ▲-Lake Dias, FL; ◆-Kilmer Pond, SC; ■-Mississippi Delta, LA; ✳-Tampa Bay Estuary, FL. (Reproduced with permission from Ref. 22. Copyright 1984 American Society of Limnology and Oceanography, Inc.)

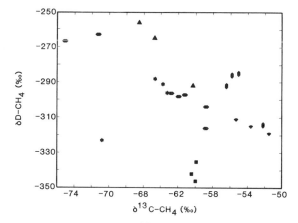

Figure 3. Plot of $\delta D-CH_4$ vs. $\delta^{13}C-CH_4$, Symbols as in Figure 2.

Table I. δD–CH4, $\delta^{13}C$–CH$_4$ and ancillary data for various freshwater systems and the Tampa Bay estuary.

LOCATION/ DATE	#	δD (°/oo) CH$_4$	δD (°/oo) H$_2$O	$\delta^{13}C$ (°/oo) CH$_4$	$\delta^{13}C$ (°/oo) CO$_2$	$\delta^{13}C$ (°/oo) SOM[a]	%C SOM	Sed. T(°C)	αC[b]	αD[c]	Conc.(mol%) CH$_4$	Conc.(mol%) CO$_2$
A 08/31/84	1	−297±1.0	−7±0.9	−61.3±0.1	−7.2			31	1.058	1.413	59.8	6.5
"	3	−304±0.5		−58.8±0.1	−6.9				1.061	1.427	55.4	2.6
10/02/84	1	−316±0.6		−58.8±0.1	−11.6			26	1.050		45.9	2.2
11/17/84	1	−298	−11±0.7	−62.0	−8.1			22	1.057	1.405	50.5	2.9
12/21/84	1	−296		−62.9	−8.8			23	1.058	1.349	46.2	2.5
"	2	−267±0.4		−75.0±0.3	−7.2				1.073		50.9	7.5
01/23/85	1	−263	−10±0.5	−71.4	−5.0	−27.7	3.3	16	1.072	1.343	59.1	4.3
"	2					−27.9	3.5					
B 12/21/84	1	−285	−3±0.4	−55.1	−2.2			21	1.056	1.394	85.3	7.2
"	2	−314		−52.2	−6.8				1.048	1.453	85.8	2.9
01/23/85	1	−292	−2±0.2	−56.4	−10.8	−22.5	0.8	14	1.048	1.410	65.0	1.4
"	2	−286		−55.8	−9.0				1.050	1.398	77.0	3.5
C 12/25/84	1	−292	+5±0.2	−60.3	−3.7	−27.5	11.1	21	1.060	1.419	60.8	8.4
"	2	−256		−66.7	−3.7	−28.0	33.6		1.068	1.351	49.0	13.1
"	3	−265±0.5		−64.8	−2.4	−26.9	0.5		1.067	1.367	74.4	8.0
D 04/23/82	2	−335	−48±0.2	−59.9±0.1						1.432	63.9	1.2
"	5	−346		−60.0						1.456	61.0	0.6
"	6	−342±0.5		−60.4±0.1	−12.2				1.051	1.447	69.4	3.6

E											
10/14/84	1	-315		-53.6	-10.9		24	1.045		41.0	2.8
"	2	-319		-51.5	-12.0			1.042		34.0	1.9
12/12/84	1	-311	-16±0.6	-55.3			9		1.428	33.6	0.7
F											
09/16/84	1	-323±1.2	+11	-70.8±0.1			29		1.493	54.9	2.0
10/11/84	2	-296	+11	-63.3			28		1.436	79.3	13.5
12/03/84	3	-288	+7±0.4	-64.7			24		1.414	70.7	7.5
12/03/84	4	-291	0±0.2	-63.8	-23.2	1.6	21		1.410	74.1	6.9

TABLE LEGEND

<u>A</u> - Crescent Lake, 12ha, fresh, St. Petersburg (SP), FL

<u>B</u> - Mirror Lake, 7ha, fresh, SP, FL

<u>C</u> - Lake Dias, 300ha, fresh, Volusia County, FL

<u>D</u> - Mississippi River Delta, fresh, LA

<u>E</u> - Kilmer Pond, fresh, near Columbia, S.C.

<u>F</u> - Tampa Bay estuary, FL
  1 - Ft. DeSoto Pk., S = 34°/oo
  2 - Bishop Harbor, S = 32°/oo
  3 - Lake Seminole, S = 20°/oo
  4 - Feather Sound, S = 6°/oo

a - SOM = Sedimentary organic matter

$$b - \alpha C = (CO_2/CH_4) = \frac{\delta^{13}CO_2 + 1000}{\delta^{13}C\text{-}CH_4 + 1000}$$

$$c - \alpha D = (H_2O/CH_4) = \frac{\delta D\text{-}H_2O + 1000}{\delta D\text{-}CH_4 + 1000}$$

are listed in Table I. Concentrations of methane and carbon
dioxide ranged from 33% to 86% and from 4% to 13%, respectively,
in the samples that we analyzed. The carbon isotopic composition
of the sedimentary organic matter (SOM) ranged from $-26.9^{\circ}/oo$ to
$-28.0^{\circ}/oo$ for all but one of the freshwater sediment samples
measured here (Table I); this implies that higher land plants (C_3
pathway) and low $^{13}C/^{12}C$ aquatic plants (44) dominate organic
matter input to these sediments. For instance, an unidentified
emergent aquatic macrophyte from Lake Dias (#3) yielded a $\delta^{13}C$ of
$-28.1^{\circ}/oo$. We determined a $\delta^{13}C$ of $-22.5^{\circ}/oo$ for the SOM of
Mirror Lake (Table I) which is substantially heavier than the
other freshwater values. Possible explanations for this include a
relatively greater contribution of isotopically heavy C_4 pathway
plant (44) material (i.e., domestic Bermuda grass), and
utilization of isotopically heavier source carbon by the submerged
aquatic plants (45). There is no obvious relationship between
either $\delta^{13}C$-SOM or %C-SOM and $\delta^{13}C$-CH_4 for the samples reported
here (Table I). For example, while the $\delta^{13}C$-CH_4 of Crescent Lake
#2 is $8-12^{\circ}/oo$ lighter than the $\delta^{13}C$-CH_4 of Crescent Lake #1,
there is no substantial difference between the two sites with
respect to $\delta^{13}C$-SOM or %C-SOM. Factors other than amount and $\delta^{13}C$
of the source organic matter, such as pathway selection (Figure
2), must also be involved in determining $\delta^{13}C$-CH_4. Fractionation
factors $\alpha C(\alpha(CO_2/CH_4))$ and $\alpha D(\alpha(H_2O/CH_4))$ of the freshwater
samples, calculated from measured isotopic compositions, ranged
from 1.042 to 1.073 and from 1.343 to 1.456, respectively (Table
I). These values are similar to those that were determined for
German lake Wurmsee (22), although the ranges reported here are
greater. The greater range (higher αC, lower αD) reflects the
larger relative contribution of the CO_2 reduction pathway in some
of our samples (Figure 2). The fractionation factors calculated
from typical deep-sea sediment isotopic compositions (1,37,38) are
significantly larger ($\alpha C \sim 1.08$) and smaller ($\alpha D \sim 1.25$) than the
factors determined here; this further implies that both isotopes
are affected by the methanogenic pathway and indicates the
potential applicability of these parameters (particularly
together) to the genetic characterization of biogenic methane
deposits.

Atmospheric Methane Isotopic Composition

As discussed earlier, recent studies (2,3) inferred a doubling of
atmospheric methane concentrations over the past few hundred
years. Furthermore, during the last 3-4 years, atmospheric
methane concentrations have increased at rates of 1-1.9% per year
(46). Decomposition of organic matter in water-covered soils,
intestinal fermentation in ruminants, biomass burning, and direct
anthropogenic input (i.e., leakage of fossil fuels) are thought to
be the major sources of methane to the atmosphere, contributing
approximately 46, 22, 5 and 7% of the total flux, respectively
(46). The major sink of atmospheric methane is thought to be
reaction with tropospheric OH radicals (46). A recent study by
Harriss et al. (47) indicated that periodically inundated soils
may also be a site of atmospheric methane removal during dry
periods.

Recent measurements (1980) indicate that the $\delta^{13}C$ of atmospheric methane is about $-47.0\pm0.3^o/oo$ (48). The average isotopic fractionation associated with the sink process is $-2.5\pm1.5^o/oo$, and there is a $+0.3^o/oo$ isotope effect resulting from the nonsteady state increasing methane concentrations (48). This implies that the average $\delta^{13}C-CH_4$ for all sources is about $-49\pm1.5^o/oo$; this is significantly less ^{13}C-depleted than most known biogenic sources (48), and is about $10^o/oo$ less negative than the average $\delta^{13}C-CH_4$ of all sources ($-61\pm3^o/oo$) calculated by Senum and Gaffney (49). Biomass burning and fossil fuels are the only known significant sources of methane more ^{13}C-enriched than $-49 /oo$ (48), and they apparently account for a relatively small fraction of the total input. Bubble ebullition is the dominant mode of methane input to the atmosphere from some shallow freshwater (50) and marine (51) aquatic environments. All of the gas samples analyzed for this work were collected as bubbles from sediments covered by very shallow (1m or less) water. Three of the freshwater lakes (Crescent Lake, Mirror Lake and Kilmer Pond) have been observed to release gas bubbles from their sediments, and it is very likely that active bubbling also occurs in the other environments studied here. There is a good deal of variation ($\sim24^o/oo$) in the $\delta^{13}C-CH_4$ data presented here (Table I); however, most of the samples are substantially lighter than $-49^o/oo$, in agreement with other data from shallow aquatic environments (48). Though limited in number, these data suggest that ebullition from these sediments is unlikely to provide an atmospheric input of methane as ^{13}C-enriched as the estimated (48) source average. Based on the results of a study by Rust (52), which demonstrated that the $\delta^{13}C$ of methane produced by ruminants depends on the $\delta^{13}C$ of the plants in their diet, Stevens and Rust (48) proposed that the anaerobic decomposition of isotopically heavy C_4 plant debris in large wetlands, such as the Sudd marshes and the Everglades, might be a significant source of isotopically heavy methane to the atmosphere. As mentioned before, no such simple relationship between the $\delta^{13}C-CH_4$ and $\delta^{13}C-SOM$ of aquatic sediments is indicated by our data (Table I). According to Wolfe (7), four well-defined groups of bacteria are involved in the conversion of complex organic matter to methane in sediments, whereas in the rumen only two of the groups are involved and there is no significant production of methane from acetate. Also, substrate input, product output, and environmental conditions are more closely controlled in the rumen than in sediments where disorder is inherently greater (53). Thus, in addition to $\delta^{13}C-SOM$, other factors such as pathway selection and possibly kinetic effects are probably important in setting the $\delta^{13}C-CH_4$ of aquatic sediments. Escape of dissolved methane across the air-water interface, rather than ebullition, may be the more likely means of transferring isotopically heavy methane from shallow aquatic environments to the atmosphere. Dissolved methane should be more readily available for aerobic methane-oxidation (39) and isotopic fractionation (41); however, assessment of the potential contributions of ebullition and dissolved methane to atmospheric $\delta^{13}C-CH_4$ is hindered by a lack of relevant data.

Using data obtained in 1960 and 1970 (54-56), Senum and
Gaffney (49) noted an apparent decrease in atmospheric methane δD
with time that they extrapolated to the present to yield an
estimate of $-104\pm4^\circ$/oo for the δD of present-day atmospheric
methane. The extrapolated δD value was corrected for kinetic
isotopic fractionation resulting from the reaction of methane with
hydroxyl radicals (the magnitude of the fractionation was
attributed to Gorden and Mulac (57)) in the troposphere to yield
an estimate of -322°/oo as the present-day average δD of all
methane sources to the atmosphere (49). Analyses reported in
(Table I) and by Woltemate et al. (22) indicate that methane as
deuterium-depleted as -322°/oo is produced in some freshwater
sediments. Organic matter decomposition in sewage sludge may also
yield methane as deuterium-depleted as -322°/oo (21). According
to the genetic characterization model of Schoell (38), which is
based on approximately 500 natural gas analyses, thermogenic
methane is generally less deuterium-depleted than -300°/oo. To
our knowledge, there are no estimates of the δD of methane
produced from either the intestinal fermentation in ruminants or
biomass burning. Although an important source of atmospheric
methane with a δD near the estimated source average (49) has been
identified (freshwater sediments), there are presently too few
measurements of the δD of atmospheric methane and its sources
available to determine whether or not this apparent agreement has
any real significance with regard to evaluating the role of
shallow aquatic sediments in the atmospheric methane budget and in
determining the δD of atmospheric methane.

Summary

Stable hydrogen and carbon isotopic compositions of biogenic
methane produced in the sediments of several freshwater and
estuarine environments have been measured and interpreted using a
previously published model. The results infer that acetate
dissimilation is the dominant methanogenic pathway in these
sediments, accounting for about 50 to 80% of the total methane
production, with CO_2 reduction responsible for the remainder. In
general, $\delta D\text{-}CH_4$ and $\delta^{13}C\text{-}CH_4$ are inversely correlated, implying
that methane oxidation alone can not explain the isotopic
variations. The inverse relationship also implies that both
isotopes are affected by methanogenic pathway selection.
Fractionation factors, αD and αC, calculated from the isotopic
data presented here are significantly larger and smaller,
respectively, than αD and αC typically found in deep-sea
sediments. Measurements performed to date indicate that the
$\delta^{13}C\text{-}CH_4$ of gas bubbles formed in shallow aquatic sediments is
generally significantly more negative than the average of all
sources to the atmosphere. There are presently too few
measurements of the δD of atmospheric methane and its sources
available to allow evaluation of the role of shallow aquatic
sediments in determining the δD of atmospheric methane.

Acknowledgments
We would like to thank Dr. Paul Carlson of the Florida Department
of Natural Resources for providing samples from the Tampa Bay
Estuary. Dr. James M. Brooks of Texas A&M University provided
shiptime which enabled us to collect the Mississippi Delta
samples. Dr. Willard S. Moore of the Univ. of South Carolina
supplied the samples from Kilmer Pond. Mr. Roger A. Burke, Sr. of
DeLand, FL provided the Lake Dias samples. This manuscript was
typed by Jodi Gray. This research was supported by NSF grants
OCE-8308945 and OCE-8417321 (Marine Chemistry Program) to W.M.S.
and R.A.B.

Literature Cited

1. Rice, D. D.; Claypool, G. E. Bull. AAPG 1981, 65, 5-25.
2. Craig, H.; Chou C. C. Geophys. Res. Lett. 1982, 9, 477-481.
3. Khalil, M. A. K.; Rasmussen, R. A. Chemosphere. 1982, 11,
   877-883. 4. Wang, W. C.; Yung, Y. L.; Lacis, A. A.; Mo, T.,
   Hansen, J. E. Science 1976, 194, 685-690.
5. Lacis, A.; Hansen, J.; Lee, P.; Mitchell, T.; Lebedeff, S.
   Geophys. Res. Lett. 1981, 8, 1035-1038.
6. Mah, R. A. Phil. Trans. R. Soc. Lond. B 1982, 297, 599-616.
7. Wolfe, R. S. In "Microbial Biochemistry (International
   Review of Biochemistry, vol. 21)"; Quayle, J. R. Ed.;
   University Park Press: Baltimore, 1979; pp. 270-300.
8. Zehnder, A. J. B. In "Water Pollution Microbiology";
   Mitchell, R., Ed.; Wiley: New York, 1978; vol. 2, pp.
   349-376.
9. Jeris, J. S.; McCarty, P. L. J. Wat. Pollut. Control Fed.
   1965, 37, 178-192.
10. Smith, P. N.; Mah, R. A. Appl. Microbiol. 1966, 14, 368-371.
11. Koyama, T. J. Geophys. Res. 1963, 68, 3971-3973.
12. Takai, Y. Soil. Sci. and Plan Nutr. 1970, 6, 238-244.
13. Cappenberg, Th. E.; Prins, R. A. Ant. van Leeuwenhoek, J.
    Microbiol. Serol. 1974, 40, 457-469.
14. Winfrey, M. R.; Zeikus, J. G. Appl. Environ. Microbiol.
    1979, 37, 244-253.
15. Lovley, D. R.; Klug, M. J. Appl. Environ. Microbiol. 1982,
    43, 552-560.16. Belyaev, S. S.; Finkel'shtein, Z. I.;
    Ivanov, M. V. Mikrobiologiya (English translation) 1975, 44,
    272-275.
17. Jones, J. G.; Simon, B. M.; Gardener, S. J. Gen. Microbiol.
    1982, 128, 1-11.
18. Sansone, F. J.; Martens, C. S. Science 1981, 211, 707-709.
19. Oremland, R. S.; Marsh, L. M.; Polcin, S. Nature (Lond.)
    1982, 296, 143-145.
20. King, G. M.; Klug, M. J.; Lovley, D. R. Appl. Environ.
    Microbiol. 1983, 45, 1848-1853.
21. Schoell, M. Geochim. Cosmochim. Acta 1980, 44, 649-661.
22. Woltemate, I.; Whiticar, M. J., Schoell, M. Limnol.
    Oceanogr. 1984, 29, 985-992.
23. Coleman, M. L.; Shepherd, T. J.; Durham, J. J.; Rouse, J. E.;
    Moore, G. R. Anal. Chem. 1982, 54, 993-995.

24. McNair, H. M.; Bonelli, E. J. "Basic Gas Chromatography";
    Varian: Palo   Alto, CA., 1969, pp. 144-147.
25. Craig, H. Geochim. Cosmochim. Acta 1953, 3, 53-92.
26. Daniels, L.; Fulton, G.; Spencer R. W.; Orme-Johnson, W. H.
    J. Bacteriol. 1980, 141, 694-698.
27. Pine, M. J.; Barker, H. A. J. Bacteriol. 1956, 71, 644-648.
28. Nakai, N.; Yashida, Y.; Ando, N. Chiku Kagaka 1974, 7/8,
    87-98.
29. Lovley, D. R.; Klug, M. J. Appl. Environ. Microbiol. 1983,
    45, 1310-1315.
30. King, G. M. Geomicrobiol. J. 1984, 3, 275-306.
31. King, G. M. Appl Environ. Microbiol. 1984, 48, 719-725.
32. Yancey, P. H.; Clark, M. E.; Hand S. C.; Bowlus, R. D.;
    Somero, G. M. Science 1982, 217, 1214-1222.
33. Cavalieri, A. J.; Huang, H. C. Oecologia 1981, 49, 224-228.
34. Blunden, G.; Gordon, S. M.; McLean, W. F. H.; Guiry, M. D.
    Bot. Mar. 1982, 25, 563-567.
35. Sackett W.; Brooks. G.; Conkright, M.; Doyle L.; Yarbro, L.
    1985, In Press.
36. Mountfort, D. O.; Asher, R. A., Mays, E. L., Tiedje, J. M.
    Appl Environ. Microbiol. 1980, 39, 686-694.
37. Claypool, G. E.; Kaplan, I. R. In "Natural Gases in Marine
    Sediments"; Kaplan, I. R., Ed.; Plenum Press: New York 1974;
    pp. 99-139.
38. Schoell, M. Bull. AAPG 1983, 67, 2225-2238.
39. Rudd, J. M. W.; Taylor, C. D. Adv. Aquat. Microbiol. 1980,
    2, 77-150.
40. Bernard, B.; Brooks, J.; Sackett W. Proc. Offshore Tech.
    Conf. 1977, OTC 2934, pp. 435-438.
41. Coleman, D. D.; Risatti, J. B.; Schoell, M. Geochim.
    Cosmochim. Acta 1981, 45, 1033-1037.
42. Zehnder, A. J. B.; Brock, T. D. Appl. Environ. Microbiol.
    1980, 39, 194-204.
43. Bernard, B. B.; Brooks, J. M.; Sackett, W. M. Earth Planet.
    Sci. Lett. 1976, 31, 48-54.
44. Smith, B. N.; Epstein, S. Plant Physiol. 1971, 47, 380-384.
45. Osmond, C. B.; Valaane, N.; Haslam, S. M.; Votila, P.;
    Roksandic, Z. Oecologia 1981, 50, 117-124.
46. Khalil, M. A. K.; Rasmussen, R. A. J. Geophys. Res. 1983,
    88, 5131-5144.47. Harriss, R. C.; Sebacher, D. I.; Day, F.
    P. Nature (Lond.) 1982, 297, 673-674.
48. Stevens, C. M.; Rust, F. E. J. Geophys. Res. 1982, 87,
    4879-4882.
49. Senum, G. I.; Gaffney, J. S. In "The Carbon Cycle and
    Atmospheric CO_2: Natural Variations Archean to Present";
    Sundquist, E. T.; Broecker, W. S., Eds.; American Geophysical
    Union: Washington, D.C., 1985; Geophysical Monograph 32, pp.
    61-69.
50. Cicerone, R. J.; Shetter, J. W. J. Geophys. Res. 1981, 86,
    7203-7209.
51. Martens, C. S.; Klump, J. V. Geochim. Cosmochim. Acta 1980,
    44, 471-490.
52. Rust, F. Science 1981, 211, 1044-1046.

53. Wolin, M. J. In "Microbiol. Interactions and Communities"; Bull, A.T.; Slater, J.H. Eds.; Academic: New York, 1982; vol.1, pp. 323-356.

54. Bainbridge, A. E.; Seuss, H. E.; Friedman, I. Nature (Lond.) 1961, 192, 648-649.

55. Begemann, F.; Friedman, I. J. Geophys. Res. 1968, 73, 1149-1153.

56. Ehhalt, D. H. In "Carbon and the Biosphere"; Woodwell, G.M.; Pecan E.V. Eds.; Office of Information Service: Washington, D.C., 1973; pp. 144-158.

57. Gorden, S.; Mulac, W. A. Int. J. Chem. Kinet. 1975, 7(Symp.#1), 289-299.

RECEIVED September 16, 1985

18

Polybromomethanes

A Year-Round Study of Their Release to Seawater from *Ascophyllum nodosum* and *Fucus vesiculosis*

Philip M. Gschwend and John K. MacFarlane

Ralph M. Parsons Laboratory for Water Resources and Hydrodynamics, Department of Civil Engineering, Massachusetts Institute of Technology, Cambridge, MA 02139

Polybromomethanes ($CHBr_3$, $CHBr_2Cl$, and CH_2Br_2) were released to seawater in laboratory incubations by Ascophyllum nodosum and Fucus vesiculosis at nanograms to micrograms each compound per gram dry algae per day throughout the year. This biological source of halogenated compounds was detectable in nearshore seawater during high tides when these littoral algae were submerged. The weak seasonality of polybromomethane formation, together with various evidence from the literature, suggest that these brominated organic compounds may arise from fungal epiphytes closely associated with these algae.

It has long been recognized that marine macroalgae contain halogenated metabolites (1 - 3). A few investigations have provided evidence that certain of the volatile halogenated compounds are released from these algae into coastal seawater (4 - 7). In particular, of all the volatile halogenated compounds observed, three polybromomethanes ($CHBr_3$, $CHBr_2Cl$, and CH_2Br_2) were released in the greatest quantities in a survey of several temperate macroalgae (7). This natural source of halogenated organic compounds to the marine environment is especially important in light of our concern for emissions of halogenated solvents and haloforms in coastal discharges (8 - 9) and to the atmosphere (10). Such natural sources must be quantified to place man's impact in perspective and to suggest natural sink mechanisms which may have evolved in response to these sources.

In this report, we describe a seasonal study of polybromomethane releases from two common fucoid algae, Ascophyllum nodosum and Fucus vesiculosis. We performed this investigation to evaluate the yearround variations in algal impact on coastal seawater chemistry. Further, we hoped to observe seasonal correlations between polybromomethane releases and factors in the life cycles of these algae to provide insights as to the genesis and function of these natural products.

0097-6156/86/0305-0314$06.00/0

METHODS

Algal Incubations. Our procedures for quantifying polybromomethane releases from macroalgae to seawater have been described previously (7). A brief description follows. Ascophyllum nodosum and Fucus vesiculosis specimens were collected approximately monthly from the rocky intertidal shoreline of the Cape Cod Canal, Sagamore MA. Algae were collected by breaking off a piece of the rock to which the plants were attached, thereby not causing any tissue lesions, and were held in a bucket of seawater until delivery to the laboratory within 2 hours. Plants supporting visible epiphytes were avoided, but it was not possible to exclude closely associated microscopic organisms. Seawater temperature and tidal stage were recorded at the time of sampling and water samples were taken for nutrients and stored frozen until analysis. Nitrate-N was determined by the cadmium reduction method and phosphate-P was assesed by the ascorbic acid or stannous chloride methods (11).

At the laboratory, the algae were placed in a 34-liter aquarium, filled with seawater and maintained without headspace by a tightly fitting glass lid. The aquarium was kept in a temperature and light-controlled incubation refrigerator set to mimic the current environmental conditions. Water motion was maintained using a two-inch magnetic stir bar. Algal incubations typically lasted one day. After incubating, seawater was withdrawn from the aquarium, spiked with a volatile halogenated organic internal standard, and analyzed for polybromomethane content.

Analysis of Polybromomethanes in Seawater. Polybromomethanes in aquarium water and Cape Cod Canal seawater samples were determined by a hybrid procedure of the "purge and trap" method (12) and the Grob closed-loop-stripping analysis (13). Briefly, headspace air was bubbled through a 1.8 liter water sample at a rate of 650 mL/min for 10 minutes. The effluent vapors pass through a condenser maintained at 10°C to reduce the water content of the gas stream, and then through a Tenax trap (65 mm long x 3 mm i.d., containing about 20 mg Tenax solid adsorbent) where the nonpolar organic compounds are collected. The "cleansed" air stream then recycles via a metal bellows pump to the water sample.

After stripping the water, the Tenax trap was immediately transferred to the hot injection port of a Hewlett Packard 5995B benchtop gas chromatograph-mass spectrometer (GC-MS). The volatiles were then thermally desorbed from the Tenax and carried with the gas stream onto the front of a glass capillary column (coated with SE54). The first loop of the capillary column was dipped in a Dewar containing liquid nitrogen and served to cryogenically focus the volatile concentrate at the column front. After 5 minutes of thermal desorption and transfer, the liquid nitrogen Dewar was removed and chromatography was begun. Data from the GC-MS was corrected for instrument response factors and stripping efficiencies to ultimately derive the amounts of polybromomethanes present in the sample. Seawater held in the aquarium without macroalgae never showed any polybromomethane formation. 1-Chloropentane, added to the seawater as an internal standard to monitor stripping efficiency, was recovered with a precision of 74 ± 18% (N=200). 1-Chlorobenzene added directly to the Tenax trap prior to thermal desorption was

recovered with a precision of 89 ± 12% (N=120). Given the conditions
of our algal incubations and analytical capabilities, our lower limit
of detection for algal releases was about 50 ng polybromomethane/g
dry algae·d. Polybromomethanes could be detected in Canal seawater
down to approximately 10 ng/L.

RESULTS AND DISCUSSION

The seasonal variations in temperature and nutrient concentrations in
the Canal where the algae were growing are shown in Table I. Water
temperature varied by about 20°C being lowest in January and highest
in August. Nitrate and phosphate were detectable yearround,
undoubtedly reflecting the influence of nutrient-rich Cape Cod Bay
seawater. In light of the literature and given these environmental
conditions, these fucoids grew optimally in the spring, summer, and
fall, and in the winter grew slower due to decreased light and cooler
temperatures (14 - 17).
    Polybromomethanes were released from both A. nodosum and F.
vesiculosis to lab aquarium seawater throughout the year (Figure 1).
Tribromomethane (or bromoform) was emitted by both algae at rates
between 0.1 and 10 μg CHBr$_3$/g·d. Release was high in fall of 1983,
but did not show any subsequent seasonality. Dibromochloromethane
was always observed from A. nodosum, but was undetectable in several
F. vesiculosis incubations. Dibromomethane was only produced and
released by A. nodosum, although not during the winter. This
dihalo-compound was released at generally lower rates (note scale
difference in Figure 1) than the two haloforms. Dibromochloromethane
and dibromomethane appear to be released somewhat more extensively
during the summer and fall.
    Cape Cod Canal seawater samples collected at high tide (i.e.,
when both fucoid populations would be submerged) always contained
detectable bromoform, although no seasonal pattern could be seen
(Figure 2). Low tide water samples generally showed undetectable
bromoform, except for a few spring and summer samples. Clearly these
fucoid populations imparted substantial bromoform to the surrounding
seawater, and this polybromomethane typically had no other detectable
sources as reflected by most low tide samples.
    Dibromochloromethane and dibromomethane were only observed in
two Canal water samples each. Since dibromochloromethane is released
by these algae at comparable rates to bromoform, it appears that some
sink is relatively more important for this compound. Kaczmar et al.
(18) report that CHBr$_2$Cl volatilizes about 50% faster than CHBr$_3$, and
possibly it is this mechanism which maintains CHBr$_2$Cl concentrations
below our detection limits in high tide Cape Cod Canal seawater.
Dibromomethane was only released from A. nodosum and then at several
times lower rates than bromoform, thus its nondetection in the Canal
water is not surprising.
    A simple calculation suggests that the algal incubation release
rates we observed in the laboratory could readily account for the
Canal water concentrations. If A. nodosum and F. vesiculosis
biomasses are about 10^{3}g dry algae per m^2 (17), they will release
about 100μg of bromoform/m^2·hr. Thus, a 20–50 cm deep layer of water
covering the algae at high tide will take only several minutes to
build up the concentrations we observe. Thus our field observations
of 20–80 ng bromoform/L seem consistent with the laboratory results.

TABLE I.   Temperature and nutrient data for Cape Cod Canal seawater
(September 1983-December 1984).

		Cape Cod Canal Seawater	
Date	Temp(°C)	NO_3 (µM)	PO_4 (µM)
9.10.83	17.5	–	–
9.20	15.0	–	–
9.25	15.8	–	–
9.29	15.0	–	–
10.05	14.5	–	–
10.18	14.5	–	–
10.26	11.0	3.5	0.16*
11.01	10.0	1.1	0.99†
11.08	9.7	11.2	1.00†
11.15	9.2	3.4	0.93†
11.22	9.0	–	–
12.01	8.5	2.4	0.95†
12.06	6.5	4.3	0.93†
12.13	7.0	4.5	1.12†
12.22	–	4.6/2.6	0.79†
1.04.84	1.5	8.0	1.68†
3.01	3.5	–	–
3.07	3.7	2.4	0.25*
3.15	1.5	4.0	0.22*
3.21	3.8	3.7	–
4.03	5.0	5.2	0.12*
5.03	8.0	3.2	0.16*
5.10	10.0	0.7/1.4	–
5.15	9.2	3.5	–
5.22	10.3	5.5	–
6.04	12.0	4.9	–
6.18	13.5	7.8	0.44/0.13*
7.10	16.5	2.2	0.19/0.15*
7.24	14.5	3.3	0.15/0.19*
8.07	21.0	1.4	0.18±0.01*
8.22	21.5	–	0.19/0.34*
9.12	14.3	–	–
10.04	14.4	–	–
10.24	14.1	–	0.34/0.35*
11.15	9.5	–	0.34*
12.12	7.2	–	0.34/0.34*

– no data available for this sample date.

* PO_4^{3-} analyzed by the Stannous chloride method (Standard Methods, APHA, 1980).

† PO_4^{3-} analyzed by the Ascorbic acid method (Standard Methods, APHA, 1980).

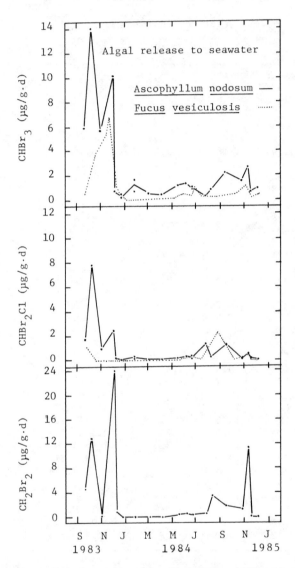

Figure 1. Seasonal release rates of $CHBr_3$, $CHBr_2Cl$, and CH_2Br_2 to seawater by the brown algae, Ascophyllum nodosum and Fucus vesiculosis. Open symbols indicate no detectable release.

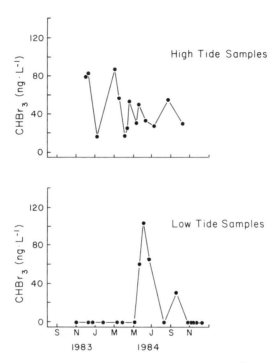

Figure 2.   Seasonal CHBr$_3$ concentrations (ng·L^{-1}) in nearshore seawater of the Cape Cod Canal, Sagamore, MA.

Algal versus Epiphytic Source. Our observations may also provide
some insight to the genesis of the polybromomethanes. Foremost in
this regard is whether the algae or some attached microorganisms
synthesize these halogenated compounds. Gschwend et al. (7)
previously reasoned that algae were most likely responsible; however,
the absence of a strong seasonality in the polybromomethane releases
from A. nodosum and F. vesiculosis indicated that this conclusion may
be wrong, since nearly every other process associated with these
macrophytes does vary yearround. First, numerous studies of
temperate fucoid algae have demonstrated major changes in their
growth rates and reproductive physiology throughout the year (15,
17). Investigations of algal metabolites, including the halogen I,
also show regular seasonality (16, 19, 20). Further algal peroxidase
enzymes, which could reasonably be proposed to oxidize halides prior
to their addition to organic substrates, have also been found to vary
greatly in activity in both A. nodosum and F. vesiculosis (21). Both
species exhibited the greatest activity in the spring and summer; A.
nodosum peroxidase activity diminished to one tenth maximal levels by
September and F. vesiculosis showed more than fifty times decrease in
this enzymatic activity by July-September. Clearly we did not see
these magnitudes of changes in polybromomethane release on a seasonal
basis. Further, Hewson and Hager (3) and Vilter (22) report that
peroxidase preparations from A. nodosum and F. vesiculosis could not
oxidize bromide (a prerequisite step to polybromomethane formation).
Thus it appears that these algae do not biosynthesize and release the
polybromomethanes we observe.

    We have attempted to support this conclusion by inoculating
seawater with surface material gently scraped off the surfaces of A.
nodosum and F. vesiculosis plants and monitoring for polybromomethane
formation. In both cases bromoform concentrations in the water were
observed to increase (Figure 3). Mercuric chloride poisoning, 0.2 μm
filtering, and autoclaving the inoculated seawater prevented this
bromoform production. Thus, some component of the epiphytic
community may be involved.

    The epiphytic communites of A. nodosum and F. vesiculosis
include bacteria, yeasts, other fungi, and phytoplankton. Most of
these microorganisms appear to undergo yearround variations in their
abundance. Using viable count techniques, Chan and McManus (23)
found the fewest bacteria on A. nodosum during the summer months of
July and August. Sieburth and Tootle (24) extended this result by
examining both A. nodosum and F. vesiculosis with scanning electron
microscopy (SEM). They found bacterial colonization of these algae
was greatest in November and April and diminished to virtually clean
in May to July. Seshadri and Sieburth (25) also enumerated the
yeasts growing on these fucoid algae and again discovered a strong
seasonal variation. Further, these same yeast species were most
abundant on a red alga, Chondrus crispus, which exhibits little or no
polybromomethane production (7). Finally, some microalgae grow as
epiphytes on fucoids (24); but no bromination capabilities have ever
been demonstrated in a microalgae (7). Although culturing and SEM
approaches to evaluating microorganisms on fucoids may be somewhat
crude for our purpose, these assays consistently demonstrate strong
seasonal trends in the dominant microbial subpopulations living on
A. nodosum and F. vesiculosis. Consequently, we suspect these
microorganisms do not make the polybromomethanes.

Finally, Vilter et al. (21) have suggested that an ascomycete, Mycosphaerella ascophylli, growing symbiotically on A. nodosum may be responsible for the bromoperoxidase activity attributed to that fucoid. This fungus is apparently always present on this brown alga (26), consistent with the continuous production of polybromomethanes. M. ascophylli has been isolated and cultured by Pederson and Fries (27) and production of brominated phenols by the culture was observed. Although the ascomycete is thought to be rare on F. vesiculosis (26), the same bromophenols are observed from that algae (28), suggesting a similar source. Since bromoperoxidases capable of forming bromophenols would also be able to yield bromoform from the appropriate organic precursors, M. ascophylli ubiquitous on A. nodosum, or similar fungi on other algae, appears to be the most likely source of the polybromomethanes. As we have noted previously (7), if these or other epiphytes are involved, the microorganisms must be quite host-specific to account for the different combinations and rates of polybromomethane releases observed from between various algal species.

If ascomycetes living in mycophycobiotic relationships with fucoids (25) are indeed the answer, it is tempting to speculate on the function of their bromoperoxidase activity. Such a strong oxidative enzymatic capability is common to fungi (29), and it could be used by the ascomycete to brominate the plant surface chemicals (e.g., forming bromophenols such as lanosol). These metabolites may then protect the algae from herbivores and microbial colonization, in return for macrophyte supply of nutrition to the fungus.

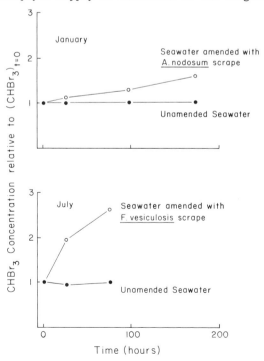

Figure 3. Production of $CHBr_3$ in seawater with and without macroalgae surface material added.

SUMMARY

Polybromomethanes ($CHBr_3$, $CHBr_2Cl$, CH_2Br_2) were found to be released yearround from two common rockweeds, A. nodosum and F. vesiculosis, or their closely associated microflora. This biological source appears sufficient to support parts per trillion levels of these halogenated organic compounds in nearby seawater. Several lines of evidence indicate that epiphytic fungi may actually accomplish the polybromomethane biosynthesis. Future research clarifying the genesis and functions of these volatile brominated compounds is surely warranted.

Literature Cited

1. Fenical, W. J. Phycol. 1975, 11, 245.
2. Moore, R.E. Acc. Chem. Res. 1977, 10, 40.
3. Hewson, W.D.; L.P. Hager. J. Phycol. 1980, 16, 340.
4. Lovelock, J.E.; R.J. Maggs; R.J. Wade. Nature 1973, 256, 193.
5. Lovelock, J.E. Nature 1975, 256, 193.
6. Dryssen, D.; E. Fogelquist. Oceanol. Acta 1981, 4, 313.
7. Gschwend, P.M.; J.K. MacFarlane; K.A. Newman. Science 1985, 227, 1033.
8. Helz, G. In "Hydrocarbons and Halogenated Hydrocarbons in the Aquatic Environment"; B.K. Afghan and D. Mackay, Eds.; Plenum Press: New Jersey, 1980; pp. 435-444.
9. Fogelquist, E.; B. Josefsson; C. Roos. Environ. Sci. Technol. 1982, 16, 479.
10. Cicerone, R.J. Rev. Geophys. Space Phys. 1981, 19, 123.
11. "Standard Methods for the Examination of Water and Wastewater," American Public Health Association, 1980.
12. Bellar, T.A.; J.J. Lichtenberg. J. Amer. Water Works Assoc. 1974, 66, 739.
13. Grob, K.; F. Zurcher. J. Chromatogr. 1976, 117, 285.
14. Conover, J.T. Publ. Inst. Mar. Sci. 1958, 5, 97.
15. Mathieson, A.C.; J.W. Shipman; J.R. O'Shea; R.C. Hasevlat. J. Exp. Mar. Biol. Ecol. 1976, 25, 273.
16. Ragan, M.A.; A. Jensen. J. Exp. Mar. Biol. Ecol. 1978, 34, 245.
17. Chock, J.S.; A.C. Mathieson. Bot. Marina 1983, 26, 87.
18. Kaczmar, S.W.; F.M. D'Itri; M.J. Zabik. Environ. Toxicol. Chem. 1984, 3, 31.
19. Black, W.A.P. J.S.C.I. 1948, 67, 355.
20. Black, W.A.P. J.S.C.I. 1949, 68, 183.
21. Vilter, H.; K.-W. Glombitza; A. Grawe. Bot. Marina 1983, 26, 331.
22. Vilter, H. Bot. Marina 1983, 26, 429.
23. Chan, E.S.C.; E.A. McManus. Can. J. Microbiol. 1969, 15, 409.
24. Sieburth, J. McN.; J.L. Tootle. J. Phycol. 1981, 17, 57.
25. Seshadri, R.; J. McN. Sieburth. Mar. Biol. 1975, 30, 105.
26. Kohlmeyer, J.; E. Kohlmeyer. Bot. Marina 1972, 15, 109.
27. Pederson, M.; N. Fries. Z. Pflanzenphysiol. 1977, 82, 363.
28. Pederson, M.; L. Fries. Z. Pflanzenphysiol. 1972, 74, 272.
29. Siuda, J.F.; J.F. DeBernardis. Lloydia 1973, 36, 107.

RECEIVED September 16, 1985

ORGANOSULFUR COMPOUNDS

19

Biogeochemical Cycling of Sulfur
Thiols in Coastal Marine Sediments

Kenneth Mopper and Barrie F. Taylor

Division of Marine and Atmospheric Chemistry, Rosenstiel School of Marine and Atmospheric Science, University of Miami, Miami, FL 33149-1098

Thiols are major intermediates in the microbial cycling of sulfur and, because of their high reactivity, they may also play important roles in geochemical processes. Preliminary studies using a new, highly sensitive HPLC assay revealed that thiols are present at concentrations up to 100μM in intertidal marine sediments from Biscayne Bay (FL). Methanethiol (MeS) and 3-mercaptopropionate (MP) were the major thiols found. The presence of the latter compound suggests that, in addition to protein degradation, anaerobic decomposition of dimethylpropiothetin (DMPT), a major sulfur compound of marine algae and higher plants, may be an important source of thiols and a significant degradation pathway for DMPT in the environment. Acrylic acid produced from this pathway readily adds HS⁻ across the double bond, via Michael addition, to form 3-mercaptopropionate. Alternatively, this thiol may be formed directly from DMPT by successive anaerobic demethylations; however the biochemical feasibility of this pathway is presently not known. Addition of a specific disulfide cleaving reagent to sediments revealed that thiols are dominantly present in bound forms. Binding of thiols to sediment particles may be an important mechanism for the incorporation of organic sulfur into geopolymers.

Diagenesis of organic matter in the water column and sediments results in the production of a wide variety of organosulfur compounds. Most studies involving these compounds in the marine environment have focussed on gaseous and hydrophobic species (1). In contrast, information regarding reduced, hydrophilic sulfur organics, in particular the thiols (general formula R-SH, where R is an organic group), is almost nonexistent.

0097-6156/86/0305-0324$06.00/0
© 1986 American Chemical Society

Thiols, or sulfhydryl compounds, play important biochemical roles in maintaining macromolecular structures, binding metals at active sites of enzymes and electron transport components, capturing and detoxifying metals, and serving as coenzymes. In aquatic sediments these compounds arise during microbial processes of sulfate reduction and lithotrophic oxidation (oxic and anoxic), and from biodegradation of organic matter (2,3). Abiotic sources of thiols include reactions of dissolved organic matter with H_2S and elemental sulfur present in pore water and on particle surfaces (4,5,6).

Functional group, spectrophotometric, electrochemical and elemental analyses have shown the presence of significant concentrations of thiols and other organo sulfur compounds in anoxic marine waters and sediments (5,7-10). For example, using polarographic techniques, Luther et al. (10) reported that thiols were the major reduced sulfur species, either inorganic or organic, in the porewaters of saltmarsh sediments (Great Marsh, Lewes, DE) with total concentrations up to 2.4 mM. The high concentrations and the high chemical reactivity of thiols strongly suggest that these compounds play a major role in the early diagenesis of organic matter in sediments, as well as in the incorporation of sulfur into organic geopolymers. For example, thiols readily react to form disulfide and polysulfide bridges (5), which may enhance the crosslinking and, hence, the molecular weight of organic matter in sediments. In addition, thiols form strong complexes with metal ions, especially transition metals (5), and may promote the mobilization of metals, e.g. arsenic (11). It is likely that through complexation with metals on particle surfaces that thiols also become strongly bound.

Despite the biogeochemical significance of thiols, relatively little is known about the nature and distribution of these species in sediments. Therefore, with the aid of a newly developed analytical method, a study was initiated to explore the role that thiols play in the marine sedimentary sulfur cycle. Some of the questions addressed were: (1) What thiols are present in sediments? (2) What are the possible formation pathways; that is, what is the relative importance of biological (e.g. microbial) versus nonbiological (e.g. chemical reactions) sources for thiols in sediments? (3) Are thiols bound to sediment particles and (4) What is the nature of the binding? The major findings and conclusions of this initial study are presented in this report.

Experimental

Thiols, as well as sulfide and sulfite, were determined in porewater samples by reversed phase high performance liquid chromatography (HPLC). The technique is based on precolumn derivatization with an o phthalaldehyde/amine reagent (Figure 1) followed by HPLC and fluorometric detection. Derivatized porewater samples were injected directly into the HPLC system; the detection limit is 0.1 nM (for 100 ul injection). Details of the method are given in Mopper and Delmas (12).

Intertidal Biscayne Bay (FL) sediments were periodically collected (using glass jars) during June through September of 1984. A total of 27 samples were taken during this period. Slurries were

prepared using two parts sediment and one part deaerated seawater
(V/V). The slurries ($\sim$ 500ml) were stored at 25°C under argon in
glass jars. Background data for typical sediment samples are given
in Table I.

Table I. Typical Background Data on Intertidal
Biscayne Bay Sediments

Date	Porewater pH	Size Fraction μm	% Dry Sediment	CaCO$_3$ %	Organic C %
Aug. 1, 1984	7.4	ϕ > 300	72.3	64.8	N.A.
		300 < ϕ > 63	20.1	39.9	1.53
		63 < ϕ	7.6		
Aug. 8, 1984	7.4	ϕ > 300	74.1	64.1	N.A.
		300 < ϕ > 63	16.1	37.7	1.34
		63 < ϕ	9.8		
Aug. 24, 1984	7.5	ϕ > 300	69.7	58.1	N.A.
		300 < ϕ > 63	22.3	41.7	1.76
		63 < ϕ	8.0		

N.A. = not analyzed; ϕ = medium grain size.

Prior to thiol analysis, slurries were allowed to settle and a
2 ml aliquot of the supernatant was filtered (0.2 μm, Nucleopore)
and derivatized. Some sediment slurries were treated with a specific
S-S cleaving reagent, tributylphosphine (13), in order to evaluate
the degree to which thiols were bound via disulfide linkages. The
reagent was added to a final concentration of 0.5-1.0 ml per liter
slurry.

Results and Discussion

Identification of Thiols in Marine Sediment Porewaters. Thiols were
present at significant levels (up to about 100 μM) in anoxic,
intertidal Biscayne Bay sediments during the entire sampling period.
The concentrations found were similar to those reported for other
low molecular weight organics in sediment porewaters (Table II).
More than 30 individual thiols were detected of which 13 were
positively or tentatively identified (Figure 2 and Table III).
Peaks were identified by co-injection with authentic compounds under
different chromatographic and derivatization conditions, as outlined
in Figure 3. For example, altering the pH of the mobile phase was
particularly effective for the identification of carboxylated
thiols, such as 3-mercaptopropionate, because protonation of the
carboxy group selectively enhanced the retention of these compounds.

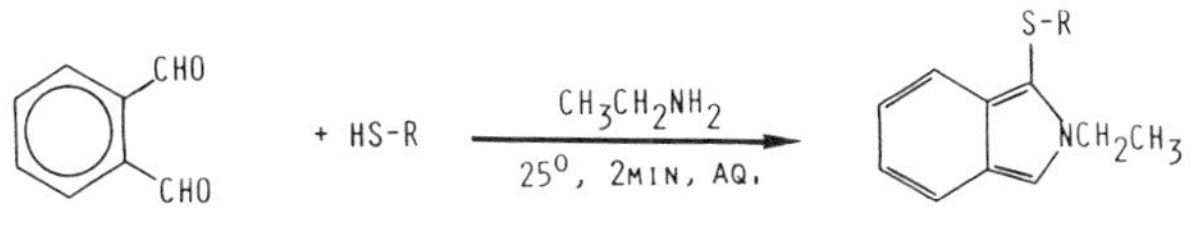

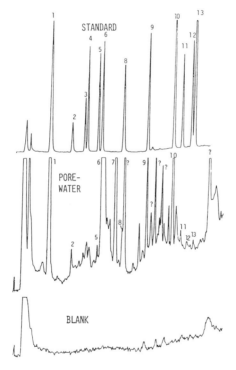

O-Phthalaldehyde   Thiol   Isoindole

Figure 1. Fluorescence derivatization of thiols; the excitation and emission maxima of the fluorescent isoindole product are at 340 nm and 450 nm respectively (12).

Figure 2. **Upper:** Gradient separation of o‑phthalaldehyde derivatives of 11 thiols and sodium sulfite according to Mopper and Delmas (12). Peaks: (1) sodium sulfite (100 pmol); (2) glutathione (7 pmol); (3) thioglycollate (200 pmol); (4) N‑acetylcysteine (7 pmol); (5) 2‑mercaptoethanesulfonate (Co‑M) (10 pmol); (6) 3‑mercaptopropionate (10 pmol); (8) monothioglycerol (10 pmol); (9) 2 mercaptoethanol (10 pmol); (10) methanethiol (15 pmol); (11) ethanethiol (10 pmol); (12) 2‑propanethiol (15 pmol); (13) 1 propanethiol (15 pmol). **Middle:** Thiols in porewater in reducing sediment slurry from Biscayne Bay. Porewater water was filter‑sterilized prior to derivatization. Peak 7: sulfide (Note: response factor is about 200 times lower than for thiols). **Lower:** reagent blank in porewater matrix.

Table II. Concentration Range of Thiols in Comparison to
Other Low Molecular Weight Organics in Sediment Porewaters

Compounds	Typical Concentrations	Reference
Thiols	0.1 - 100 µM; 0.1 - 2.4 mM	This work and 10, respectively
Sugars	0.2 - 2 µM	33, 34
Amino Acids	1 - 200 µM *	35, 36
†LMW Carboxylic Acids	1 - 30 µM	37, 38
LMW Carbonyls	0.1 - 5 µM	Mopper (unpublished)

* Note: The highest value was measured in Thioploca mats underlying
the Peru upwelling. Maximum amino acid concentrations found in
typical marine sediments are 10 - 20 uM.
† LMW = Low molecular weight.

Table III. Thiols Identified in Slurries of Marine Sediments

Major Thiols (0.5 - 20 µM)
3-Mercaptopropionate (dominant)
Methanethiol (dominant)
Ethanethiol
Monothioglycerol
2-Mercaptoethanol
2-Mercaptopyruvate

Minor Thiols (<0.5 µM)
Mercaptoacetate (thioglycollate)
Glutathione
Coenzyme M (2-mercaptoethanesulfonate)
1-Propanethiol
2-Propanethiol
N-Acetylcysteine
2-Mercaptopropionate

Unknowns
At least 20 unknown thiols, including 4 or 5 which are major.

Methanethiol (MeS) and 3-mercaptopropionate (MP) were by far
the dominant thiols (Figure 2). Biochemical formation pathways are
feasible for these thiols, although the importance of these pathways
relative to abiotic pathways is presently unknown, especially for
3-mercaptopropionate, as summarized in Figure 4.   The relative
importance of these pathways are discussed in the following
sections.

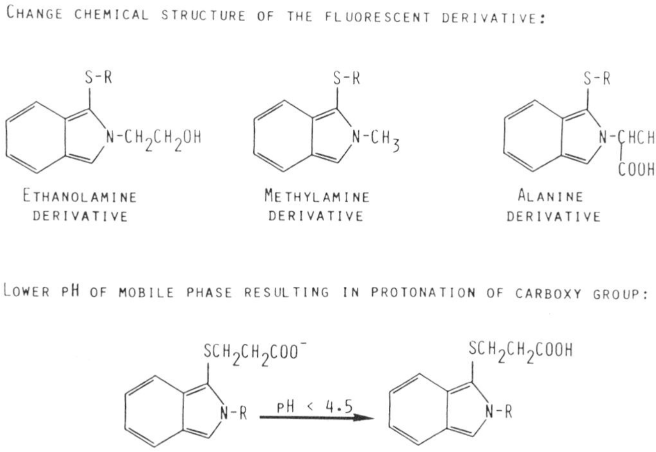

CHANGE CHEMICAL STRUCTURE OF THE FLUORESCENT DERIVATIVE:

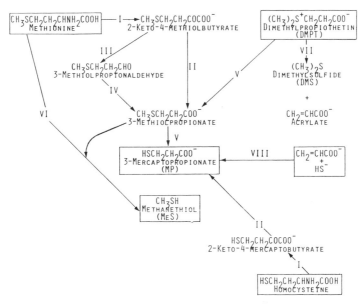

ETHANOLAMINE
DERIVATIVE

METHYLAMINE
DERIVATIVE

ALANINE
DERIVATIVE

LOWER pH OF MOBILE PHASE RESULTING IN PROTONATION OF CARBOXY GROUP:

Figure 3. Identification of thiols in sediments by altering chromatographic selectivity.

Figure 4. Possible routes for the formation of 3‑mercaptopropionate and methanethiol. I = Oxidative deamination or transamination; II. Oxidative decarboxylation; III. Nonoxidative decarboxylation; IV. Oxidation; V. Demethylation; VI. Demethiolation; VII. Aerobic/anaerobic catabolism; VIII. Michael addition.

Microbial Production of Thiols. Sulfur containing amino acids have
been shown to be important precursors of thiols in many
environments, including sediments (14-17). Biscayne Bay sediments
are not likely to be an exception. Thus, methionine degradation
might contribute to the production of methanethiol and 3-mercapto-
propionate (the major thiols detected in the porewaters, Table III).
Microbes in anoxic environments, specifically the rumen and lake
sediments, convert methionine to methanethiol (14-17). Oxidative
deamination or transamination of methionine, followed by oxidative
decarboxylation, yields 2-keto-4-methiolbutyrate and then
3-methiolpropionate (18); subsequent demethylation would produce
3-mercaptopropionate (Figure 4). Similar mechanisms, but without
the necessity for a final demethylation reaction, could convert
homocysteine into 3-mercaptopropionate via 2-keto-4-thiolbutyrate
(Figure 4). The pathway from homocysteine would be analogous to
that from cysteine to 2-mercaptoacetate via 3-mercaptopyruvate (18).
Ethanethiol (Table III) might be derived from S-ethylcysteine or
ethionine (19). Alkylated thiols, other than methanethiol and
ethanethiol, could be derived from either alkylated analogs of
methionine or via mechanisms involving reductive deaminations and
nonoxidative decarboxylations.

While amino acids are undoubtedly important precursors of
thiols, in coastal marine habitats dimethylpropiothetin (DMPT) is
probably also an important precursor because of its high
intracellular concentrations (up to 300 mM) in algae and seagrasses
(20,21). Dimethylpropiothetin is degraded under oxic conditions by
bacterial and algal enzymes yielding principally acrylic acid and
dimethylsulfide (DMS) (19,22,23):

$$(CH_3)_2\overset{+}{S}CH_2CH_2COO^- \longrightarrow CH_3SCH_3 + CH_2{=}CHCOOH$$

          DMPT   DMS   acrylic acid

DMS is also generated during anaerobic fermentation of dimethylpropiothetin (24) and was detected as the principal volatile sulfur
compound evolved from salt marsh flats (25). Therefore, it is
reasonable to hypothesize that DMPT, is also a major precursor for
other organosulfur compounds, e.g. thiols, in anoxic marine
sediments. However, no studies have been reported on the microbial
production of thiols from this compound. Hypothetical formation
pathways for methanethiol and 3-mercaptopropionate from DMPT are
given in Figure 4. Two successive demethylations of DMPT would
yield 3-methiolpropionate and 3-mercaptopropionate. The first
demethylation of dimethylpropiothetin is biochemically feasible; for
example, homocysteine can accept a methyl-group from DMPT, in a
reaction analogous to that involving dimethylthetin (26), to yield
methionine and 3-methiolpropionate (27). The second demethylation,
that of 3-methiolpropionate, has not been demonstrated but it might
be catalyzed by methanogenic and/or acetogenic bacteria. Methyl
transfer reactions from methylated sulfur substrates could operate
and involve cobalamine-containing enzymes as shown for methanol (28)
and postulated for methylamines and methoxy-aromatic molecules
(29,30). Finally, 3-methiolpropionate is a known precursor of
methanethiol (18).

Abiotic Production of Thiols. While the formation pathway of
3-mercaptopropionate from successive demethylations of DMPT remains
to be proven, DMPT may nonetheless be an important precursor of this
thiol by a less speculative pathway involving established reactions.
The first reaction is the anaerobic cleavage of DMPT to DMS and
acrylic acid (reaction VII, Figure 4). If this reaction occurs in
the sulfate reducing zone, acrylic acid will then rapidly react with
HS⁻ (a powerful nucleophile) by adding it across the double bond via
the well known Michael addition reaction (31) (Figure 4). In the
past, this reaction has been studied at high concentrations of
reactants and in the absence of water (32). Therefore, the
environmental relevance of this reaction is questionable.

Several experiments were performed to examine this possible
reaction. Acrylic acid and sodium sulfide were added to deaerated,
aged Gulf Stream seawater to final concentrations of 0.1 mM and 1.0
mM, respectively. Controls consisted of seawater alone, seawater
plus acrylic acid, and seawater plus sodium sulfide. The reaction
was run at pH 8.2-8.4 under argon at 25°C and 60°C. After two
hours, aliquots were removed for thiol analysis by HPLC. Only two
thiols were detected, of which 3-mercaptopropionate, the expected
product, was one. The identity of the other thiol has not been
established. Controls showed negligible thiol production (Figure
5). The apparent yield of the reaction (% acrylic acid converted to
3-mercaptopropionate) was about 1-2% at 60°C and 0.3-0.4% at 25°C.
Addition of tributylphosphine, a disulfide cleaving reagent, to the
reaction mixture either before or after the 2 hour incubation,
increased the thiol yield by about a factor of two. These results
suggest that abiotic reactions may indeed be responsible for the
formation of some thiols in the environment and that thiols oxidize
rapidly to form disulfide compounds even under reducing conditions.
A kinetic study of the Michael reaction in seawater is currently
being undertaken.

In order to provide additional evidence in support of the
proposed Michael addition reaction, a sediment study was also
performed. Acrylic acid was added directly to slurries of reducing
sediments from Biscayne Bay and the formation of 3-mercaptopro-
pionate in the porewater, relative to unspiked controls, was
monitered by HPLC. The concentration of added acrylic acid was 0.1
mmol per liter slurry (0.2 mM in the porewater) and the slurries
were incubated under argon at 37°C for 2 hours prior to thiol
analysis. Figure 6 clearly shows that addition of acrylic acid to
reducing sediment gives rise to 3-mercaptopropionate, the main
product expected from the Michael addition of HS⁻ to acrylic acid.

The addition of H_2S, traces of which will be present at the pH
of the porewater (Table I), to acrylic acid probably follows the
Markownikoff addition rule to yield 2-mercaptopropionate:

$$H_2C=CHCOO^- + H_2S \longrightarrow H_3C-\underset{\underset{SH}{|}}{C}HCOO^-$$

In fact, 2-mercaptopropionate was tentatively identified (at trace
concentrations) in some Biscayne Bay sediments (Table III).

It is tempting to conclude from these studies that abiotic
reactions play a major role in the formation of thiols in sediments.

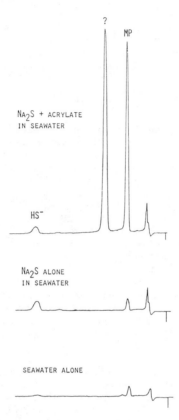

Figure 5. Upper: Abiotic production of 3-mercaptopropionate (MP) and an unknown thiol (?) in deaerated seawater from reaction of acrylic acid (0.1 mM) with sodium sulfide (1.0 mM) at 60°C for 2 hours under argon; detection by fluorescence derivatization and HPLC. Middle: Control = sodium sulfide (1.0 mM) alone in seawater under identical reaction conditions as above. Lower: Control = seawater alone under identical reaction conditions as above.

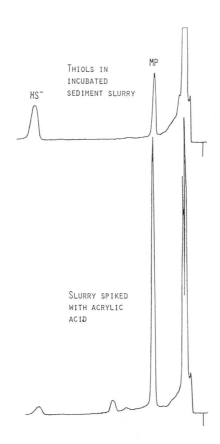

Figure 6. <u>Upper</u>: 3-mercaptopropionate (MP) and hydrogen sulfide (HS⁻) present in aqueous phase of Biscayne Bay sediment incubated at 37°C for 2 hours under argon; detection by fluorescence derivatization and HPLC; the large initial peaks are probably humic substances. <u>Lower</u>: Aliquot of same sediment slurry sample incubated under the same conditions as above but with added acrylic acid (10 μmole per 100 ml slurry); note increase in 3-mercaptopropionate and decrease in HS⁻.

While this may be true for 3-mercaptopropionate, the actual
importance of these reactions relative to microbial production of
thiols can only be properly assessed with tracer experiments using
radiolabelled substrates and reactants.

Release of Bound Thiols. Tributylphosphine quantitatively and
rapidly reduces disulfides to thiols and, by maintaining reducing
conditions, prevents their reoxidation (13):

$$R'-S-S-R + (C_4H_9)_3P + H_2O \longrightarrow R'SH + RSH + (C_4H_9)_3P=O$$

Although this reagent had only been used in the past to preserve
standard thiol solutions, it was felt that it could also be employed
to quantify the relative amounts of -S-S- bound thiols in sediments.
However, since tributylphosphine had not been previously used in
seawater media, a preliminary test was performed to evaluate its
chemical behavior under these conditions. Dimethyldisulfide and
oxidized glutathione were added to seawater to a final concentration
of 1 µM each. Tributylphosphine was added (0.5 ml per liter
seawater) and the samples were incubated at 25°C in the presence of
air. Aliquots were periodically removed for thiol analysis. After
20 min, 95% of the disulfides were reduced to thiols (Figure 7).
The thiols were stable for three days even in the presence of air.
After this period, reoxidation occurred probably due to loss of the
excess and protective phosphine by reaction with oxygen. From these
tests it was concluded that the phosphine reagent could be used to
study the relative abundance of free versus -S-S- bound thiols in
reducing sediments.

Sediment slurries were incubated at 30°C (approximate in situ
temperature) with and without tributylphosphine. Aliquots of
porewater were periodically removed for thiol analysis over the
following 2-4 days. During the course of this study, a total of 7
such incubations were performed on freshly collected sediment.
Results were similar in all cases and typically showed that
tributylphosphine induced a dramatic and rapid release of bound (or
oxidized) thiols (Table IV and Figure 8). Bound thiols were present
at approximately 20 times greater concentrations than free thiols
(i.e., ~95% of all thiols released from sediment were initially
bound). If air is not excluded during the incubation, released
thiols become reoxidized after several days (Figure 8) probably due
to the oxidation of the protective phosphine. Addition of fresh
tributylphosphine regenerated the thiols.

Addition of tributylphosphine to extracted porewater (particle
free) resulted in only a minor increase in thiol concentrations.
This result indicates that the dramatic increases obtained with
slurries (Table IV and Figure 8) are probably due to release of
thiols bound to particle surfaces, as opposed to release from
disulfides dissolved in the interstial water.

Surface binding is most likely through -S-S- linkages, but it
is also conceivable that some fraction of the released thiols may be
due to displacement of thiols from metal complexes on particle
surfaces by the phosphine nucleophile. However, when an equivalent
amount of a strong metal complexing agent (EDTA) was substituted for
the phosphine, only negligible releases of thiols were observed,
suggesting that, for the sediments studied, binding by metal

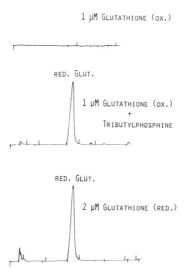

Figure 7. Cleavage of oxidized glutathione in seawater by tributylphosphine to yield reduced glutathione (red. Glut.); detection by fluorescence derivatization and HPLC. <u>Upper:</u> Oxidized glutathione (1µM) alone. <u>Middle:</u> Oxidized glutathione (1µM) plus tributylphosphine (50 µl/100ml seawater) reacted for 20 min at 25°C; <u>Lower:</u> Reduced glutathione standard (2µM) in seawater.

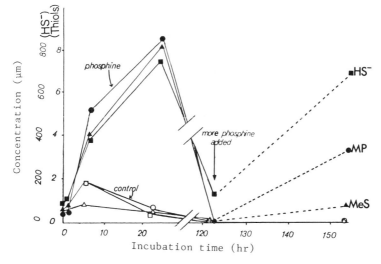

Figure 8. Release of thiols (3-mercaptopropionate - MP and methanethiol - MeS) and HS⁻ upon addition of tributylphosphine (50µl per 100 ml slurry) to a Biscayne Bay sediment slurry; incubated at 25°C; control slurries received no tributylphosphine.

complexation may not be as important as binding by −S−S− bond formation. Specific methods for releasing metal bound thiols are being explored.

Table IV. Release of Bound Thiols from Sediment

Incubated Sediment*	3-Mercaptopropionate (μM)	Methanethiol (μM)	HS$^-$ (mM)	SO$_3^=$ (mM)
Control Slurry*	0.19	0.05	4.24	0.071
Slurry with Tributylphosphine*	6.53	1.31	6.16	0.033
% Thiol Bound*	97%	96%	31%	0%

* 8/24/84; Biscayne Bay intertidal sediment incubated for 23h at 25°C; thiols detected in the "control slurry" (no tributylphosphine added) are interpreted as being in the unbound (or free) state; thiols detected in "slurry with tributylphosphine" are interpreted as free plus bound species; the "% thiol bound" was calculated from: [(thiols in tributylphosphine treated slurry) − (thiols in control)] x 100%/(thiols in tributylphosphine treated slurry).

Summary and Conclusions

Thiols are present at significant levels in reducing, intertidal sediments and apparently arise as a result of interacting biotic (microbial) processes and abiotic reactions (Figure 9). Over 30 thiols were detected, of which methanethiol and 3-mercaptopropionate were present in the highest concentrations throughout the entire 4 month sampling period. Methanethiol can readily arise through a number of known anaerobic pathways; however, no such pathways have been reported for the formation of 3-mercaptopropionate. It can be speculated that this compound is produced by successive anaerobic demethylations of dimethylpropiothetin (DMPT), a major sulfur compound of marine algae and plants. On the other hand, the known anaerobic breakdown pathway of DMPT is via enzymatic cleavage to yield dimethylsulfide and acrylic acid. Acrylic acid is a highly reactive species and, in zones of active sulfate reduction, it will readily undergo a Michael addition with HS$^-$ to yield 3-mercaptopropionate (Figure 6). If this reaction is in fact occurring in sediments and in the water column of anoxic basins, a number of important geochemical implications can be inferred. For every mole of DMPT hydrolyzed, up to two moles of organosulfur compounds (dimethylsulfide and 3-mercaptopropionate) are produced. If direct biological sources are indeed negligible for the latter compound, then its concentration and turnover may be used to estimate that of acrylic acid and indirectly that of DMPT hydrolysis

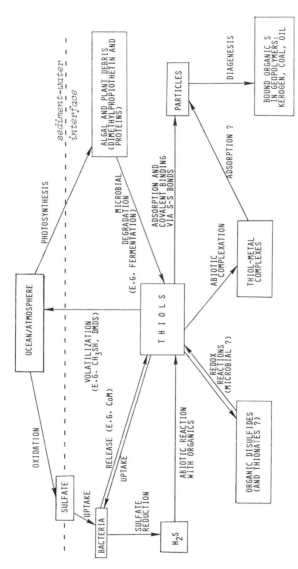

Figure 9. Cycling of thiols in sediments and potential role of thiols in the biogeochemistry of sulfur.

in the environment. More generally, the results imply that a major
chemical pathway for the incorporation of sulfur into organic
geopolymers is by reaction of HS⁻ with reactive sites, e.g. olefinic
double bonds, displaceable halogens (39), within sedimentary organic
matter. The Michael addition reaction of HS⁻ to acrylic acid may be
used as a model case of such interactions.

Addition of a specific –S–S– cleaving reagent,
tributylphosphine, to reducing marine sediments resulted in a
dramatic and rapid release of thiols into the porewater. The
results showed that in the sediments studied bound thiols are
present in at least 20 times greater concentration than freely
dissolved thiols or disulfides. These results imply that another
important route for the incorporation of sulfur into organic
geopolymers may be binding of thiols to reactive sites (e.g., R–SH
groups and metal ions) on particles.

The results of this initial study indicate that thiols play an
active role in the biogeochemical cycling of sulfur in marine
sediments (Figure 9). Many questions remain to be addressed. In
particular, how fast is the turnover of thiols in sediments and what
organisms are involved? What fraction of sedimentary sulfur passes
through the thiol pool? What are the precursors of the thiols? How
do thiol–metal interactions affect the geochemistry (e.g. migration
and binding) of heavy metals and thiols in sediments? How is biotic
and abiotic production of thiols in porewaters related to the sulfur
content and sulfur speciation within organic geopolymers?

Acknowledgments

We would like to thank R. Cuhel, G.R. Harvey, and G.E. Luther for
valuable input regarding possible formation routes for thiols.
Financial support was provided in part by a grant from the National
Institutes of Health (Biomedical Research Support Grant to K.M.) and
from the National Science Foundation (OCE–8516020).

Literature Cited

1. Balzer, W.   In "Marine Organic Chemistry"; Duursma, E.K.;
    Dawson, R., Eds.; Elsevier: Amsterdam, 1981; Chap. 13.
2. Trudinger, P.A.   Phil. Trans. Roy. Soc. 1982, B298, 563–581.
3. Zinder, S.H.; Doemel, W.N; Brock, T.D.   Appl. Environ.
    Microbiol. 1977, 34, 859–860.
4. Altschuler, Z.S.; Schnepfe, M.M.; Silber, C.C.; Simon, F.O.
    Science 1983, 221, 221–227.
5. Boulegue, J.; Lord, C.J.; Church, T.M.   Geochim. Cosmochim.
    Acta 1982, 46, 453–464.
6. Martin, T.H.; Hodgson, G.W.   Chem. Geol. 1973, 12, 189–208.
7. Adams, D.D.; Richards, F.A.   Deep–Sea Res. 1968, 15, 471–481.
8. Nissenbaum, A.; Kaplan, I.R.   Limnol. Oceanogr.   1972, 17,
    570–582.
9. Dyrssen D.; Haraldsson, C.; Westerlund, S.; Aren, K.   Mar.
    Chem., submitted.
10. Luther, G.W.; Church, T.M.; Giblin, A.E.; Howarth, R.W.   In
    "Organic Marine Geochemistry"; Sohn, M., Ed.; ACS Symposium
    Series No.   American Chemical Society: Washington, D.C., 1986;
    in press.

11. Cullen, W.R.; McBride, B.C.; Reglinski, J. J. Inorg. Biochem. 1984, 21, 174-194.
12. Mopper, K.; Delmas, D. Anal. Chem. 1984, 56, 2558-2560.
13. Humphrey, R.E.; Potter, J.L. Anal. Chem. 1965, 37, 164-165.
14. Bird, R.P.; Moir, R.J. Aust. J. Biol. Sci. 1972, 25, 835-848.
15. Salsbury, R.L.; Merricks, D.L. Plant Soil 1975, 43, 191-209.
16. Zikakis, J.P.; Salsbury, R.L. J. Dairy Sci. 1969, 52, 2014-2019.
17. Zinder, S.H.; Brock, T.D. Appl. Environ. Microbiol. 1978, 35, 344-352.
18. Cooper, A.J.L. Ann. Rev. Biochem. 1983, 52, 187-222.
19. Bremner, J.M.; Steele, C.G. Adv. Microbial Ecol. 1978, 2, 155-201.
20. White, R.H. J. Mar Res. 1982, 40, 529-536.
21. Vairavamurthy, A.; Andreae, M.O.; Iverson, R.L. Limnol. Oceanogr. 1985, 30, 59-70.
22. Cantoni, G.L.; Anderson, D.G. J. Biol. Chem. 1956, 222, 171-177.
23. Kadota, H.; Ishida, Y. Ann. Rev. Microbiol. 1972, 26, 127-138.
24. Wagner, C; Stadtman, E.R. Arch. Biochem. Biophys. 1962, 98, 331-336.
25. Aneja, V.P.; Overton, J.H., Jr.; Cupitt, J.T.; Durham, J.L.: Wilson, W.E. Tellus 1979, 31, 174-178.
26. Lehninger, A.L. "Biochemistry"; Worth: New York, 1975; 2nd Edition.
27. Maw, G.A.; du Vigneaud, V. J. Biol. Chem. 1948, 176, 1037-1045.
28. Wood, J.M.: Moura, I.; Moura, J.G.G.; Santos, M.H.; Xavier, A.V.; LeGall, J.; Scandellari, M. Science 1982, 216, 303-305.
29. Naumann, E.; Fahlsbuch, K.; Gottschalk, G. Arch. Microbiol. 1984, 138, 79-83.
30. Tschech, A.; Pfennig, N. Arch. Microbiol. 1984, 137, 163-167.
31. Noller, C.R. In "Textbook of Organic Chemistry"; W.B. Saunders: Philadelphia, 1966; 3rd Edition; p. 619.
32. Dahlbom, R. Acta. Chem. Scand. 1951, 5, 690-698.
33. Mopper, K.; Dawson, R.; Liebezeit, G.; Ittekkot, V. Mar. Chem. 1980, 10, 55-66.
34. King, G.M.; Klug, M.J. Appl. Environ. Microbiol. 1982, 44, 1308-1317.
35. Henrichs, S.M.; Farrington, J.W.; Lee, C. Limnol. Oceanogr. 1984, 29, 20-34.
36. Jørgensen, N.O.G.; Lindroth, P.; Mopper, K. Oceanolog. Acta 1981, 4, 465-474.
37. Hordijk, K.A.; Cappenberg, T.E. Appl. Environ. Microbiol. 1983, 46, 361-369.
38. Ansbaek, J.; Blackburn, T.H. Microb. Ecol. 1980, 5, 253-264.
39. Schwarzenbach, R.P.; Giger, W.; Schaffner, C.; Wanner, O. Environ. Sci. Technol. 1985, 19, 322-327.

RECEIVED September 16, 1985

20

Speciation of Dissolved Sulfur in Salt Marshes by Polarographic Methods

George W. Luther III[1], Thomas M. Church[2], Anne E. Giblin[3], and Robert W. Howarth[3]

[1] Chemistry-Physics Department, Kean College of New Jersey, Union, NJ 07083
[2] College of Marine Studies, University of Delaware, Newark, DE 19716
[3] Ecosystems Center, Marine Biological Laboratory, Woods Hole, MA 02543

Polarographic techniques are capable of measuring a variety of sulfur species. These species include thiosulfate, sulfite, polythionates, sulfide, organic sulphydryl groups and inorganic and organic polysulfides. All of these can be produced in salt marsh porewaters as a result of sulfate reduction and the consequent oxidation of sulfide and sulfide minerals. Other wet and instrumental methods generally are not capable of measuring all of these sulfur species or distinguishing between these species.
    Sulfur speciation in the porewaters of two salt marshes (Great Marsh, Delaware and Great Sippewissett, Cape Cod, Massachusetts) are presented. A major finding is that the porewaters from Great Sippewissett salt marsh are dominated by inorganic sulfur species throughout the depth profile. However, Great Marsh, Delaware porewaters are dominated by organic sulphydryl species at millimolar concentrations from the surface to 12 cm and by inorganic sulfur species below 12 cm. Glutathione may be the thiol principally responsible for the field observations. The physical and geochemical characteristics of the two salt marshes differ widely and appear to be related to the differences in sulfur chemistry. There appears to be an interconversion between the inorganic and organic forms of sulfur in Great Marsh. Pyrite oxidation may be important to the formation of thiols.

Sulfur can form a number of inorganic and organic species with oxidation states varying from -2 to $+6$ and as a result undergoes a variety of biogeochemical transformations in the estuarine environment. Target compounds or ions for study in the dissolved phase

0097-6156/86/0305-0340$06.00/0

are sulfate, sulfide, polysulfide [S(-2) and S(0)], thiosulfate, sulfite, polythionates and organic compounds such as thiols. Most research over the last several years has shown that sulfur species with intermediate oxidation states are present in porewaters in substantial quantities and are important to many biogeochemical processes(1-8).

In particular, the partially oxidized forms of sulfur may be produced by a variety of reactions including the oxidation of hydrogen sulfide(1,2), the reductive dissolution of goethite(3), the oxidation of sulfide minerals(4) and microbial processes(5). Inorganic polysulfides are important to metal cycling(6) and to pyrite formation(7). Thiosulfate as well as polysulfides have been linked to energy export in salt marsh ecosystems(8-10).

The formation of organic sulfur compounds in salt marsh porewaters is not as well understood(11) as the inorganic moieties although progress has been made in coastal marine sediments(12). Their presence as thiols and alkyl sulfides has been determined but generally in trace amounts(12-16). The importance of these trace organic sulfur compounds to atmospheric sulfur emissions is presently an area of intense research(14,15). Less appears to be known about organic sulfur compounds in salt marsh porewaters.

Traditional techniques to measure dissolved sulfur in marine porewaters are iodometric titrations and colorimetric methods based on methylene blue color development. However, these methods(17-19) are not capable of measuring a wide variety of sulfur species or distinguishing between sulfur species. Major sulfur species are defined as those with concentrations typically greater than 10 uM. Iodometric titrations can determine sulfur in -1 and -2 oxidation states as shown in equations (1) through (4). However, these titrations cannot distinguish between all of the various forms.

$$2 \; S_2O_3^{2-} \quad + \quad I_2 \quad \longrightarrow \quad 2 \; I^- \quad + \quad S_4O_6^{2-} \tag{1}$$

$$S^{2-} \quad + \quad I_2 \quad \longrightarrow \quad 2 \; I^- \quad + \quad S \tag{2}$$

$$S_x^{2-} \quad + \quad I_2 \quad \longrightarrow \quad 2 \; I^- \quad + \quad x \; S \tag{3}$$

$$2 \; RSH \quad + \quad I_2 \quad \longrightarrow \quad 2 \; I^- \quad + \quad RSSR \quad + \quad 2 \; H^+ \tag{4}$$

Thus, iodometric titrations are limited to the analysis of pure standards. The popular methylene blue colorimetric techniques of Cline(17) and Gilboa-Garber(18) only measure inorganic sulfide and not organic sulfur compounds. UV-VIS spectroscopy can measure a number of organic and inorganic sulfur compounds but many of these compounds have overlapping peaks and thus cannot be distinguished. Inorganic and organic sulfur can be determined by a Raney nickel reduction of many sulfur compounds followed by colorimetric

measurement of the inorganic sulfide formed(19). However, this
method cannot distinguish the forms of sulfur readily. Therefore,
quantitative analysis of all major sulfur compounds is difficult at
best with these methods.

To date only two methods have been able to give a rather
complete speciation of major sulfur species in salt marsh
porewaters. First, Boulegue et al(20,21) have used electrochemical
titrations with UV-VIS spectrophotometry to determine organic and
inorganic sulfur species in Great Marsh, Delaware. The titrations
yield specific information on inorganic sulfide S(-2), sulfite,
thiosulfate and polysulfides and general information on organic
compounds. The major drawback to these titrations is that they are
tedious. The method also uses UV-VIS spectrophotometry to
determine thiols and polysulfides and to check on inorganic species
which are better determined by electrochemical titrations. The
analysis of inorganic and organic compounds by UV-VIS spectrophotometry is complicated by overlapping peaks and is less quantitative.

Secondly, Luther et al(22) have used polarographic techniques
to measure inorganic sulfur speciation in a New England salt marsh.
Thiosulfate, sulfite, inorganic S(-2) (as polysulfide and sulfide)
and S(0) polysulfides were measured. Polythionates and thiols were
not determined although each react at the mercury electrode(22,23).
The objectives of this paper are (1) to show the usefulness of
polarographic techniques in determining organic sulfur and its
likely structure(s), (2) to show that polarographic techniques can
provide routine field data and (3) to compare the chemistry of
major sulfur species in salt marsh porewaters from two different
salt marshes - Great Marsh, Delaware and the Great Sippewissett
Marsh, Cape Cod, Massachusetts.

Experimental

Reagents. Iodine and thiosulfate standard solutions were purchased
from Fisher Scientific Co.. Sodium sulfide was purchased from Alfa
Products and was checked for purity by differential scanning
calorimetry (DSC), X-ray diffraction (XRD) and polarographic
methods. Final standardization was by iodometric methods. Pure
sodium polysulfides were prepared and standardized as previously
described(22). Glutathione(GSH), oxidized glutathione, L-cysteine
* HCl, DL-penicillamine, L-cystine, D-penicillamine disulfide,
2-mercaptoethylamine, 3-mercaptoproprionic acid and mercaptosuccinic acid were purchased from Sigma Chemical Co.. Methanethiol, ethanethiol, 1,2-ethanedithiol and 2-mercaptoethanol were
purchased from Eastman Kodak Co.. The thiols were standardized by
iodometric methods. In the case of mercaptosuccinic acid,
iodometric titrations yielded 1.3 thiol groups per formula weight
rather than the expected 1.0 indicating an impure sample. The
disulfides were used without further purification. All solutions
were prepared with deoxygenated and deionized water in a nitrogen
filled glove bag or by using Schlenk type apparatus.

Apparatus. Polarograms were recorded with a Princeton Applied

Research (PAR) model 174A or 364 system with a model 303 static mercury drop electrode (SDME). The SDME cell assembly was modified for a SCE reference electrode as described previously(22). Prepurified argon was used to purge the cell assembly. Linear Sweep Voltammetry (LSV) was performed at a 50 or 100 mV/s scan rate from 0 to -1.5 V using the SDME in the hanging mercury drop electrode mode (HMDE). Differential pulse polarography (DPP) and sampled DC polarography (SDC) were performed with a 2 or 5 mV/s scan rate from 0 to -1.5V and a one s drop time for the SDME in the dropping mercury electrode mode.

Field sampling. Marsh cores from Great Sippewissett Marsh, Massachusetts were taken near the eastern most site and in the short Spartina zone as described in Giblin and Howarth(24). Cores were taken in August, 1984 at the beginning of senescence. Marsh cores from Great Marsh, Delaware were taken in June and August 1984 at the site described in Lord and Church(25). All cores were handled in a nitrogen filled glove box or glove bag. Sediment cores were sectioned in 3 to 5 cm intervals, pressed wih a Reeburgh squeezer and filtered as previously described(10,25). The filtered samples were placed into gas tight syringes until ready for splitting and analysis. Measurements of pH were made on aliquots of these samples.

Sulfur analytical scheme. The polarographic analysis of many of the major sulfur species which are found in porewaters has been described in detail by Luther et al(22). Tables I and II show the relevant reactions of sulfur species at the mercury electrode. Sulfur species are quantified by the method of standard additions. Briefly, the samples are split in the following manner.

The first subsample (5 mL) is analyzed by LSV or DPP to determine thiosulfate, tetrathionate, and sulfite. All samples are acidified with 1M HCL (degassed) to a pH of 5 and then purged to remove sulfide. This step is necessary to determine sulfite. Replicate measurements are always precise to ±5% and are frequently precise to ±1%.

The second subsample (2 to 5 mL) is placed in an ampoule and spiked with an excess of freshly prepared 0.01 M sodium sulfite and covered with a septum. The ampoule is later sealed under nitrogen and heated to 60°C for 2 hours. Sulfite reacts with inorganic zerovalent polysulfide, S(0), to form thiosulfate. Thiosulfate is measured again by LSV or DPP. The difference between the values of thiosulfate in this subsample and the first subsample equals the total S(0) from polysulfides.

A third subsample (25 uL to 500 uL) is added to 5 mL of a pH 10 buffer of 0.2 M nitrate/0.2 M bicarbonate as described previously(22,23). This solution is analyzed by SDC for thiols, disulfides, inorganic S(-2) and inorganic polysulfide S(0). Figure 1 shows the anodic wave for a thiol and inorganic S(-2) (negative current) and the cathodic wave for inorganic polysulfide S(0) (positive current) in two separate field samples from Great Marsh, Delaware. Figure 2 shows the anodic wave for glutathione and the cathodic wave for its disulfide in the pH 10 buffer.

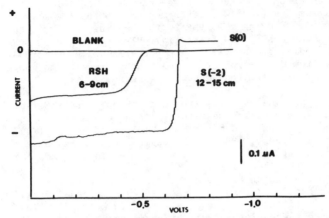

Figure 1.   SDC polarographic waves for two field samples from
Great Marsh, Delaware.

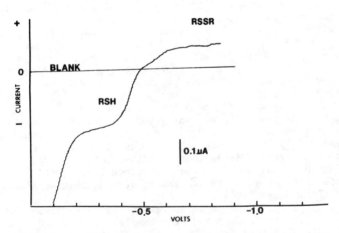

Figure 2.   SDC polarogram of glutathione and oxidized
glutathione prepared by adding iodine to glutathione.

Sulfate and chloride ion were measured by ion chromatography using a Dionex model 2010I system on samples from Great Sippewissett Marsh. On samples from Great Marsh, sulfate ion was determined gravimetrically as $BaSO_4$ and chloride ion was determined by an argentometric Mohr titration.

Results and Discussion

Methods. In a recent study([22]), polarography was used to determine inorganic sulfur species in marine porewaters as in Tables I and II. In that study, no thiols or organic polysulfides were detected

Table I. Oxidation reactions at the mercury electrode.

---

(1) $2 S_2O_3^{2-}$ + Hg $\longrightarrow$ $Hg(S_2O_3)_2^{2-}$ + $2e^-$ $E_{1/2}$ = -0.12 V

(2) $2 SO_3^{2-}$ + Hg $\longrightarrow$ $Hg(SO_3)_2^{2-}$ + $2e^-$ $E_{1/2}$ = -0.60 V

(3) SH^- + Hg $\longrightarrow$ HgS + H^+ + $2e^-$ $E_{1/2}$ = -0.68 V

(4) S_x^{2-} + Hg $\longrightarrow$ HgS + $(x-1)S$ + $2e^-$ $E_{1/2}$ = -0.68 V

(5) 2 RSH + Hg $\longrightarrow$ $(RS)_2Hg$ + $2 H^+$ + $2 e^-$ $E_{1/2}$ < -0.68 V

$E_{1/2}$ for (1) and (2) are for 0.01 mM solutions in 0.60 M NaCl.

$E_{1/2}$ for (3-5) are for 0.1 mM solutions in pH = 10 buffer.

For reaction (5) $E_{1/2}$ varies with R.

---

when the porewaters were prepared according to the third subsample described in the experimental section. In this study, a one electron reversible polarographic wave representing a thiol was observed in porewaters from the surface to 12 cm zone in Great Marsh, Delaware. This wave occurred at a half wave potential of -0.45 V (Figure 1) with an n value for the electrode process of 0.997. In addition, two samples (0-3 cm and 3-6 cm) from Great Marsh gave broad irreversible reduction waves at half wave

potentials of -0.59 V and -0.75 V. The wave at -0.59 V is similar
to organic disulfide standards (Figure 2). The wave at -0.75 V has
not yet been assigned but may be indicative of polysulfides of
composition RS_x^-. Thus, we were not able to calculate the
concentration of the species responsible for this latter wave.

Table II. Reduction reactions at the mercury electrode.

(1) $S_x^{2-} + 2(x-1)\ e^- + x\ H_2O \longrightarrow x\ OH^- + x\ SH^-$	$E_{1/2} = -0.68$ V
(2) $S_4O_6^{2-} + 2e^- \longrightarrow 2\ S_2O_3^{2-}$	$E_{1/2} = -0.32$ V
(3) $RSSR + 2e^- \longrightarrow 2\ RS^-$	$E_{1/2} = -0.54$ V

$E_{1/2}$ for (2) is for 0.01 mM solutions in 0.60 M NaCl.

$E_{1/2}$ for (1) and (3) are for 0.1 mM solutions in pH = 10 buffer.

$E_{1/2}$ for (3) is for cystine. $E_{1/2}$ and sensitivity vary with R.

In previous studies of Great Marsh, Delaware(25), the Cline
method was used to measure inorganic sulfide, and no dissolved
sulfur was detected in the 0-12 cm zone. The Cline(17) and the
Gilboa-Garber(18) methods depend upon methylene blue formation to
determine free inorganic sulfide. Because thiols have strong C-S
bonds and no free sulfide, the colorimetric methods should not
measure thiol compounds. This hypothesis was tested using cysteine
and glutathione. Neither of these compounds reacted with the
reagents to develop the methylene blue color used for inorganic
sulfide analysis. Thus, these colorimetric methods are not useful
for the study of thiols in porewaters from the marine environment.
As part of this study, an intercalibration was performed using the
colorimetric methods and the polarographic techniques for inorganic
sulfide. The results were generally satisfactory. In seven
samples from Great Sippewissett salt marsh, the colorimetric and
polarographic methods agreed within 5%. However, the polarographic
technique has a wider linear range.

Polarography of thiols. In an effort to duplicate the field
results, several thiols which are commercially available were
studied. The compounds were prepared in the pH 10 matrix which is
used for the porewater analysis. The half-wave potentials are
given for 0.1mM solutions of each compound in pH 10 buffer. Our

results for these compounds are listed in Table III. The compounds range from simple thiols and amino acids to the tripeptide glutathione, which is commonly found in plant and animal tissue at intercellular concentrations ranging from 0.1 to 10 mM(26).

Table III. Polarographic data for selected sulfur species.

Compound	$E_{1/2}$ (Volts)	µA/M
	Thiols	
1,2-Ethanedithiol	−0.759	3490
Sodium sulfide	−0.680	4960
2-Mercaptoethanol	−0.557	3327
Mercaptosuccinic Acid	−0.533	1964
D-L Penicillamine	−0.530	1743
2-Mercaptoethylamine	−0.526	2878
L-Cysteine	−0.525	2000
3-mercaptoproprionic acid	−0.492	2800
Glutathione	−0.462	1731
	Disulfides	
Cystine	−0.540	800
Penicillamine disulfide	−0.550	96
Oxidized glutathione	−0.582	3615

The half wave potentials for each compound or class is indicative of its relative structure. The reaction of each of these thiols with the mercury electrode is generally a reversible one electron oxidation of the mercury electrode (anodic depolarization of the electrode). Large negative values of the half-wave potential indicate a stronger interaction between the thiol and the mercury electrode. The highest negative potential for the compounds studied is for 1,2-ethanedithiol and shows that both sulphydryl groups form a complex with the mercury electrode over a wide range of potentials. Inorganic sulfide which forms a stable HgS film at the electrode is the next highest. As reported before(22,27), simple mercaptans with only the thiol functional group form stable mercury complexes and yield half-wave potentials near −0.58 V. Amino acids and low molecular weight thiols with at least one other functional group yield more positive half-wave potentials indicating overall weaker interactions with the mercury electrode. Glutathione yields the most positive half wave potential and the value corresponds to that of the field data from

Great Marsh, Delaware. The n value for glutathione equals 0.965.
Because the pK values of the sulphydryl groups in cysteine,
penicillamine and glutathione are similar(28), the large relative
size of the tripeptide should make the mercury complex $(RS)_2Hg$ weak
because of steric hindrance. Because glutathione is the only
commonly found tripeptide containing SH groups in nature(26), it is
likely that it is the thiol principally responsible for our field
observations.

The relative slopes from current versus concentration plots
(Table III) for these thiols show that they are analytically useful
and similar in sensitivity. Figure 3 and Table III show that the
slopes of most of these thiols is lower than that of inorganic
sulfide. The principal reason for this difference in sensitivity
is due to the n value of the electrode process for inorganic
sulfide versus thiols. The n value for inorganic sulfide is 2.0
and the n value for thiols is typically 1.0 per thiol group. As a
result, thiols have broader polarographic waves than inorganic
sulfide as shown in Figures 1 and 2. Inorganic polysulfide, [S(-2)
and S(0)], yields a similar slope per sulfur atom as that of
inorganic sulfide, S(-2). The thiols show two distinct classes of
sensitivity. The thiols with straight chain structures (e.g.
mercaptoethanol) have higher sensitivities (2800 - 3200 uA/M) than
amino acids with branched chain structures (1700 - 2000 uA/M).
Sterically hindered structures should have lower sensitivity
because of poor electron transfer at the mercury electrode. This
behavior has been noted for certain organic disulfides(29).

Minimum detectable limits (MDL) for these thiols using SDC are
about 1-2 uM and were calculated according to the method of Turner
et al(30). MDL values for thiols can be lowered an order of
magnitude if DPP is used. Because our field samples are diluted in
a buffer matrix to a maximum of 500 uL in 5 mL of buffer, the MDL
for field samples is about 10 uM. The linear range for analysis of
these thiols is from MDL to about 0.4 mM.

Polarography of organic disulfides. Organic disulfides are reduced
but react irreversibly at the mercury electrode in a manner as per
equation (3) in Table II. However, Hall(29) has shown that the n
value for a number of disulfides is typically much less than 1.0.
Values for n of 0.6 or less were found for cystine, penicillamine
disulfide and oxidized glutathione. Only in the case of oxidized
glutathione was the slope large enough to be suitable for
analytical work (Table III). In fact, oxidized glutathione is the
most soluble of these disulfides. This lowered sensitivity for
penicillamine disulfide and cystine may be related to low
solubility as well as steric hindrance(29). For penicillamine
disulfide, which has the lowest sensitivity, each carbon bound to
sulfur has two methyl groups rather than two protons, as cystine
and oxidized glutathione have. These methyl groups would prevent
electron transfer at the electrode. A better method to determine
organic disulfides would be to reduce them chemically to form the
more sensitive thiols and then to determine the increase in thiol
concentration after the reduction. Unfortunately it was not

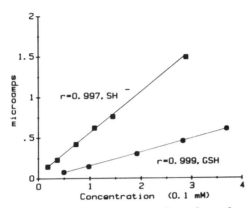

Figure 3. Current versus concentration plots from SDC polarographic data for bisulfide ion and glutathione in pH 10 buffer solution.

possible to perform this experiment because of the small volumes obtained from the porewaters.

As shown in Figures 1 and 2 and Tables I and II, polarographic methods can distinguish between many of the organic and inorganic sulfur species by the shape and the potential of each wave. However, they cannot measure organic sulphinic acids such as taurine and sulfides (R_2S), which are likely to be present in porewaters.

Field Results

Organic sulfur species. Tables IV and V show results for the field samplings from Great Marsh, Delaware and Great Sippewissett Marsh, Cape Cod, Massachusetts respectively. The major discovery is the presence of thiols in the surface to 12 cm zone of the Delaware marsh. In addition, organic disulfides are present from 0-6 cm.

Table IV. Dissolved sulfur speciation in Great Marsh, Delaware, June 1984

Depth (cm)	$SO_4^{2-}/\Delta SO_4^{2-}$ (mM)	$S_2O_3^{2-}$ (μM)	$S(0)$ (μM)	SH^- (μM)	RSH (μM)	pH
0-3	35.3/+4.63	6.6	<1	<5	25	6.20
3-6	42.0/+12.0	n.m.	<1	<5	336	5.04
6-9	54.1/+22.5	18.6	<1	<5	2411	4.87
9-12	–	30.6	<1	<5	209	6.66
12-15	18.3/-7.65	37.4	326	948	<10	6.80
18-21	19.2/-7.78	104	243	1406	<10	6.96
30-33	–	16.0	162	3360	<10	6.83

Table V. Dissolved sulfur speciation in Great Sippewissett, Massachusetts, August 1984

Depth (cm)	$SO_4^{2-}/\Delta SO_4^{2-}$ (mM)	$S_2O_3^{2-}$ (μM)	$S(0)$ (μM)	SH^- (μM)	RSH (μM)	pH
0-4	21.0/-1.10	624	<1	<5	<10	6.60
4-8	17.0/-0.75	157	104	2220	<10	7.10
8-13	15.7/-1.81	186	83	3630	<10	6.90
23-28	14.7/+0.26	172	64	2610	<10	7.00

Assuming oxidized glutathione as the disulfide responsible for the polarographic wave at -0.59 V, the disulfide concentrations were calculated as 5 uM and 139 uM for the 0-3 and 3-6 cm depths respectively. Below 12 cm, the pore water profile is dominated by inorganic sulfide. Inorganic sulfide is not present when thiols are present. The thiol concentration increases with depth until just above the onset of the inorganic sulfide zone which shows increased sulfide content with depth. The highest concentration of thiol corresponds to the pH minimum and to sulfate production as evidenced by the excess sulfate values relative to open ocean sulfate chloride ratios (ΔSO_4^{2-} in Tables IV and V).

In the Massachusetts marsh, only inorganic sulfide and thiosulfate occur as the major reduced forms of sulfur over the depth profile. Although this could be related to a lack of sulfide oxidation, sulfate depletion is not very high in this core. Therefore, neither process is the more dominant biogeochemical process. Previous studies(10,22) have not found evidence for thiols in this marsh even when sulfide oxidation was prevalent. In addition, cores were prepared from this marsh for controlled microcosm experiments. In cores in which oxidation was induced, thiols were not detected in the porewaters.

The formation of thiols appears related to the oxidaton of sulfide or sulfide minerals in Great Marsh, Delaware because of the high sulfate excesses whenever thiols are present. Conversely, the reaction of bisulfide ion with labile organic compounds such as halides and alcohols via nucleophilic displacement reactions is not likely because of the absence of bisulfide ion in the porewaters. A number of mechanisms for thiol formation using other sulfur species are plausible. Two nucleophilic displacement type mechanisms appear most promising. However, these mechanisms may not occur in each marsh because of differing physical and geochemical characteristics which are described later.

First, thiosulfate is capable of acting as a nucleophile to displace hydroxyl groups in labile compounds such as amino acids. This reaction forms thiols, hydrogen sulfate and leads to increased acidity(31). Although thiosulfate is pervasive in the Delaware core, it never reaches particularly high values in the thiol zone, indicating a possible steady state concentration. Thiosulfate could be produced when hydrogen sulfide diffuses upward into the oxidized zone (0-12 cm) where thiols are forming or by sulfide mineral oxidation. When diagenetically modeled(25), the rate of reoxidation of dissolved sulfides and sulfide minerals is greater than or equal to sulfate reduction rates in this zone. Therefore, any inorganic sulfide which is formed in, or which diffuses into this zone, should be readily oxidized.

Secondly, the zone of greatest thiol production occurs at 6-9 cm which is also the interval of optimal pyrite formation in that marsh during the winter season(25). Thus, pyrite is a possible starting material for thiol production because pyrite formation and its eventual oxidation during the growing season are important geochemical processes in most salt marshes(25,32,33). Pyrite formally contains zero and -2 valent sulfur. In the case of pyrite oxidation, the S(-2) may be transformed to a thiol and the S(0) to the fully oxidized sulfate ion. This process would be rapid if mediated by microbial catalysis(34). Roy and Trudinger(31) propose that glutathione or some other membrane bound thiol is the true enzymatic intermediate for sulfide and elemental sulfur oxidations to sulfate. Because inorganic sulfide concentrations were not detected in the 0-12 cm zone in Great Marsh, our data suggests that a thiol such as glutathione may serve as a link between the inorganic and organic sulfur pools in some salt marsh wetlands.

Inorganic sulfur species. Other differences in major sulfur species are evident between the two systems as shown in Tables IV and V. Thiosulfate concentrations in Great Sippewissett were higher by 5 to 100 times than those in Great Marsh, Delaware at similar depths. Polysulfide S(0) sulfur concentrations in Great Marsh, Delaware were higher than those in Great Sippewissett. These sulfur species were present even in the strongly reducing zone. As described above, thiosulfate can form from dissolved sulfide(1,2) and sulfide mineral oxidation(4). Also, it can form from polysulfide decomposition(1).

Polysulfides can be generated via two major pathways. First, polysulfides can be formed by the oxidation of dissolved sulfide and sulfide minerals(1,2). Second, they can be formed by the reaction of elemental sulfur with bisulfide ion(35). Polysulfide levels can be predicted for the second process as described in previous studies(22,36-38). Equilibrium calculations as described in a previous study(22) were performed for the polysulfide levels in these samples. The ratio of S(0) experimental to S(0) calculated for all samples from Great Sippewissett were 0.145 (4-8 cm), 0.137 (8-13 cm) and 0.128 (23-28 cm). Because these ratios are less than 1.0, these results indicate that polysulfides should form primarily from the reaction of bisulfide ion with elemental sulfur(5) rather than sulfide oxidation. This data set is

$$SH^- \ + \ x \ S^0 \ \longleftrightarrow \ S_x^{2-} \ + \ H^+ \eqno(5)$$

different than that on which we reported previously(22). In the previous study, the ratios were much greater than 1.0 demonstrating that polysulfides were forming primarily by sulfide oxidation reactions. The different results are attributed primarily to differences between sites in the Massachusetts marsh. The present results were observed in the eastern site where dissolved sulfide levels are typically above 1mM and pH values are near 7.0. Our previous results(22) were from the western site where dissolved sulfide levels are typically lower than 1 mM and pH values are below 6.5, further demonstrating a more oxidizing environment. For Great Marsh, Delaware the ratio was 2.74 (12-15 cm), 1.02 (18-21 cm) and 0.362 (30-33 cm). These results demonstrate that polysulfide formation via sulfide oxidation can occur at the 12-15 cm zone but at greater depths polysulfides likely form from elemental sulfur reacting with bisulfide ion.

Sulfite concentrations were observed only at depths of 20 cm

or more in both systems. In samples from Great Marsh, Delaware,
sulfite concentrations were 8.3 uM at depths of 18–21 cm and 30–33
cm. In the core from Great Sippewissett, sulfite was determined
only at the 23–28 cm depth (37.8 uM). Incomplete sulfate reduction
may be responsible for the formation of sulfite with depth.

Marsh characteristics. In the upper zone, Great Marsh appears to
act as an organic sulfur producing marsh which may be similar to
the production of organic sulfur compounds in paper pulp mills(27).
The formation of thiols and organic disulfides in Great Marsh,
Delaware but not in Great Sippewissett, Massachusetts may be a clue
to any differences between biogeochemical processes in different
salt marshes. These processes in turn may be related to a number
of marsh characteristics. First, Great Marsh is inundated by tides
only 3 days a month, creating dessicated sediments, whereas Great
Sippewissett is flooded twice daily. There is significant
groundwater intrusion in Great Sippewissett but not in Great Marsh.
Thus exchange of water and the removal of soluble labile materials
is more likely in Great Sippewissett than in Great Marsh. Second,
Great Marsh has higher salinity in the upper zone during the summer
than Great Sippewissett. Third, solid material in cores from Great
Sippewissett approaches 80% peat content and 20% inorganic content
with very little clay(10). Great Marsh contains 20% organic and
80% inorganic content which is essentially all aluminosilicate
clay(39). Clays are important natural catalysts for the synthesis
of labile organic compounds because of their high surface area and
reactive surfaces(40). Great Marsh appears to be a more saline,
closed system with higher clay content and potentially important
labile organic content. These conditions appear to be significant
for the formation of organic sulfur compounds.

Conclusions

     The half wave potentials, sign of the current for the
electrode reactions and the shape of the polarographic waves allow
inorganic and organic sulfur species to be distinguished readily.
Thus, polarographic methods are capable of providing useful field
data in a routine manner for many of the major inorganic and
organic sulfur species present in salt marsh waters. The data
derived from these methods show that differences between salt
marshes can be substantial. We believe that the polarographic
methods rather than the conventional colorimetric methods should be
used to study major sulfur species in porewaters from other salt
marshes as well. With their use, the overall importance of organic
sulfur to the biogeochemistry of salt marshes should become better
understood.

Acknowledgments

     We thank M. Cosman, J. Ortega and R. Varsolona for
polarographic measurements of organic sulfur standards. J.
Scudlark and G. Banta assisted in field sampling, core sectioning

and sulfate - chloride measurements. This work was funded by NSF grants DEB-8104701, DEB-8216376 and OCE-8201056. We also thank the Donors of the the Petroleum Research Fund, administered by the American Chemical Society, for partial support of this research.

Literature Cited

1.   Chen, K.Y.; Morris. J.C. Environ. Sci. Technol., 1972, 6, 529-537.
2.   Hoffmann, M. R. Environ. Sci. Technol., 1977, 11, 61-66.
3.   Pyzik, A.J.; Sommer, S.E. Geochim Cosmochim. Acta, 1981, 45, 687-698.
4.   Goldhaber, M. B. Am. J. Sci., 1983, 283, 193-217.
5.   Goldhaber, M. B.; Kaplan, I. R. In "The Sea"; Goldberg, E. D., Ed. Wiley: New York, 1974; Vol. 5, p.569-655.
6.   Gardner, L.R. Geochim. Cosmochim. Acta, 1974, 38, 1297-1302.
7.   Rickard, D. T. Am. J. Sci., 1975, 275, 636-652.
8.   Howarth, R. W. Biogeochemistry, 1984, 1, 5-27.
9.   Howarth, R. W.; Teal, J. M. Am. Nat.,1980, 116, 862-872.
10. Howarth, R.W.; Giblin, A. E.; Gale, J.; Peterson, B. J.; Luther, G. W. Environ. Biogeochem. Ecol. Bull.(Stockholm), 1983, 35, 135-152.
11. Balzer, W. In "Marine Organic Chemistry, Dursma, E. K.; Davison, R; Eds.; Elsevier Sci.: New York, 1983, p. 395-414
12. Mopper, K.; Taylor, B. In "Organic Marine Geochemistry", M. Sohn, Ed.; ACS SYMPOSIUM SERIES this volume, Amer. Chem. Soc.; Washington, D.C., 1986.
13. Wakeham, S. G.; Howes, B. L.; Dacey, J. W. H. Nature, 1984, 310, 770-772.
14. Steudler, P. A.; Petersen, B. J. Nature, 1985, 311, 455-457.
15. Adams, D. F.; Farwelll, S. O.; Pack, M. R.; Robinson, E. APCA Journal, 1981, 31, 1083-1089.
16. Mopper, K.; Delmas, D. Anal. Chem., 1984, 56, 2557-2560.
17. Cline, J. D. Limnol. Oceanogr., 1969, 14, 454-458.
18. Gilboa-Garber, N. Anal. Biochem., 1971 43, 129-133.
19. Kijowski, W.; Steudler, P.A. Limnol. Oceanogr., 1982, 27, 975-978.
20. Boulegue, J.; Ciabrini, J. P.; Fouillac, C.; Michard, G.; Ouzounian, G. Chem. Geol., 1979, 25, 19-29.
21. Boulegue, J.; Lord, C. J.; Church, T. M. Geochim. Cosmochim. Acta, 1982, 46, 453-464.
22. Luther III, G.W.; Giblin, A.E.; Varsolona, R. Limnol. Oceanogr., 1985, 30, 727-736.
23. Jordan,J.; Stahl, J.; Yakupkovic, J. E. "Instrumental methods of analysis of sulfur compounds in synfuel process streams". DOE/PC/40783-T9 and DOE/PC/40783-T10, NTIS, U.S. Department of Energy, 1983.
24. Giblin, A. E.; Howarth, R. W. Limnol. Oceanogr., 1984, 29, 47-63.
25. Lord III, C. J.; Church, T. M. Geochim. Cosmochim. Acta, 1983, 47, 1381-1391.

26.   Smith,E.; Hill, R. L.; Lehman, I. R.; Lefkowitz, R. J.;
      Handler, P.; White, A. "Principles of Biochemistry."
      McGraw-Hill: New York, 1983, pp. 886.
27.   Renard, J.J.; Kubes, G.; Bolker, H. I. Anal. Chem., 1975,
      47, 1347-1352.
28.   Friedman, M.; Cavins, J.F.; Wall, J.S. J. Am. Chem. Soc.,
      1965, 87, 3672-3682.
29.   Hall, M. E. Anal. Chem., 1953, 25, 556-561.
30.   Turner, J.A.; Abel, R.H. Osteryoung, R. A. Anal. Chem., 1975,
      47, 1343-1347.
31.   Roy, A.B.; Trudinger, P.A. "The biochemistry of inorganic
      compounds of sulphur." Cambridge University Press: London,
      1970, pp. 400.
32.   Howarth, R.W.; Teal, J. M. Limnol. Oceanogr., 1979, 24,
      999-1013.
33.   Luther III, G. W.; Giblin, A. E.; Howarth, R. W.; Ryans, R. A.
      Geochim. Cosmochim. Acta, 1982, 46, 2665-2669.
34.   Ehrlich, H. L. "Geomicrobiology". Marcel Dekker, Inc.: New
      York, 1981, pp. 393.
35.   Giggenbach, W. Inorg. Chem., 1972, 11, 1201-1207.
36.   Emerson, S.; Jacobs, L.; Tebo, B. In "Trace metals in sea
      water", Wong et al, Ed.; Plenum: New York, 1983, p. 579-608.
37.   Jacobs, L.; Emerson, S. Earth Planet. Sci. Lett., 1982, 60,
      237-252.
38.   Boulegue, J.; Michard, G. In "Chemical modeling in aqueous
      systems", E.A. Jenne, Ed.; ACS SYMPOSIUM SERIES No. 93, Amer.
      Chem. Soc.; Washington, D. C., 1979, p.25-50.
39.   Lord III, C. Ph. D. Thesis, University of Delaware, 1980.
40.   Pinnavaia, T.J. Science, 1983, 220, 365-371.

RECEIVED September 16, 1985

ORGANIC–INORGANIC INTERACTIONS

21

Chemical Speciation in High-Complexation Intensity Systems

Robert H. Byrne and William L. Miller

Department of Marine Science, University of South Florida, St. Petersburg, FL 33701

Theoretical and experimental developments indicate that mixed ligand complexes are important in multi-ligand systems. Consequently, the formation of ternary and higher order mixed complexes must be incorporated in the structure of multi-ligand metal speciation models. Our experimental investigation of Cu(II) complexation by oxalate and glycine is in accord with previous work in demonstrating that mixed-ligand stability constants can be reasonably assessed from theoretical considerations:

β_{11}(experimental) = $(5.0 \pm 0.5) \times 10^{12}$ and

β_{11}(theoretical) = $2(_G\beta_2 \cdot _{Ox}\beta_2)^{\frac{1}{2}}$ = 5.4×10^{12} where

β_{11} = $[CuOxG^-]/[Cu^{2+}][Ox^{2-}][G^-]$ and

$_x\beta_2$ = $[CuX_2]/[Cu^{2+}][X]^2$.

Mixed ligand Cu(II) complexes of probable importance in natural systems include ligand number two inorganic, organic and mixed organic/inorganic species. Due to the probable existence of ternary organic/inorganic metal species, trace metal complexation cannot be cleanly divided into organic and inorganic components.

Equilibrium models are widely used in assessments of trace metal bioavailability, toxicity, and transport through the environment. Properly applied, equilibrium models are powerful tools in such assessments. Due to a variety of factors, however, equilibrium modeling often falls short of its full potential. One problem, of special importance in equilibrium characterizations, is simplistic modeling. The use of simplistic chemical models is particularly important because it affects not only the modeling of complex natural systems, but also modeling of relatively simple chemical media used to generate primary thermodynamic data.

A common starting point in metal speciation schemes is the construction of a model in the form:

0097–6156/86/0305–0358$06.00/0
© 1986 American Chemical Society

$$\frac{M_T}{[M]} = 1 + \Sigma \ _j\beta_1[L_j] + \Sigma \ _j\beta_2[L_j]^2 + \Sigma \ _j\beta_3[L_j]^3 + \ . \ . \ . \ . \ . \tag{1}$$

where M_T is the total concentration of metal M, [M] denotes the free concentration of metal, $[L_j]$ denotes the free concentration of each ligand L_j, and $_j\beta_n$ denotes the cumulative formation constant for each trace metal complex $M(L_j)_n$:

$$_j\beta_n = \frac{[M(L_j)_n]}{[M][L_j]^n} \tag{2}$$

The chemical model embodied in equation 1 is flawed due to the omission of important chemical species. In chemical systems containing two or more ligand types, it is important to include the possibility of mixed ligand complex formation. As an example, appropriate descriptions of two ligand systems must include the following terms:

$$\frac{M_T}{[M]} = 1 + \ _x\beta_1[X] + \ _x\beta_2[X]^2 + \beta_{11}[X][Y] + \ _Y\beta_1[Y] + \ _Y\beta_2[Y]^2 + \ .. \tag{3}$$

$$\text{where} \quad \beta_{11} = \frac{[MXY]}{[M][X][Y]} \tag{4}$$

and brackets denote the concentration of each chemical species. Since the existence of mixed complexes is well documented (1) and the potential significance of mixed ligand species increases greatly in high ligand variety systems (2), it is quite important that the potential significance of mixed complexes is explicitly acknowledged in quantitative modeling efforts.

Statistical considerations are widely used in predictions of mixed complex stabilities (1, 3, 4). The formation of a ternary complex, MXY, from the ligand number 2 complexes MX_2 and MY_2 can be described using a stability constant of the form:

$$\frac{[MXY]^2}{[MX_2][MY_2]} = \chi \tag{5}$$

Since MXY is inherently twice as probable as MX_2 or MY_2, statistical considerations alone predict that $\chi = 4$ (1, 3, 4, 5). Accordingly, the statistically modeled mixed ligand formation constant, β_{11} is given as (1, 3, 5):

$$\beta_{11} = \frac{[MXY]}{[M][X][Y]} = 2(_x\beta_2 \cdot _Y\beta_2)^{\frac{1}{2}} \tag{6}$$

Using statistical relationships, it can be shown that mixed ligand complexes are of great potential importance even in media containing only two ligand types (5). In high ligand variety media, the vast number of possible mixed ligand complexes compels the incorporation

of statistical assessments in the general framework of metal
speciation models.

In high ligand variety systems, equation 1 can be replaced with
an equation which implicitly accounts for the formation of ternary
and higher order mixed complexes (2).

$$\frac{M_T}{[M]} = 1 + \Sigma_j \beta_1 [L_j] + (\Sigma_j \beta_2^{\frac{1}{2}} [L_j])^2 + \ldots \qquad (7)$$

The important difference between equation 1 and equation 7 is seen
in the treatment of ligand number two and higher order species found
in high complexation intensity systems. Equation 1 provides an
account of ligand number two complexes by obtaining a weighted
summation of squared free ligand concentrations. Equation 7
provides an account of ligand number two complexes by obtaining a
weighted summation of free ligand concentrations and subsequently
squaring the result.

It is important to note that although equation 7 can provide
much greater calculated $M_T/[M]$ values than equation 1, it is,
nevertheless, expected that equation 7 will provide underestimates
of the total to free metal ratio. This conclusion is based on the
observation (1) that equation 6, in the vast majority of cases,
underestimates the magnitude of mixed ligand stability constants.
In general, mixed ligand stability constants are larger, and often
considerably larger, than predictions based on single ligand type
stability constants, $_j\beta_n$ (1, 6). A simple example of this occurs
when steric factors greatly diminish formation of the complex MX_2
where X is a high molecular weight ligand, but do not limit the
formation of MXY where Y is a low molecular weight ligand. In
another example, for a four coordinated metal such as Cu^{2+}, ligand
number two MX_2 complexes with tridentate ligands, X, may be of
little importance, while the significance of mixed complexes, MXY,
involving monodentate and tridentate ligands, can be considerable.
Due to these factors and others, experimentally determined values
of χ are commonly observed between the statistically predicted
value, 4, and values of 10^3 or more (1, 6). Accordingly, while
equation 7 should be expected to provide speciation assessments
which are superior to those provided by equation 1, the assessments
provided by equation 7 should consistently underestimate the
complexation of metals as mixed ligand species.

Trace Metal Complexation in Natural Systems

Equation 7 can be used to provide insights about the nature of trace
metal speciation in high complexation intensity systems. Using
copper as an example, equation 8 provides speciation predictions in
a system of inorganic ligands, L_i, and organic ligands, L_k.

$$\qquad (8)$$

$$\frac{Cu_T}{[Cu^{2+}]} = 1 + \Sigma_i \beta_1 [L_i] + \Sigma_k \beta_1 [L_k] + (\Sigma_i \beta_2^{\frac{1}{2}} [L_i] + \Sigma_k \beta_2^{\frac{1}{2}} [L_k])^2 + \ldots$$

Expanding this equation produces the result:

$$\frac{Cu_T}{[Cu^{2+}]} = 1 + \Sigma_i \beta_1[L_i] + \Sigma_k \beta_1[L_k] + \tag{9}$$

$$(\Sigma_i \beta_2^{\frac{1}{2}}[L_i])^2 +$$

$$2(\Sigma_i \beta_2^{\frac{1}{2}}[L_i])(\Sigma_k \beta_2^{\frac{1}{2}}[L_k]) +$$

$$(\Sigma_k \beta_2^{\frac{1}{2}}[L_k])^2$$

The terms $\Sigma_i \beta_1[L_i] + \Sigma_k \beta_1[L_k]$ provide the sum contributions of ligand number one inorganic and organic complexes to the sum $Cu_T/[Cu^{2+}]$ ratio. The term $(\Sigma_i \beta_2^{\frac{1}{2}}[L_i])^2$ provides the sum contribution of all inorganic ligand number two complexes to the sum $Cu_T/[Cu^{2+}]$ ratio. The term $(\Sigma_k \beta_2^{\frac{1}{2}}[L_k])^2$ provides the contribution of all organic ligand number two complexes. The expression $2(\Sigma_i \beta_2^{\frac{1}{2}}[L_i])$ $(\Sigma_k \beta_2^{\frac{1}{2}}[L_k])$ is the complexation contribution provided by ligand number two mixed, organic/inorganic complexes. In assessing the nature of this term it is noteworthy that the magnitude of the term can be increased by increasing either inorganic or organic ligand concentrations. Accordingly, it is possible to increase the contribution of organic complexation to the $Cu_T/[Cu^{2+}]$ ratio by increasing only the concentration of inorganic ligands. The only exception to this conclusion should be obtained when ligand number one type species are of exclusive importance. In the general case where $Cu(L_i)_2$ and $Cu(L_k)_2$ species are of some significance, however slight, increasing $[L_k]^2$ will increase the concentration ratio $[CuL_iL_k]/[Cu^{2+}]$, and will thereby increase the contribution of organic ligands to the $Cu_T/[Cu^{2+}]$ ratio. In such a case, equation 9 indicates that the fraction of metal complexes involving organic ligands will decrease. However, in contrast to the predictions of equation 9, when mixed ligand stability constants are much larger than those based on statistical arguments, increasing the concentration of inorganic ligands can increase both the fraction of metal complexes involving organic ligands as well as the contribution of organic complexes to the $Cu_T/[Cu^{2+}]$ ratio.

Since the inorganic and organic chemistry of Cu(II) has been the object of considerable scrutiny, it is useful to consider in some detail the application of equation 9 to the chemistry of Cu(II) in organic rich systems. Recent work on inorganic Cu(II) speciation is sufficient to provide a reasonable assessment of the terms $_i\beta_1[L_i]$ and $_i\beta_2^{\frac{1}{2}}[L_i]$. Our assessment of the inorganic speciation terms in equation 9 is based on the formation constant results of Byrne and Miller (7) and Paulson and Kester (8) and is appropriate to 35 °/oo seawater at 25°C. Under these conditions the inorganic speciation scheme of Cu(II) in seawater at pH 8.2 can be expressed as follows:

$$1 + \Sigma_i \beta_1[L_i] = 20.03 = ([Cu^{2+}] + [CuCl^+] + [CuSO_4^o] + \tag{10}$$

$$[CuOH^+] + [CuHCO_3^+] + [CuCO_3^o])/[Cu^{2+}]$$

$$(\Sigma_i \, _{i}\beta_2^{\frac{1}{2}}[L_i])^2 \; = \; 4.42 \; = \; ((1.95)^{\frac{1}{2}} + (0.5)^{\frac{1}{2}})^2 \; = \tag{11}$$

$$([Cu(CO_3)_2^{2-}]^{\frac{1}{2}} + [Cu(OH)_2^{\circ}]^{\frac{1}{2}})^2/[Cu^{2+}]$$

The interactions of Cu(II) and dissolved organic matter are considerably more complex and consequently are less well defined. However, an assessment of the nature of organic copper complexation can be provided by treating organic complexation as an independent variable. The two unknown terms $\Sigma_k \, _{k}\beta_1[L_k]$ and $\Sigma_k \, _{k}\beta_2^{\frac{1}{2}}[L_k]$ can be reduced to one by considering the behavior which typifies stepwise complex formation (6,9):

$$\frac{K_2}{K_1} \; = \; \frac{\beta_2}{\beta_1^2} \; = \; R \quad \text{and} \tag{12}$$

$$0.2 \; \leq \; R \leq 0.3 \quad \text{monodentate ligands}$$
$$0.01 \; \leq \; R \leq 0.1 \quad \text{bidentate ligands}$$

We will take as a reasonable estimate, $R \simeq 0.03$. Using this estimate, it follows that $_{k}\beta_2^{\frac{1}{2}} = 0.173 \, _{k}\beta_1$. Consequently, equation 9 can be written as:

$$\tag{13}$$

$$\frac{Cu_T}{[Cu^{2+}]} \; = \; 24.45 + \Sigma_k \, _{k}\beta_1[L_k] + 0.727 \; \Sigma_k \, _{k}\beta_1[L_k] + 0.03 \; (\Sigma_k \, _{k}\beta_1[L_k])^2$$

Equation 13 partitions the $Cu_T/[Cu^{2+}]$ ratio into four components: (a) ligand number one and two inorganic complexes, (b) ligand number one organic complexes, (c) mixed inorganic-organic complexes, and (d) ligand number two organic complexes. The results of calculations using equation 13 and assumed values of $\Sigma_k \, _{k}\beta_1[L_k]$ between 0 and 50 are shown in Table I.

Table I. The influence of organic ligands on Cu(II) speciation in seawater.

$Cu_T/[Cu^{2+}]$ =	Inorganic (N=1, N=2)	+	Organic (N=1)	+	Mixed (N=2)	+	Organic (N=2)
24.45	24.45		0		0		0
44.8	24.45		10		7.3		3
71.0	24.45		20		14.5		12
103.3	24.45		30		21.8		27
141.6	24.45		40		29.1		48
185.9	24.45		50		36.4		75

Examination of Table I shows that as organic N=1 complexes become comparable in concentration to inorganic complexes, a large percentage of Cu(II) is present as ligand number two complexes. In addition, Table I shows that mixed organic/inorganic complexes are,

under all conditions, comparable in concentration to N=1 organic
copper complexes. The latter result highlights the conclusion that,
due to the formation of mixed ligand species, complexation by
inorganic and organic ligands should be viewed as a cooperative
rather than a competitive process.

The results shown in Table I are contingent upon the
statistical model embodied in equation 7, and the assumption $K_2/K_1 \cong$
0.03. It should be noted that a significant fraction of organic
ligands in natural aqueous systems have high molecular weights and,
as a consequence of unfavorable steric effects, K_2/K_1 for
complexation by such ligands may be very much less than the value
assumed. As such, it is reasonable to view the term $\Sigma_k \beta_1 [L_k]$ as a
summation over a low molecular weight subset of the entire suite of
organic ligands. The complexation properties of the remaining high
molecular weight system may have an interesting characteristic which
should be noted. While steric effects may greatly limit the
formation of N=2 organic complexes, such effects should be greatly
reduced for mixed complexes involving a small inorganic ligand.
Consequently, in systems of inorganic ligands and high molecular
weight organic ligands, the relative importance of mixed
organic/inorganic complexes may actually be somewhat enhanced.

Because of the substantial potential significiance of mixed
ligand species in trace metal complexation, it is important to
evaluate the quality of theoretically predicted mixed-ligand
constants for a wide variety of ligand types. Toward this
objective, we have examined the complexation characteristics of
systems containing Cu(II), glycine and oxalate ions. Systems
containing Cu(II) and bidentate ligands are particularly useful in
investigations of mixed ligand complexation because the coordination
chemistry of Cu(II) effectively precludes the formation of ligand
number three and higher order complexes. Glycine was used in this
investigation because of the occasional importance of amino acids in
natural systems (10) and because of the high affinity of amino acids
for Cu(II) (11). Oxalate was used as the second ligand type in our
investigation because oxalate is a useful analog in assessments of
the complexation properties of carbonate ions (12, 13), and because
of the abundance and importance of oxalate in the environment (14,
15).

Experimental Examination of Mixed Ligand Stability Constants

Our formation constant determinations were conducted using a copper
ion selective electrode and closely follow the potentiometric
methods outlined by Byrne and Miller (7) in investigations of cupric
carbonate complexation. The total copper concentration used in our
work was 1×10^{-4} M. Total glycine and oxalate concentrations were
0.02 M. The ionic strengths of our media were maintained at
1.0±0.02 M or 0.7±0.02 M using $NaClO_4$ as a supporting electrolyte.
Each experiment was conducted at 25°C. Free copper concentrations
were monitored using an Orion model 942900 copper specific ion
electrode, and a Ross type reference electrode (Orion No. 800500).
Measurements of pH were conducted on the free hydrogen ion
concentration scale (16, 17, 18) using an Orion model 810200
combination pH electrode. Measurements in perchlorate solutions
necessitated the use of 3 molar NaCl filling solutions rather than

KCl in order to preclude KClO$_4$ precipitates at the reference
electrodes' liquid junctions.[4] The use of two Corning model 130 pH
meters permitted simultaneous measurement of pH and free copper
concentrations. Electrode calibrations demonstrated a Nernstian
response (within ± 1% of theoretical) for both pH and copper
electrodes. Free ligand concentrations were varied by titrating our
test solutions with 1 N HClO$_4$. All measurements were obtained under
constant and subdued lighting conditions.

Free ligand concentrations in our experimental media were
calculated using equations of the form:

$$(Ox)_T = [Ox^{2-}] (1 + (_{Ox}K_2)^{-1} [H^+] + (_{Ox}K_2 \, _{Ox}K_1)^{-1} [H^+]^2) \quad (14)$$

$$G_T = [G^-] (1 + (_GK_2)^{-1} [H^+] + (_GK_2 \, _GK_1)^{-1} [H^+]^2) \quad (15)$$

where

$$_{Ox}K_2 = \frac{[Ox^{2-}] \, [H^+]}{[HOx^-]} \quad \text{and} \quad _{Ox}K_1 = \frac{[HOx^-] [H^+]}{[H_2Ox]} \quad (16)$$

$$_GK_2 = \frac{[G^-][H^+]}{[HG]} \quad \text{and} \quad _GK_1 = \frac{[HG][H^+]}{[H_2G^+]} \quad (17)$$

Oxalic acid dissociation constants were selected from the critical
compilation of Martell and Smith (19). At 25.0°C and 1 molar ionic
strength the selected results are:

$$-\log \, _{Ox}K_2 = 3.55 \pm 0.02M \quad (18)$$

$$-\log \, _{Ox}K_1 = 1.04 \pm 0.04M \quad (19)$$

Aminoacetic acid dissociation constants were directly determined
through titrametric analysis. At 25.0°C we obtained the results
shown in Table II.

Table II. Dissociation Constants of Aminoacetic Acid at 25°C

Ionic Strength	$-\log_G K_2$	$-\log_G K_1$
1.0	9.68±0.01	2.50±0.01
0.7	9.63±0.01	2.49±0.01

Two types of glycine complexation experiments were performed in
our study. Work at 0.70 M ionic strength was performed at moderate
complexation intensities within the pH range 4.6 to 2.5. Cu(II)
glycine complexation constants were determined by non-linear least
squares fits using the equation:

$$\frac{Cu(II)_T}{[Cu^{2+}]} = 1 + {}_{HG}\beta[HG] + {}_{G}\beta_1[G^-] + {}_{G}\beta_2[G^-]^2 \tag{20}$$

where $Cu(II)_T/[Cu^{2+}]$ is treated as the dependent variable, [HG] and [G$^-$] are treated as independent variables, and

$$_{HG}\beta = \frac{[CuHG^{2+}]}{[Cu^{2+}][HG]} \tag{21}$$

$$_G\beta_1 = \frac{[CuG^+]}{[Cu^{2+}][G^-]} \tag{22}$$

$$_G\beta_2 = \frac{[CuG_2^o]}{[Cu^{2+}][G^-]^2} \tag{23}$$

In a second set of experiments, results were obtained at one molar ionic strength and pH between 10.5 and 2.3. Free copper concentrations in these experiments were as low as 10^{-15} M. Due to the extreme range of copper complexation intensity in these experiments ($1 \leq Cu_T/[Cu^{2+}] \leq 10^{11}$), use of equation 20 in our data analysis produced a very strong implicit weighting of our high pH data. Under these conditions the derived β_2 value was quite well defined while the derived constants $_{HG}\beta$ and $_G\beta_1$ were so poorly defined as to be useless. This problem was alleviated by analyzing our one molar ionic strength data using the equation

$$(Cu_T) = [Cu^{2+}] (1 + {}_{HG}\beta_1 [HG] + {}_G\beta_1 [G^-] + {}_G\beta_2[G^-]^2). \tag{24}$$

Cu_T was thereby treated as the dependent variable in our analysis and the independent variables were [Cu^{2+}], [HG], and [G$^-$]. Analysis of our 0.7 M ionic strength data ($1.0 \leq Cu_T/[Cu^{2+}] \leq 100$) using equation 24 produced essentially the same results as were obtained using equation 20.

Cupric oxalate complexation experiments were analyzed using an equation identical in form to equation 24:

$$Cu(II)_T = [Cu^{2+}] (1 + {}_{HOx}\beta [HO\bar{x}] + {}_{Ox}\beta_1 [Ox^{2-}] + {}_{Ox}\beta_2 [Ox^{2-}]^2) \tag{25}$$

where

$$_{HOx}\beta = \frac{[CuHOx^+]}{[Cu^{2+}][HOx^-]} \tag{26}$$

$$_{Ox}\beta_1 = \frac{[CuOx^o]}{[Cu^{2+}][Ox^{2-}]} \quad \text{and} \tag{27}$$

$$_{Ox}\beta_2 = \frac{[Cu(Ox)_2^{2-}]}{[Cu^{2+}][Ox^{2-}]^2} \tag{28}$$

Using equation 25, our oxalate complexation data were analyzed
within the range $1.3 \leq pH \leq 4.8$. The results obtained in our analyses
of Cu(II) complexation by oxalate and glycine are shown in Table
III. The uncertainties provided with our 1.0 M glycine and oxalate
complexation results at 1.0 M ionic strength reflect, in each case,
the range of parameter estimates obtained in two experiments. The
uncertainties provided with our 0.7 M results reflect the range of
parameter estimates obtained in six experiments.

Table III. Cu(II) - glycine and Cu(II) - oxalate stability
            constants at 25.0°C.

Ionic Strength	$\log_{HG}\beta$	$\log_G\beta_1$	$\log_G\beta_2$	Source
1.0 M	1.4±0.1	8.4 ±0.1	15.36±0.01	this work
0.7 M	1.2±0.1	8.26±0.06	15.15±0.06	this work
0.5 M	–	8.14±0.02	15.0 ±0.10	Martell and Smith (19)

Ionic Strength	$\log_{HOx}\beta$	$\log_{Ox}\beta_1$	$\log_{Ox}\beta_2$	Source
1.0 M	–	5.56±0.01	9.50±0.01	this work
1.0 M	–	5.53±1	9.54±0.5	Martell and Smith (19)

The β_2 results shown in Table III provide the following
theoretically predicted mixed ligand stability constant:

$$\beta_{11}(\text{theoretical}) = 2(_{Ox}\beta_2 \cdot _G\beta_2)^{\frac{1}{2}} = 5.4 \times 10^{12} \qquad (29)$$

Our direct experimental determinations of this constant were
conducted at 25°C and 1.0±0.02 M ionic strength by methods nearly
identical to those used in our oxalate and glycine stability
constant determinations. The solutions used in our β_{11}
determinations were 0.02 M in glycine plus 0.02 M in oxalate. The
results of our titrations between pH 10.4 and 2.2 were analyzed
using the equation

$$(Cu_T) = [Cu^{2+}](1 + _{HG}\beta[HG] + _G\beta_1[G^-] + _G\beta_2[G^-]^2 + \qquad (30)$$
$$\beta_{11}[G^-][Ox^{2-}] + _{Ox}\beta_1[Ox^{2-}] + _{Ox}\beta_2[Ox^{2-}]^2)$$

The single ligand type formation constants in equation 30 were taken
from our Table III results. The CuOxG⁻ formation constant obtained
using our methods is:

$$\beta_{11} = (5.0\pm0.5) \times 10^{12} \qquad (31)$$

where the uncertainties provided encompass the β_{11} results obtained in two analyses.

Discussion

Under proper conditions, Cu(II) ion selective electrodes are well suited to extreme complexation conditions. Our glycine complexation experiments at low pH and at high pH produced very similar results in spite of extremely different extents of Cu(II) complexation between the two types of experiments.

Our experimental design entailed the investigation of Cu(II) speciation under very intense complexation conditions. Investigation of mixed ligand systems at high pH is potentially very complicated due to the probable significance of hydroxide complexes and mixed ligand hydroxide complexes. According to our experimental design, oxalate and glycine concentrations were sufficiently high to render insignificant all species involving OH^-. Consequently, the only significant complex species in our solutions were $CuOx^{\circ}$, $Cu(Ox)_2^{2-}$, CuG^+, GuG_2° and $CuOxG^-$. The relative simplicity of our experimental system was an important factor in our ability to sensitively examine the formation of the ternary complex. A second important factor was the substantial importance of $CuOxG^-$ in our experimental system. In our titrations, performed over a wide range of pH, conditions were obtained whereby $CuOxG^-$ was the dominant solution species.

Our mixed ligand formation constant result is in very good agreement with theoretical predictions. The value of χ obtained in our study is $\chi = 3.5 \pm 0.7$. Spectrophotometric determinations of this constant at 30.0°C and 0.25 M ionic strength (20), produced the estimate $\chi = 21.4$. Taken together, these studies are in good accord with the observation that experimentally determined mixed ligand stability constants are as large or larger than theoretically predicted formation constants (1, 6, 21). Our observations support the contention that mixed ligand complexation should be included in the structure of trace metal speciation assessments (e.g. equation 7). Oxalate has been successfully used as a carbonate ion analog in assessments of trace metal–carbonate ion interactions (12, 13). Consequently, our theoretical and experimental developments indicate that carbonate may play a significant role in the interactions of metal ions and organic ligands.

Summary and Conclusions

The results of our analyses reaffirm the importance of incorporating mixed ligand complexes in the structure of metal speciation models. Failure to account for such complexes, even in the relatively simple solutions used in complexation analyses, can result in very large formation constant misestimations. In high ligand variety media, failure to consider mixed complexes can result in substantial underestimation of a metal's degree of complexation. In natural media, failure to consider mixed complex formation may produce a flawed assessment of trace metal bioavailability, toxicity, and interactions with organic matter.

Participation of organic ligands in trace metal complexation may be enhanced by cooperative complexation with inorganic ligands.

Since high ligand number complexes involving hydroxide and carbonate are important in the marine speciation schemes of many metals, these ligands may be of particular importance in organic as well as inorganic complexation.

Few, if any, investigations have produced useful mixed ligand complexation data involving carbonate ions. However, to the extent that oxalate serves as a useful carbonate ion analog, the results of this work indicate that the formation of mixed carbonate complexes involving simple bidentate organic ligands is well predicted from statistical considerations.

Acknowledgments

This work was supported by the National Science Foundation (OCE-81-10162). The authors wish to thank Dr. Mark S. Shuman for his constructive criticism of this work.

Literature Cited

1. Martin, R.P.; Petit-Ramel, M.M.; Scharff, J.P. In "Metal Ions in Biological Systems"; Sigel, H., Ed.; Dekker: New York, 1973; Vol. II, Chap. 1.
2. Byrne, R.H. Mar. Chem. 1983, 12, 15-24.
3. Watters, J.I.; DeWitt, R. J. Am. Chem. Soc. 1960, 82, 1333-9.
4. Dyrssen, D.; Jagner, D.; Wengelin, F. "Computer Calculation of Ionic Equilibria and Titration Procedures"; Almquist and Wiksell Stockholm, 1968.
5. Byrne, R.H. Mar. Chem. 1980, 9, 75-80.
6. Sigel, H. In "Metal Ions in Biological Systems"; Sigel, H. Ed.; Dekker: New York, 1973; Vol. II, Chap. 2.
7. Byrne, R.H.; Miller, W.L. Geochim. Cosmochim. Acta 1985, 49, 1837-44.
8. Paulson, A.J., Kester, D.R. J. Sol. Chem. 1980, 9, 269-277.
9. Sillen, L.G.; Martell, A.E. "Stability Constants of Metal Ion Complexes"; Special Pub. No. 17, The Chemical Society; Burlington: London, 1964.
10. Henrichs, S.; Farrington J.W. Nature. 1979, 279, 319-21.
11. Martell, A.E.; Smith, R.M. "Critical Stability Constants"; Plenum: New York, 1974; Vol. 1.
12. Langmuir, D. In "Chemical Modeling in Aqueous Systems; Jenne, E.A., Ed.; American Chemical Society: Washington, D.C., 1979; Chap. 18.
13. Turner, D.R.; Whitfield, M.; Dickson, A.G. Geochim. Cosmochim. Acta 1981, 45, 855-881.
14. Graustein, W.C. Science 1977, 198, 1252-54.
15. Zutic, V; Stumm, W. Geochim. Cosmochim. Acta 1984, 48, 1493-1503.
16. McBryde, W.A.E. Analyst 1969, 94, 337-46.
17. McBryde, W.A.E. Analyst 1971, 96, 739-40.
18. Byrne, R.H.; Kester, D.R. J. Sol. Chem. 1978, 7, 373-83.
19. Martell, A.E.; Smith, R.M. "Critical Stability Constants"; Plenum: New York, 1982; Vol. 5.
20. Ramanujam, V.V.; Krishnan,U. Ind. J. Chem., 1980, 19A, 779-82.
21. Byrne, R.H.; Young, R.W. J. Sol. Chem. 1982, 11, 127-36.

RECEIVED October 11, 1985

The Adsorption of Organomercury Compounds from Seawater onto Sedimentary Phases

Cristie Dalland, Eva Schumacher, and Mary L. Sohn

Department of Chemistry, Florida Institute of Technology, Melbourne, FL 32901

Adsorption isotherms and conditional adsorption constants (Kads) were determined for the adsorption of diphenyl mercury and phenylmercuric ion (introduced as phenylmercuric acetate) from a seawater matrix onto several solid phases. Diphenyl mercury was found to adsorb onto humic acid, but no adsorption was detected onto bentonite, amorphous $Fe(OH)_3$ or hydrous MnO_2. The value of Kads was found to decrease with increasing ionic strength for the diphenyl mercury - humic acid system in seawater. As the concentraction of suspended humic acid increased, the nonlinearity of the diphenyl mercury - humic acid system became more pronounced at an increasingly lower diphenyl mercury concentration. Phenylmercuric ion adsorbed most strongly onto humic acid although adsorption onto $Fe(OH)_3$ and MnO_2 was detected and Kads values were determined. Kads values for PMA with humic acid, MnO_2 and $Fe(OH)_3$ were found to decrease with decreasing salinity.

The speciation, concentrations and residence times of dissolved substances in natural waters are dependent on many factors and processes. Important factors include temperature, pH, redox potential, ionic strength and the concentrations of other dissolved species such as organic and inorganic ligands as well as the presence of suspended particulate and colloidal matter. Important processes in addition to rate of input, and biochemical cycling include precipitation, complexation, coagulation and adsorption onto suspended particulate matter.

The focus of this study is the process of adsorption. The adsorption of substances onto solid suspended phases is a primary mechanism for the removal of dissolved matter from the water column and the consequent accumulation of many substances in sediments. Of major concern is the value of the adsorption constant (Kads) which is a measure of the strength of the adsorption interaction. Knowledge of the variation in Kads with changes in experimental condi-

tions can provide valuable insight into the mechanism of adsorption
processes.

The majority of adsorption studies deal with the interactions
between metal ions and either real or model sediment phases. Amorphous $Fe(OH)_3$, hydrous MnO_2, clays (bentonite, illite, kaolinite),
SiO_2, Al_2O_3 and organic matter (often humic substances) are the
solid phases most frequently studied.

Equilibration of hydrous oxide solid phase adsorbents with a
metal ion adsorbate is normally accomplished by mechanical shaking
of the sample over a time period extending from several hours to
several days when working with model solid phases which were prepared in the laboratory. An initial equilibration is usually
reached within one or two hours and is followed by a period of
much less extensive adsorption which may extend over a much longer
time period (1-3) However, sediments and suspended matter
isolated from natural systems seem to require longer time periods
of seven to twenty days before an accurate determination of Kads
can be made (4,5). In addition, when working with natural sediments, the major role played by adsorbed organic coatings on solid
phases must be considered (2,6).

The effect of pH on metal ion adsorption is usually quite
dramatic with the percent of metal adsorbed onto a surface increasing from near zero to a maximum, over a range of several pH units
(1-4,7) as pH increases. Increased ionic strength of the
supporting medium typically results in a decrease in Kads due to
increased competition for available surface active sites (7,8).
However, the value of Kads for some metals with natural suspended
sediments has been found to increase with increasing salinity
(conditions similar to estuarine mixing), suggesting removal of
metal ions from solution by coagulation (9). Solid phase concentration is also a critical parameter affecting Kads values. The
inverse relationship between the concentration of suspended adsorb-
ing solids and Kads is most readily explained by a solid-solid
interaction which decreases the availability of the strongest
binding sites (9). The introduction of ligands into a metal
ion-solid adsorbent equilibrium consideration can lead to
either an increase or decrease in Kads. Ligands which are not
adsorbed will decrease metal adsorption by competing with the
adsorbent for the metal, while enhancement of metal adsorption
can occur if the ligand is adsorbed and additional complexing
functional groups are still available on the adsorbed ligand
for metal ion interaction (2). The relative importance of the
various solid phases to trace metal distributions in natural
aquatic systems is dependent on the metal ion of interest. However, the fraction of metal adsorbed typically increases as the
percentage of organic matter in the sediment increases (9). The
sorption behavior of metal ions onto solid surfaces is often
described by a Langmuir isotherm. The value of Kads can be
determined from the slope of the isotherm which is usually linear
in the range of low metal ion concentrations (3).

Studies involving the adsorption of organic substances onto
solid phases have largely centered around organic pesticides
because of the environmental significance of these toxic substances. The extent of adsorption of Bromacil onto freshwater

sediments was found to correlate strongly with the percent of organic matter in the sediments while weaker adsorption onto clays and amorphous $Fe(OH)_3$ surfaces was indicated (10,11,12). In many cases, the extent of adsorption of neutral organic compounds is inversely related to solubility (13). With respect to the effect of the concentration of solid phase on Kads, results seem to be similar to those obtained in metal ion solid interactions. Kads values for the partitioning of DDT, Kepone, Lindane, and Heptachlor decrease with increasing sediment concentration. The effect of pH on the adsorption of organic substances is dependent on the pK_a of the substance considered (14,15).

The sorptive behavior of organic substances is usually best described by a Freundlich isotherm (10,13,14,16). The value of Kads can be determined from the y-intercept of a log-log plot of the modified Freundlich equation:

$$S_{ads} = K_{ads} \; C^{1/n} \tag{1}$$

S represents the concentration of the adsorbate in the solid phase, while C represents the equilibrium concentration in the bulk solution. The adsorption isotherm of diphenyl mercury and humic acid is illustrated in Figure 1.

Nature of This Study. The following represents preliminary results of an initial study on the adsorptive behavior of two organometallics. Numerous organometallics comprise the active ingredient of various pesticides. Examples include phenylmercuric salts, diphenyl mercury, triphenyl, tributyl, and tricyclohexyltin salts, as well as organoarsenic compounds (Table 1). Because of the toxicity of these substances their behavior and associations in natural water systems is a matter of concern.

Table I. Commercial and Agricultural Uses of Some Organomentallics

Compound	Use
Diphenyl mercury	Fungicide
Phenylmercury acetate	Eradicant fungicide
Phenylmercury borate	Paint mildewstat
Phenylmercury chloride	Seed treatment fungicide
Phenylmercury hydroxide	Lumber treatment fungicide
Phenylmercury iodide	Fungicide
Ethylmercury chloride	Seed treatment fungicide
Methylmercury chloride	Seed treatment fungicide
Methylmercury iodide	Mercury analysis reagent
Fentin acetate	Fungicide
Fentin hydroxide	Fungicide

Materials and Methods. Laboratory glassware was soaked in 50% nitric acid overnight and then rinsed repeatedly with distilled water. All liquid solutions were stored in polyethylene containers. Stock solutions of diphenyl mercury and phenylmercuric acetate were prepared by dissolving weighed portions of the pure solids (obtained from the U.S. Environmental Protection Agency, Pesticides and Industrial Chemicals Repository) in distilled water and were

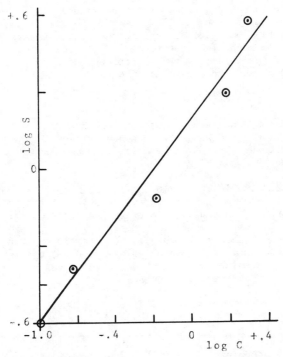

Figure 1. Freundlich isotherm for the adsorption of DPM
onto 150 ppm humic acid in 50% seawater.

stored under refrigeration for a maximum of 100 days. Both diphenyl
mercury (DPM) and phenyl mercuric acetate (PMA) are fungicides used
in limited nonagricultural areas only (17,18).

Hydrous manganese oxides and amorphous iron oxides were
prepared in the laboratory according to the methods described by
Oakley et al (3). The addition of manganese sulfate solution to a
slightly basic potassium permanganate solution produces a
suspension of hydrous MnO_2. A suspension of $Fe(OH)_3$ is produced by
simply adjusting a ferric nitrate solution to a pH of 8.0 with a
dilute sodium hydroxide solution. Both suspensions were washed
repeatedly with seawater and stored in seawater for several days.

Bentonite was obtained from Fischer Scientific. Humic acid
was obtained from Aldrich. An additional humic acid sample (BV)
which was extracted from Chesapeake Bay sediments, was used in one
set of adsorption experiments. The extraction and purification of
this sample has been described in detail elsewhere (19). Further
characterization of BV humic acid is also presented elsewhere (20).

The adsorption of diphenylmercury (DPM) and phenylmercuric ion
(PM) was studied on the solid phases described above (hydrous
manganese oxides, amorphous iron oxides, humic acid and bentonite
clay). The solid phase (5-15 mg) was added to 25-50 ml of filtered
seawater yielding solid phase concentrations of approximately 100
to 400 ppm suspended matter. The concentration of seawater was
also varied in order to study the variation of adsorptive behavior
with changes in ionic strength. The suspension was then spiked
with either DPM or PMA to yield concentrations of organometallic
which varied from 0.10 to 3.5 ppm. The range in organometallic
concentration used for this study was determined by the sensitivity
of the detection method and the solubility of DPM and PMA in
seawater.

The suspensions were then agitated on a shaker table for 48
hours. Adsorption constants which were measured after 10 days were
identical to constants measured after a 48 hour equilibration
period. Thus all adsorption measurements were made after a 48 hour
equilibration period.

Samples were then removed from the shaker table and centrifuged. Initially, filtration was employed to separate liquid and
solid phases, but the organometallics were strongly adsorbed onto
the filter paper, so this procedure was abandoned and centrifugation was adopted. The concentration of DPM or PMA remaining in
solution was measured on a Perkin-Elmer Model 460 atomic absorption
spectrophotometer employing the Mercury-Hydride System (MHS-10).
The reduction of phenylmercuric cations to elemental mercury has
previously been reported (21, 22). The average relative precision
of the method was 1.8%. Linear calibration curves for DPM and PMA
were obtained for almost the entire solubility ranges of these
compounds in seawater (Figure 2). The calibration curve for DPM
became nonlinear near the solubility limit (3.5 ppm for DPM and
>5.0 ppm for PMA). The detection limits for DPM and PMA were 0.10
ppm and 0.04 ppm respectively, in seawater. Because of the
relatively high detection limits for DPM and PMA, it was not
possible to work in the ppb or lower concentration range which
would be environmentally more relevant.

The amount of organometallic adsorbed by the solid phase was

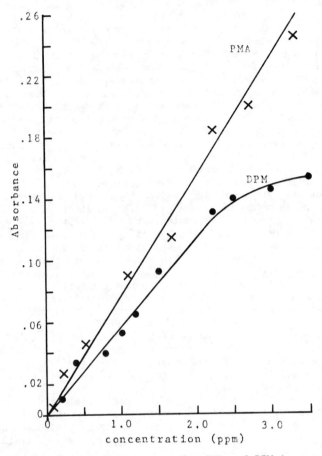

Figure 2. Calibration curves for PMA and DPM in seawater.

calculated as the difference between the spike and the amount
remaining in solution after the equilibration period of 48 hours.
Values of adsorption constants were determined graphically, as
described below.

Results and Discussion

Diphenyl Mercury Adsorption. Adsorption of DPM from seawater onto
amorphous iron hydroxide, manganese oxide and bentonite clay was
not detected in this study. A comparison of standard diphenyl
mercury solutions in seawater with identical solutions to which
sediment phase had been added and shaken for 48 hours was routinely
performed as part of the isotherm determination. There was no
significant difference in the concentration of dissolved diphenyl
mercury for standard versus standard plus solid phase for any of
the suspensions of amorphous, $Fe(OH)_3$, MnO_2, or bentonite in
seawater, implying no significant adsorption of DPM from seawater
onto these phases under the concentrations studied. If lower
concentrations of DPM could have been used (ppb or lower) it is
possible that adsorption might have been detected.

Adsorption of DPM from seawater by humic acid was recorded and
the value of Kads was determined by a simple least squares analysis
of the linear portion of the isotherm to be 1.11 for a 150 ppm
suspension of humic acid in seawater (Table II). This low value of
Kads is not unexpected, since one would expect the interaction
between the neutral nonpolar DPM molecule and the suspended humic
acid to be due to a rather weak molecular attraction.

Table II. Values of Adsorption Constants for DPM onto Humic Acid
(150 ppm) as a Function of Seawater Concentration

Kads (L/G)	% Seawater
1.11 (1.07, BV)	100
1.25	50
3.10	25
1.28	0

The effect of changing ionic strength on the value of Kads for
the DPM - humic acid system was investigated by diluting seawater
with distilled water. The pH of these systems varied from 7.0 to
7.6 and depended only on the relative amounts of seawater and
freshwater. This small variation in pH should not significantly
affect the adsorptive behavior of DPM. Schwartz (14) found no pH
dependence for the sorptive behavior of organics when these substances are predominantly in their molecular forms. The slight
decrease in pH with freshwater addition mirrors the changes which
would occur in natural brackish systems where salt and fresh water
mix. The effect of decreasing ionic strength on the value of Kads
is shown in Table II. The progression from seawater to a mixture
of 25% seawater (75% distilled water) led to an increase in Kads
for the DPM-humic acid system. However, at 100% distilled water,
the value of Kads decreased.

The trend of decreasing Kads with increasing ionic strength is

the same trend typically seen when studying metal ion adsorption
(7, 8). It is somewhat suprising that DPM, a neutral molecule,
exhibits this same trend. It had been expected that increasing
ionic strength would have had a salting out effect on the DPM,
increasing the tendency of the DPM to associate with the organic
humic phase. However, the trend from 100% to 25% seawater was just
the opposite, and suggests that DPM and the ions of seawater may be
competing for the same type of adsorption sites.

The reversal of this trend, or the decrease in Kads when one
progresses from 25% seawater to distilled water can be explained if
one considers the solubility behavior of the adsorbent, humic acid.
Humic acid was found to be fairly insoluble in solutions varying
from 25% to 100% seawater. These solutions are characterized by a
faint coloration even after the humic acid has been centrifuged
out. As mentioned earlier, filtration of these samples resulted in
retention of DPM by the filter paper, probably due to adsorption.
The coloration of the centrifugate was considerably darker for the
distilled water experiments, suggesting that the humic acid was
more soluble in this medium. Thus the increased concentration of
dissolved humic acid in distilled water led to greater solubilization of DPM and its sorptive behavior was thus modified by the
greater concentration of dissolved humic acid.

Table II lists two values for Kads at 100% SW. The bracketed
value is that evaluated using a humic acid extracted from estuarine
sediments. The sedimentary sample was obtained from the Chesapeake
Bay near the town of Bivalve (BV). Because the values of Kads were
so similar for the estuarine humic acid and the commercially prepared soil humic acid, use of the estuarine sample was discontinued
and the soil humic acid was used consistently thoughout the study.

The effect of concentration of suspended adsorbent on sorptive
behavior in a seawater matrix was studied by determining adsorption isotherms for DPM and 94, 150, 200,400, and 1000 ppm humic
acid. The corresponding values of Kads (Table III) show no
definite trend with respect to increased adsorbent concentration.
This is not totally unexpected in view of O'Connor and Connolly`s
observation that systems with low Kads values do not show dramatic
changes of Kads with changes in suspended matter concentrations
(9). Only systems with high Kads values typically show definite
decreases in Kads with increases in adsorbent concentration.

Table III.   Values of Adsorption Constants for DPM onto Humic
             Acid as a Function of Suspended Humic Acid Concentration

Kads (L/g)	Concentration of Humic Acid (ppm)
2.2	94
1.1	150
3.6	200
1.5	400
2.3	1000

Although least squares analysis of the linear portions of the adsorption isotherms do not indicate a systematic change in Kads with changes in suspended humic acid concentration, the corresponding isotherms (Figure 3) show a definite and meaningful pattern. As the concentration of suspended humic acid increases, the non-linear behavior of the isotherm is initiated at a lower and lower DPM concentration. This suggests that as the concentration of adsorbent increases, nonideal (nonlinear) behavior is attained at a correspondingly lower concentration of adsorbate (DPM). This in turn implies fewer energetically favorable adsorption sites available at higher adsorbate concentration, implying solid-solid interaction resulting in the partial elimination of the availability of certain adsorption sites (9). Thus, inspection of the adsorption isotherms (Figure 3) demonstrates a trend consistent with that seen in other studies (9) although Kads values do not mirror this.

Phenylmercuric Adsorption. Phenylmercuric ion was chosen for this study primarily because of its structural similarity to DPM, the major difference, other than the loss of one benzene ring, being the charge. The presence of the (+1) charge greatly modifies the adsorption behavior of the species with respect to DPM.

Phenylmercuric ion was introduced to the various solid phase-seawater suspensions as an aqueous solution of phenylmercuric acetate (PMA). Adsorption of phenylmercuric ion onto all solid phases studied, except bentonite clay, was noted at solid concentrations of 150 ppm (suspended). Adsorption isotherms for humic acid, hydrous MnO_2 and amorphous $Fe(OH)_3$ are shown in Figure 4. Values of Kads for phenylmercuric ion and these solid phases are listed in Table IV.

Table IV. Values of Adsorption Constants for PMA in Seawater

Kads (L/g)	Solid Phase
4.5	MnO_2
5.3	$Fe(OH)_3$
50	HA

Phenylmercuric ion showed a marked preference for humic acid (Kads = 50) although adsorption onto two of the three inorganic phases was pronounced with respect to the behavior of DPM. The charge of the phenylmercuric ion allows interactions to occur with adsorption sites on MnO_2 and $Fe(OH)_3$, which were not reactive towards DPM.

The effect of changing ionic strength on the value of Kads for PMA adsorption was also investigated. Decreasing the salinity of the humic acid suspension, led to a very pronounced decrease in Kads. In 100% seawater, Kads was evaluated as 50 L/g (Table IV) while at 25% seawater Kads dropped to a value of 1.2. Decreasing salinity of the MnO_2 and $Fe(OH)_3$ systems yielded similar results. At 25% and 50% seawater, no adsorption of PMA onto either phase was detected. These results are similar to the results of Li et al. for mercuric ion adsorption (23). In that study, a decrease in the adsorption constant for mercuric ion onto sedimentary phases with

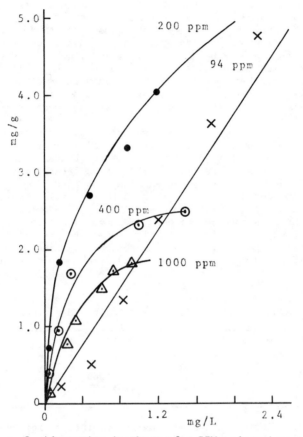

Figure 3. Adsorption isotherms for DPM and various concentrations of suspended humic acid in seawater.

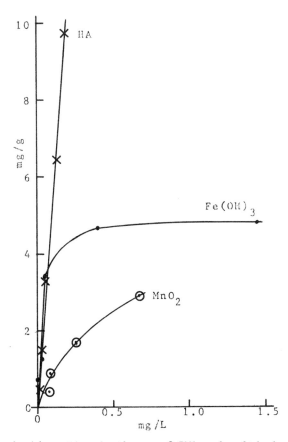

Figure 4. Adsorption isotherms of PMA and solid phases in seawater.

decreasing salinity was described. Thus the adsorptive behavior of
PMA (a charged ionic species) with respect to salinity changes, was
found to be somewhat similar to that of mercuric ion.

Conclusions. Diphenyl mercury, a neutral organometallic compound,
which does not contain any markedly acidic or basic functional
groups was found to adsorb only onto humic acid. No sorptive
behavior could be detected with respect to bentonite, MnO_2 or
$Fe(OH)_3$. The rather low value of Kads indicates a simple molecular
attraction. However, the decrease in Kads with an increase in
ionic strength indicates that DPM is competing with the metal ions
in seawater for adsorption sites. The increasing nonlinearity of
adsorption isotherms with increasing suspended humic acid concentration is also similar to results obtained for metal ion adsorption.
     Phenylmercuric ion was found to associate most strongly with
humic acid, although adsorption onto MnO_2 and $Fe(OH)_3$ was appreci-
ciable. Adsorption onto bentonite was not detectable for either
organometallic.
     Although further studies are needed on the behavior of organo-
metallic pesticides, the results of this study indicate that
neutral organometallics can be expected to associate with organic
phases such as humics. Charged organometallics, such as phenylmercuric ion associate most strongly with organic phases, although
adsorption onto some inorganic phases does occur. Obviously,
additional research is needed to generalize about the adsorptive
behavior of organometallics other than organomercurials. Adsorption studies on organotin compounds are currently in progress.

Acknowledgments

Acknowledgement is made to the Donors of the Petroleum Research
Fund, administered by the American Chemical Society, for the
support of this research. The authors also wish to thank Rolf Sohn
for preparation of figures.

Literature Cited

1. Benjamin, M.M.; Leckie, J.O. J. Colloid Interface Sci 1981, 79,
   209-221.
2. Davis, J.A.; Leckie, J.O. Environ. Sci. Technol. 1978, 12, 1309
   1315.
3. Oakley, S.M.; Nelson, P.O.; Williamson, K. J. Environ. Sci.
   Technol. 1981, 15, 474-480.
4. Balistrieri, L.S.; Murray, J. W. Geochim. Cosmochim. Acta
   1984, 48, 921-929.
5. Li, Y.; Burkhardt, L., Buchholtz, M.; O'Hara, P., Santschi,
   P.H. Geochim. Cosmochim. Acta 1984, 48, 2011-2019.
6. Balistrieri, L. S.; Brewer, P.G.; Murray, J.W. Deep-Sea Res.
   1981, 28A, 101-121.
7. Davies-Colley, R. J.; Nelson, P.O.,; Williamson, K.J. Environ.
   Sci. Technol. 1984, 18, 491-499.
8. Gaudette, H.E.; Grim, R.E.; Metzger, C. F. Amer. Mineral. 1966,
   51, 1649-1656.

9. O'Connor, D. J.; Connolly, J.P. Water Res. 1980, 14, 1517-1523.
10. Corwin, D. L.; Farmer, W. J. Environ. Sci. Technol. 1984, 18, 507-514.
11. Lambert, S.M.; Porter, P.E.; Schieferstein, R.H. Weeds 1965, 13, 185-190.
12. Lambert, S.M. J. Agr. Food Chem. 1968, 16, 340-343.
13. Hague, R.; Freed, V.H. Residue Rev. 1974, 52, 89-116.
14. Schwartz, H.G. Environ. Sci. Technol. 1967, 1 332-337.
15. Weber, J.B. Amer. Mineral. 1966, 51, 1657-1670.
16. Lotse, E.G.; Graetz, O.A.; Chesters G.; Lee, G.B.; Newland,L.W. Environ. Sci. Technol. 1968, 2, 353-357.
17. "Analytical Reference Standards and Supplemental Data," Environmental Protection Agency, 1984.
18. Meister, T.T. "Farm Chemicals Handbook 1977"; Meister Publishing Co.; Willoughby, OH, 1977; p. D 212.
19. Sohn, M.L.; Hughes, M.C. Geochim.Cosmochim.Acta 1981, 45, 2393-2399.
20. Sohn, M.L. Org.Geochem. 1985, 8, 203-206.
21. Toffaletti, J; Savory, J. Anal. Chem. 1975, 47, 2091.
22. Braman, R.S. In "The Hydrolysis of Cations"; Baes, C.F., Jr.; Mesmer, R.E., Eds.; Wiley-Interscience: N.Y., 1976; p. 37.
23. Li,Y.; Burkhardt, L.; Teraoaka, H. Geochim.Cosmochim.Acta. 1984, 48, 1879-1884.

RECEIVED September 23, 1985

23

Effects of Humic Substances on Plutonium Speciation in Marine Systems

G. R. Choppin, R. A. Roberts[1], and J. W. Morse[2]

Department of Chemistry, Florida State University, Tallahassee, FL 32306-3006

The dominant oxidation state of plutonium in the dissolved phase of seawaters has been shown to be Pu(V). Data are presented that indicate a significant role by humic materials which cause rapid reduction of Pu(VI) to Pu(IV). The latter leaves the solution phase via hydrolysis. The humic material also seems to reduce Pu(V) but at a much slower rate and, in sunlight, this reduction may be negated by an oxidation process of unknown origin at this time. The role of humics sorbed on suspended particulate and sedimentary matter is discussed.

<u>Sources and Amounts of Plutonium in the Environment.</u> Since 1945 approximately 3300 kg of plutonium has been injected into the environment, mostly (>90%) from atmospheric explosions of nuclear weapons. This corresponds to about 380 kCi total alpha radioactivity. The addition to this amount by releases from nuclear power operations is much smaller; the major continuing addition is ca. 0.1 kCi per month released to the Irish Sea from the British nuclear reprocessing plant at Windscale. About 2/3 of the plutonium from nuclear explosions would be formed into highfired oxides which would be rather inert chemically. However, the remainder, created during the explosion as single atoms via the $^{238}U(n, \gamma)^{239}U(2\beta^-)^{239}Pu$ reaction sequence, should be more reactive and behave similarly to that released from reprocessing plants or nuclear waste repository sites.

Sampling of filtered water samples of the Pacific Ocean indicates a concentration of ca. $2 \times 10^{-17}M$ (i.e. 10^{-3} dpm L^{-1}) (<u>1</u>). These values, however, are open to question as the plutonium associated with suspended particulates may be more than an order of magnitude greater than that in true solution. For example, for the

[1]Current address: Mallinckrodt Medical Products Research and Development, St. Louis, MO 63134.

[2]Current address: Department of Oceanography, Texas A&M University, College Station, TX 77843.

Mediterranean Sea, the plutonium activity in unfiltered water samples was twenty five times greater than that of filtered samples (2). In experiments in which plutonium was added to seawater, the solubility of ionic species after 30 days was found to be 5×10^{-12}M (3). Addition of relatively large quantities of humic material in these experiments increased the plutonium solubility sixfold.

Properties of Plutonium. As nuclear power utilization with its associated reprocessing and waste disposal operations expands, the fate of any released plutonium assumes greater importance. However, there is another rationale for studying the environmental behavior of plutonium. Because of its solution chemistry, it is an element with rather unique qualities as a probe of environmental properties.

Plutonium has four oxidation states, VI, V, IV, and III, all of which can exist in aqueous solution within the E_h and pH range found in nature. Under the proper conditions, all four states can coexist, although one or more of the states is usually favored. The III and IV oxidation states generally exist as hydrated or complexed cations. The V and VI states, on the other hand, exist as PuO_2^+ and PuO_2^{2+} - the dioxo cations - also hydrated or complexed. Normally, acidic conditions stabilize the lower oxidation states while more basic conditions favor the higher states.

Distribution of plutonium between its different oxidation states is, of course, dependent not only on pH. Complexation, for example, can change the relative stabilities of the different states. The various oxidation states differ in their ability to form complexes, with Pu(V) exhibiting the weakest complexation. Pu(VI) and Pu(III) are rather similar in complexation strength although Pu(VI) is usually stronger than Pu(III) for most ligands. Pu(IV) forms the strongest complexes. Hydrolysis, including hydrolytic polymerization, can also play an important role in stabilizing one oxidation state over another. The effects of redox potentials, complexation, hydrolysis, etc. can combine to give the possibility of more than one oxidation state coexisting under the same conditions which complicates the study of the environmental behavior of plutonium but also can serve to define the E_h, etc. of systems. The relative tendencies of various species of the different oxidation states to form colloids or to sorb to particulates add further dimensions to the chemical behavior of plutonium.

A topic which requires further study is the effect of organics on the behavior of plutonium in the marine environment. Based on a bioassay method, Fisher, et al. (4) have suggested that marine humic and fulvic acids produce no substantial complexation of transuranic elements in the oceans. However, Dahlman, et al. (5) support the idea of complexation by humic and fulvic acids and provide experimental evidence of reduction of Pu(VI) to Pu(IV) by fulvic acid. Complexation is also supported by the work of Pillai and Mathew (3), although the concentrations of organics used in their studies were unrealistically high.

Previous Studies. Studies from the Irish Sea (6) and the Pacific
Ocean (7) show that the soluble form of plutonium is predominantly
in the oxidized forms (V and VI states) with evidence that V is
the more important state. However, modeling calculations with experimental complexation and redox data led Aston (8) to conclude
that Pu(VI) is the dissolved form (no effect of organic material
was included). In a recent extensive review of the geochemistry
of plutonium in water environments, Sholkovitz (9) concluded the oxidation state distribution in waters requires more investigation.
In this paper we report the result of studies designed to clarify
the dominant oxidation state of plutonium in sea water and the role
of humic material.

Experimental

The stability of Pu(VI) and Pu(V) was studied in solutions of standard artificial (same mineral constitution) seawater and of true
seawater from the Gulf of Mexico (collected ca. 5 miles from the
Florida coastline below Tallahassee). All solutions were buffered
to pH 8.0 by addition of trishydroxymethylaminomethane ("tris") and
filtered by vacuum through 0.45 micron Millipore filters. The glassware, pipets, etc. used were treated to reduce adsorption losses of
Pu at pH 8 by a method developed in this laboratory (10).
        The humic acid was recovered from Bahamian marine sediments
obtained from sites in less than 5 m water depth and at least 0.5
miles from shore. The isolation and purification procedures have
been described earlier (11).
        Pu(VI) was prepared by oxidation of acidic stock ^{238}Pu or ^{239}Pu
tracer solutions with $KMnO_4$. The oxidized tracer solution was neu-
tralized with NaOH prior to addition to the experimental solution.
        Pu(V) was prepared by photolysis of a thenoyltrifluoracetone
(TTA) solution as described previously (12). The Pu(V) was stripped
from the organic-TTA solution directly into the experimental aqueous
solution. The concentration of ^{239}Pu was ca. 10^{-7}M in all experiments while that of ^{238}Pu was ca. 10^{-9}M.
        Ten ml of the test solutions were kept in treated screw cap
glass vials and 0.500 ml aliquots withdrawn periodically. Plutonium
was separated by oxidation state by a solvent extraction method (13)
and the alpha activity counted with a liquid scintillation counter
using "Handifluor" (Mallinckrodt) as the cocktail.

Results

When Pu(V) was added to either artificial seawater or to real seawater, essentially no change in oxidation state was observed over
the 120 hours of these experiments (Figure 1). Preliminary data
from longer term experiments which have continued over 800 hours
indicate a slow reduction of Pu(V) in seawater when kept in the dark
but little or no reduction when exposed to normal daily sunlight
(14).
        Figure 2 shows data for addition of Pu(VI) to artificial sea-
water with different amounts of humic acid. In all solutions, the
Pu(VI) was reduced rapidly to Pu(V) + Pu(IV) with the ratio of the
V/IV decreasing with humic acid concentration. The Pu(IV) is not
found in solution but can be recovered quantitatively from the walls

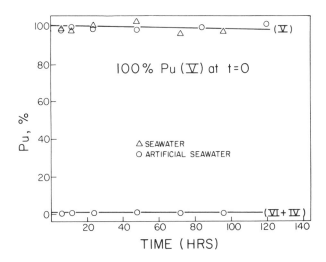

Figure 1.   Variation of concentration of Pu(V) with time in artificial and true seawater at pH 8.0 and 25°C; at t = 0, [Pu(V)] $\tilde{} \ 10^{-7}$M.

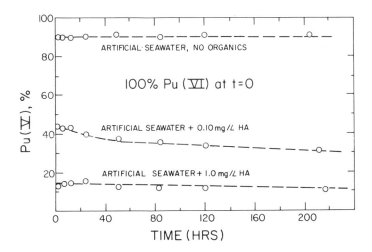

Figure 2.   Percentage of Pu(V) of original ca. 10^{-7}M Pu(VI) added to artificial seawater with 0, 0.1 and 1.0 mg/L humic acid at pH 8.0 and 25°C.

by washing the empty vials with acid. Very similar results were
obtained when 0.7M NaCl solution at pH = 8.0 was used instead of
artificial seawater. In true seawater, the Pu(VI) was reduced
rapidly to ca. 30% Pu(V). Humic material is reported to be 6-30%
of the DOC of seawater, (15), indicating a range of 0.1-1.0 mg/L as
reasonable. The reduction to 30% Pu(V) in seawater compares well
with the reduction observed in artificial seawater for such a range
of humic concentration.

Discussion

The data in Figures 1 and 2 are consistent with the existence of
Pu(V) as the dominant oxidation state in solution in seawater. It
is not possible to state whether this represents thermodynamic equilibrium or some balance of opposing redox conditions which results in
a steady state concentration of Pu(V). Pu(VI) can be rapidly reduced to Pu(V). However, in the presence of humic materials a significant fraction of the Pu(VI) is reduced directly to Pu(IV) which
hydrolyzes and sorbs to the walls and to particulate matter. Apparently there is competition between reduction of Pu(VI) to Pu(V) by
seawater and complexation of Pu(VI) by humic acid (16). The latter
results in rapid reduction of Pu(VI) to Pu(IV) with subsequent hydrolysis. Complexation of Pu(V) by humic acid should be much weaker
and may account for the slow reduction of Pu(V) in the dark. Photo-
lysis of the humic material apparently results in some oxidation of
Pu(IV) to Pu(V), to provide a metastable Pu(V) concentration.

In an experiment with pore water (pH 7.3) squeezed from the
Bahamian carbonate sediments from which the humic acid was extracted,
we observed a significant reduction of Pu(V) with time. This observation would be consistent with a higher humic acid concentration in
such pore water. It is in agreement with the observation that reduced forms (i.e. Pu(IV)) are predominant on particulate and sedimentary matter (5).

Observations that particulate organic matter has strong adsorption properties and that dissolved organic compounds often are adsorbed on inorganic phases to such an extent that they alter the
surface properties of these phases have led to increased interest in
the role played by organic matter on the sorption of dissolved compounds from seawater (17-29). Recently, interesting results have
been reported for the sorptive properties of heterogeneous inorganic-
organic solids (e.g., iron oxyhydroxide-humic acid) that may closely
mimic natural particulate matter properties (19,21,30-32). The
influence of adsorbed organic compounds on the sorptive behavior of
inorganic substrates is not the same for all sorbates (e.g., adsorption of copper is often strongly influenced, while cadmium adsorption is usually not significantly influenced) (19).

These new findings have added to the complexity of understanding
and modeling the interaction of dissolved phases with particulate
matter in marine waters, but the evidence is increasing that we
are now capable of making reasonable quantitative predictions for
the behavior of many trace metals in the deep ocean (21). To date,
no direct studies have been made of the influences of organics, such

as humic and fulvic acids, on the sorptive behavior of Pu on particles coated with these compounds. However, based on the strong
interactions of Pu with dissolved humic acids, it is quite likely
that they can also significantly influence both adsorption and
surface redox reactions.

Acknowledgments

We acknowledge support of this research through a contract with the
USDOE Office of Health and Environmental Research.

Literature Cited

1. Choppin, G. R. In "Environmental Inorganic Chemistry";
    Martell, A. E.; Irgolic, K. J., Eds.; VCH Publ. Inc.:
    Deerfield Beach, Fl.; 1985; pp. 301-320.
2. Beasley, T. M.; Cross, F. A. In "Transuranium Elements
    in the Environment"; Hanson, W. C., Ed.; DOE/TIC-22800,
    National Tech. Inform. Serv.: Springfield, Va.; 1980;
    pp. 524-540.
3. Pillai, K. C.; Mathew, E. In "Transuranium Nuclides
    in the Environment"; Inter. At. Ener. Agency, STI/PUB/410:
    Vienna; 1976; pp. 25-44.
4. Fisher, N. D.; Bjerregaard, P.; Huynh-ngoe, L.; Harvey, G. R.
    Mar. Chem. 1983, 13, 45-56.
5. Dahlman, R. C.; Bondietti, E. A.; Eyman, E. D. In "Actinides
    in the Environment"; Friedman, A. M., Ed.; ACS SYMPOSIUM SERIES
    No. 35, American Chemical Society: Washington, D.C., 1976;
    pp. 47-80.
6. Nelson, D. M.; Lovett, M. B.; Nature 1978, 276, 599-601.
7. Nelson, D. M.; Carey, A. E.; Bowen, V. T. Earth Planet
    Sci. Lett. 1984, 68, 422-430.
8. Aston, S. R. Mar. Chem. 1980, 8, 319-325.
9. Sholkovitz, E. R. Earth-Sci. Rev. 1983, 19, 95-161.
10. Caceci, M.; Choppin, G. R. Radiochem. Acta 1983, 33, 113-114.
11. Bertha, E. L.; Choppin, G. R. J. Inorg. Nucl. Chem. 1978,
    40, 655-658.
12. Saito, A; Roberts, R. A.; Choppin, G. R. Anal. Chem. 1985,
    57, 390-391.
13. Bertrand, P. A.; Choppin, G. R. Radiochem. Acta 1982,
    31, 135-137.
14. Kobashi, A.; Choppin, G. R., unpublished data.
15. Stumm, W. and Morgan, J. J. "Aquatic Chemistry" 2nd Ed.;
    John Wiley & Sons, Inc.: New York, 1981, p. 514.
16. Shanbhag, P. M.; Choppin, G. R. J. Inorg. Nucl. Chem.
    1981, 43, 3369-3372.
17. Davis, J. A. In "Contaminants and Sediments"; Ann Arbor
    Science: 1980; Vol. II.
18. Davis, J. A. Geochim. Cosmochim. Acta 1982, 46, 2381-2393.

19. Davis, J. A. Geochim. Cosmochim. Acta 1984, 48, 679-691.
20. Hunter, K. A. Limnol. Oceanog. 1980, 25, 807-822.
21. Hunter, K. A. Deep-Sea Res. 1983, 30, 669-675.
22. Hunter, K. A.; Liss, P. S. Nature 1979, 282, 823-825.
23. Hunter, K. A.; Liss, P. S. Limnol. Oceanogr. 1982,
    27, 322-335.
24. Loeb, G. I.; Neihof, R. A. J. Mar. Res. 1977, 35, 283-291.
25. Neihof, R. A.; Loeb, G. I. Limnol. Oceanogr. 1972,
    17, 7-16.
26. Neihof, R. A.; Loeb, G. I. Mar. Res. 1974, 32, 5-12.
27. Tipping, E. Chem. Geol. 1981a, 33, 81-89.
28. Tipping, E. Geochim. Cosmochim. Acta 1981b, 45, 191-199.
29. Tipping, E.; Cooke, D. Geochim. Cosmochim. Acta 1982,
    46, 75-80.
30. Balistrieri, L. S.; Murray, J. W. Geochim. Cosmochim. Acta
    1984, 48, 921-929.
31. Davis, J. A.; Leckie, J. O. Environ. Sci. Technol.
    1978b, 12, 1309-1315.
32. Tipping, E.; Griffith, J. R.; Hilton, J. Croatica Chem.
    Acta 1983, 56, 613-621.

RECEIVED September 16, 1985

The Interaction of Trace Metal Radionuclides with Humic Substances

Ljerka Musani-Marazović[1], Danielle Faguet[2], and Zdenka Konrad[1]

[1] Center for Marine Research Zagreb, "Rudjer Bošković" Institute, Zagreb, Yugoslavia
[2] Centre de Recherches de Sedimentologie Marine de Perpignan, France

The fate of trace metal radionuclides in the aquatic
environment and their participation in the biogeochemical cycle depend strongly on the chemical and physico-
-chemical form in which radionuclides are introduced in
natural waters. The abundance of natural humic substances
and their ability to form metal complexes and to adsorb
on suspended matter and sediment makes these substances
especially important in transport, availability and accumulation of trace metal radionuclides in natural water
environments. In that sense complexation of di- and
tri-valent metal radionuclides with humic and fulvic
acids of different origin was studied. The sorption
properties of natural suspended matter and undissolved
humic acid for the sorption of some radionuclides was
also studied.

The toxicity and fate of metals in natural waters and their participation in biogeochemical cycles depend strongly on the physico-
-chemical forms of metals entering natural waters. Particularly important is the extent of organic complexation of trace metals. In
recent years special interest has been paid to dissolved organic matter, a part of which consists of humic substances. Humic materials
have been widely presumed to be important for organic complexation of
metals. From that point of view it is of special importance to determine and characterize the species of metals actually present in
aquatic systems. Dissolved and colloidal humic material enter the
marine environment by river input into estuaries and are produced by
biological and chemical processes occurring in the sea. According to
Mantoura and Riley (1), humic substances are formed by random condensation of breakdown products of dead organisms and of extracellular
metabolites of phytoplankton. Harvey et al. (2) propose that marine
humic substances are formed from either light-induced oxidative
cross-linking of two or more polyunsaturated fatty acids or from
polyunsaturated glycerides.
    Evidence of the formation of metal complexes with humic sub-
stances in natural waters has been reported by many authors (3-12).
The same applied to the stability constants of the above complexes

0097-6156/86/0305-0389$07.00/0
© 1986 American Chemical Society

(13-18). Tremendous work has been done on the characterization of
humic substances in the aquatic environment (2,19,20) and much analytical data on the concentration of humic substances in natural
waters is reported (2,21-25). Special attention has been paid to the
physico-chemical processes governing behaviour of metals and dissolved organic matter during estuarine mixing (6,26-28).

   In our work we concentrate on the interactions between di- and
tri-valent metal radionuclides which enter estuarine and seawater in
ionic forms and humic and fulvic acids of different origins. However,
as humic substances react with metals, not only in dissolved but also
in undissolved states, we report some data on the adsorption of trace
metal radionuclides on suspended humic acids.

Materials and Methods

Electrophoretic experiments were performed with a high voltage device
(Flat Plate Electrophoresis, Savant Instruments Inc., USA). Temperature was maintained at 20 ± 0.1 ^{o}C with an ultrathermostat. The experimental conditions for high voltage paper electrophoresis were:
basic electrolyte: 0.45 um filtered 10% seawater (sampled in Mid-
-Adriatic, original salinity 37.4%o) and estuarine water, the Krka
estuary (salinity 2%o) with or without humic or fulvic acids; pH
adjusted to 8.0; Whatman 3 MM filter paper with a free length and
width of 85×1.5 cm; voltage 1000 V; current per strip of paper:
2.0-3.9 mA (10% seawater) and 1.3-1.5 mA (estuarine water); current
per 1 cm width of the paper strip: 1.3-2.6 mA (10% seawater) and
0.9-1 mA (estuarine water); specific electric effect on the paper
strips: 0.016-0.031 VA cm^{-2} (10% seawater) and 0.010-0.012 VA cm^{-2}
(estuarine water); duration of an experimental run: 102 min. The
radionuclides ^{54}Mn, ^{55}Fe and ^{51}Cr were from N.E.N. (New England
Nuclear), USA. The specific radioactivities were as follows: ^{54}Mn:
10 mCi ml^{-1}, concentration of Mn: 1.27 ug ml^{-1}; ^{55}Fe: 10 mCi ml^{-1},
concentration of Fe: 4.1 ug ml^{-1}; ^{51}Cr: 100 uCi ml^{-1}, concentration
of Cr: 0.62 ug ml^{-1}.

   The samples for electrophoresis were prepared in the following
way. An aliquot was evaporated to dryness. The residue was dissolved
in estuarine or diluted seawater adjusting a specific radioactivity
of 0.45 uCi/20 ul (^{54}Mn); 0.22 uCi/20 ul (^{55}Fe) and 2 uCi/20 ul
(^{51}Cr). This corresponds to a Mn concentration of 5.3×10^{-8} M, an Fe
concentration of 7.9×10^{-8} M, and a Cr concentration of 1.19×10^{-5} M.
Twenty microlitres of those solutions were applied to the starting
point of the electrophoretic strips which were wetted previously with
the respective solvent containing the respective humic/fulvic acid
concentration. This caused a dilution of the respective trace metal
concentration in the liquid phase on the electrophoretic paper strip
by a factor of 3.355. For experiments in the presence of humic/fulvic
acid, the respective humic/fulvic acid concentration was adjusted in
the test solution, and from the latter, after 30 min, an aliquot of
20 ul was subjected to electrophoresis. This measurement was termed
0 day. The measurements were extended for each adjusted humic/fulvic
acid concentration over a period of 0 to at least 7 days (maximum 30
days) to detect possible aging effects. Experimental conditions for
^{65}Zn, ^{109}Cd and ^{210}Pb were described earlier (8).

   The humic and fulvic acids used in our experiments were from
different sources. Humic acid (HAL sample) was isolated from sediment

in the shallow waters of the Lim Channel (the Adriatic Sea) by a procedure reported in the literature (3,29). Humic acid was extracted
from wet sediment with a mixture of 0.2 M NaOH and 0.2 M Na_2CO_3 at
80°C for 12 h, purified by repeated precipitation with HCl, redissolved in NaOH, dialysed and finally Chelex-100 was used to remove
bound metals. Then it was again precipitated with HCl, dried at 40°C
and stored as crystalline powder in the acidic form. The HAN humic
acid sample was isolated from the bottom sediment of the deep Nor-
wegian Sea (1 m below the sediment surface). The HAM humic acid
sample was isolated from the sediment of a tropical estuary at
Mahakam (Borneo). The FAC and HAC samples of fulvic and humic acids
were isolated from a Mediterranean lagoon (Canet, near Perpignan,
France), and the HALR sample was isolated from deposit of Ruppia
maritima, Linneaus from the same lagoon. The HAN, HAM, HAC, FAC and
HALR samples were extracted from the sediment with a mixture of 0.1 M
NaOH and 0.1 M $Na_4P_2O_7$ according to the procedure of Kononova and
Balachirova (30). Those samples were purified by repeated precipita-
tion with HCl, redissolved in NaOH solution, dialysed and finally
passed through the Dowex 50W-X8 resin.

Stock solutions of all samples were prepared in diluted NaOH.
The concentrations of humic and fulvic acids varied from 3 to 200 mg
l^{-1} and the pH was adjusted to pH 8.0 by the addition of NaOH or HCl.

The behaviour of radionuclides was followed up by measuring
their electrophoretic mobility (u) ($cm^2 V^{-1} s^{-1}$) (31) and by evaluating
the amount of the respective radionuclide in the three observable
zones, i.e. cationic, anionic, and the immobile zone at the starting
point. The electrophoretic zones could be detected conveniently by
autoradiography exposing an X-ray film overnight to the electrophoretic strips. The respective zones were cut from the paper and
counted in a β-liquid scintillation counter or in a γ-counter.

Some characteristics of humic and fulvic acid samples. Unfortunately, we were not able to perform some desired analyses of the samples; in the first place because of very small quantities in which
some of the samples were available, and secondly, because some samples
were not soluble enough to do molecular size distribution analyses.

The results of the elemental analyses of the humic substances
are presented in Table I.

According to Nissenbaum and Kaplan (33) the C/N ratios from most
humic acids of marine origin fall into the range 9.4 - 14.2. Our
samples named HAL, HAN, HAC might be of marine origin.

In Table II total acidity and carboxyl groups analyses are given
for some of the samples used.

For some of the samples adsorption spectra were recorded (11).
The ratio between the optical densities at 465 and 665 nm has been
used as an indicator of the condensation degree of humic substances
(34). The lower the E_4/E_6 ratio the higher the condensation degree of
the humic acids. According to this, the condensation degree of HAN is
the highest (E_4/E_6 = 3.7), followed by HAL (E_4/E_6 = 6.7) and the lowest is HAM (E_4/E_6 = 8.3). The optical densities of the UV spectra
increase with decreasing wavelength. For the same weight concentrations the sequence of UV-light absorption is HAN > HAL > HAM at wavelengths below 240 nm (11). From the measured IR spectra Raspor et al.
(11) concluded that of the three samples (HAL, HAN and HAM) the HAN
sample showed the most pronounced aliphatic character followed by HAL.

Table I. Elementary composition of investigated humic material and
C/H and C/N ratios

Humic substances	Elementary composition (%)						C/H	C/N	Ref.
	C	H	N	S	Cl	O			
HAL-humic acid from Lim Channel (Adriatic Sea)	54.1	5.5	4.2	1.4	3.4	31.4	9.8	12.8	(11)
HAN-humic acid from Norwegian sea sediments	52.9	6.7	4.2	2.2	1.6	32.4	7.9	12.7	(11)
HAM-humic acid from estuarine sediments (Mahakam, Borneo)	50.7	4.8	1.7	1.2	ND	39.8	10.7	30.7	(11)
HAC-humic acid from lagoon sed. (Canet, France)	50.8	5.8	4.0	1.5	ND	37.7	8.8	12.7	(32)
FAC-fulvic acid from lagoon sed. (Canet, France)	38.7	5.4	2.2	2.0	ND	51.6	7.2	17.2	(32)

ND = not determined. Reproduced with permission from Ref. 11. Copyright 1984 Elsevier Science.

Table II. Total acidity and carboxyl groups for samples of some
humic substances (32)

Humic substance	Total acidity ($meq\ g^{-1}$)	Carboxyl groups ($meq\ g^{-1}$)
HAN (Norw. S. sed.)	3.73	2.21
HAM (Mahakam)	6.78	3.25
HAC (Canet)	3.65	2.89
FAC (Canet)	5.82	4.62

They also concluded that HAN and HAL samples have higher molecular
weights than the HAM sample, and that HAN and HAL samples are of
marine origin. This is also supported by the findings of Hatcher et
al. (19) showing that humic acids from marine sediments are predominantly composed of highly branched, unsubstituted aliphatic structures.

     For most of the samples the molecular size distribution was not
measured owing to the samples' incomplete solubility, which shows
their hydrophobic character. The molecular size distribution was determined (32) only for HALR humic acid sample (deposit of Ruppia
maritima, Linnaeus) using size exclusion chromatography. The carbon
content was measured by DOC and by a UV detector. Two peaks were

found. The dominant fraction has a molecular size corresponding to a molecular weight of approximately 50,000, and the smaller fraction has an apparent molecular weight of approximately 18,000.

Some of the samples were analyzed for trace metal concentration and the results are presented in Table III.

Table III. Trace metal concentrations of the humic substances investigated

Humic substance	Concentrations of trace metals (ug g^{-1})						Ref.
	Al	Cu	Cd	Pb	Zn	Fe	
HAL (Lim Channel)	170	166	0.3	3.9	56.4	ND	(11)
HAN (Nor. S. sed.)	2000	433	2.9	12.2	101.5	ND	(11)
HAM (Mahakam)	3100	188	0.6	7.9	69.7	2500	(11)
HAC (Canet)	900	1041	8.3	113.0	113.0	ND	(32)
FAC (Canet)	140	ND	ND	ND	ND	ND	(32)

ND = not determined. Reproduced with permission from Ref. 11. Copyright 1984 Elsevier Science.

Raspor et al. (11) have also measured the adsorption of HAN, HAL and HAM samples at the hanging mercury drop electrode applying out--of-phase ac-voltammetry. According to their results the molecular size and the hydrophobicity of the investigated sample increase in the order HAM < HAL < HAN.

Results

The electrophoretic results on the interaction of ^{65}Zn, ^{109}Cd and ^{210}Pb with humic acid (HAL sample) in different dilutions of seawater have been reported earlier (8). All three trace metal radio-nuclides in 10% seawater (without the addition of humic acid) are mostly in the cationic form (^{65}Zn: 98.2%; ^{109}Cd: 99.5% and ^{210}Pb: 61.8%).

The distribution of ^{109}Cd species in the 10% seawater-humic acid (HAL) system is presented in Table IV, and the binding of ^{65}Zn and ^{210}Pb with HAL is presented in Figure 1. (The percentage of the metal humates was calculated from the decrease in the amount of the cationic form of metal ion).

^{54}Manganese. In diluted seawater (10%, S = 3.7‰) and in estuarine water (S = 2‰) ^{54}Mn behaves as a cation with electrophoretic mobilities u = -3.03 x 10^{-4} (cm^2 V^{-1} s^{-1}) in diluted seawater and u = -3.22 x 10^{-4} (cm^2 V^{-1} s^{-1}) in estuarine water, but shows also some cationic tailing, i.e. moves more slowly in the electric field than the cationic zone. The distribution of ^{54}Mn species in the estuarine and diluted seawater is shown in Table V.

By the addition of humic acid to the diluted seawater (10%), the amount of the cationic zone of ^{54}Mn decreases, while the cationic tailing and the immobile zone increase (Figure 2). One hundred percent represents the total amount of ^{54}Mn in the system. The percentage of

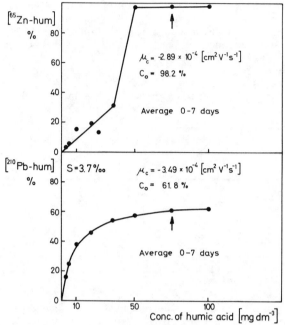

Figure 1. Dependence of Zn- and Pb-humates formation on the concentration of humic acid (HAL) in 10% seawater.(Reproduced with
permission from Ref. 8. Copyright 1980 Academic Press, London.)

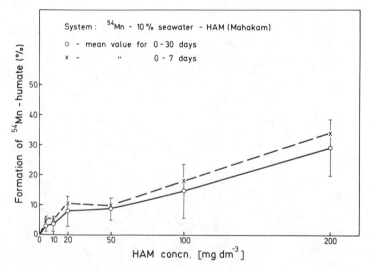

Figure 2. Dependence of Mn-humate formation on the concentration
of humic acid (HAM) in 10% seawater.

Table IV.   Distribution of ^{109}Cd species in the 10% seawater - humic
acid system (8)

Concentration of humic acid (mg dm^{-3})	Distribution of ^{109}Cd species (%)		
	C	S	A
0	99.5	0.5	0
3	98.6	1.4	0
5	98.3	1.7	0
20	98.8	1.2	0
150	4.2	95.2	0.6

C = cationic zone, $u = -2.03 \times 10^{-4}$ (cm^2 v^{-1} s^{-1}); S = zone at the
starting point; A = anionic tailing. Reproduced with permission
from Ref. 8.   Copyright 1980 Academic Press.

Table V.   Distribution of ^{54}Mn species in 10% seawater and in estuarine water

Electrolyte	Aging of the system (days)	Distribution of ^{54}Mn species (%)			
		C_{zone}	$C_{tail.}$	S	A
Estuarine water	0	93.5	5.8	0.7	0.0
	1	93.5	5.4	1.1	0.0
	7	92.3	6.9	0.7	0.1
	15	84.0	9.9	6.1	0.0
10% seawater	0	92.1	7.9	0.0	0.0
	1	95.6	4.4	0.0	0.0
	7	95.7	3.6	0.7	0.0
	15	93.7	5.9	0.3	0.1
	30	93.2	6.5	0.3	0.0

C = cationic zone; $C_{tail.}$ = cationic tailing; S = starting point
zone; A = anionic zone.

^{54}Mn bound to the humic acid was calculated from the decrease of ^{54}Mn
in the cationic zone, or, in other words, the ^{54}Mn bound to humic
acid represents the sum of the increase in the cationic tailing, the
immobile zone and in the anionic tailing.

    The aging of the system plays an important role in the binding
of ^{54}Mn to the humic acid. The maximum of binding was reached at 1-7
days of aging. For older systems the percentage of the bound ^{54}Mn to
the humic subtances decreased, which is partially seen when the 0-30
day and 0-15 day curves are compared (Figure 2).
    Figure 3 shows the change in the distribution of the ^{54}Mn zones
when 100 mg of HAM (Mahakam humic acid) is added to the seawater. It
is well demonstrated that the maximum of binding to the humic acid is
seen in the 1 day old system; older systems show relatively smaller
decreases of the cationic ^{54}Mn zone.

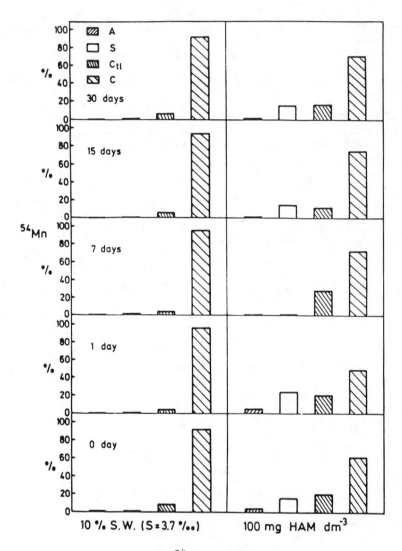

Figure 3. Distribution of ^{54}Mn zones in the 10% seawater - 100 mg HAM system.

Figure 4 presents the distribution of [54]Mn zones in the presence of 100 mg of different humic acids and fulvic acid samples in estuarine water. As seen in Figure 4, in older systems, the binding of [54]Mn to humic substances is stronger. The maximum for HAL and FAC samples is reached in the 7 day old systems and for HAM at 1 day of aging. If we calculate the mean binding of [54]Mn to humic substances for all system aged from 0 to 15 days in the same way as demonstrated in Figure 2, the strongest binding of [54]Mn is with the HAM sample, followed by the FAC and HAL samples.

[55]Iron. In diluted seawater (10%) [55]Fe introduced as Fe(III) is present almost completely in the immobile electrophoretic zone (98.7%), only a small part in the cationic tailing (1.2%) and a negligible amount in the anionic tailing zones (0.1%). However, the addition of humic substances to diluted seawater produces not only anionic but also cationic zones or the tailing of [55]Fe. In Table VI, VII and VIII the distribution of [55]Fe zones in diluted seawater - humic acid systems (HAM, HAN and HAC samples) is presented.

As seen from the tables in addition to the concentration and the origin of humic acid, the aging of the system is also very important in the distribution of the [55]Fe zones.

Comparing these three tables it is obvious that the HAN sample produces more of the cationic [55]Fe zone/tailing than the other samples, and that HAC produces more of the anionic [55]Fe zone, but that the highest amount of the anionic [55]Fe is found in the 24 day old HAM - 10% seawater system.

In Figure 5 we have tried to demonstrate the influence of both the concentration of fulvic acid and of the aging of the system on the distribution of electrophoretic zones of [55]Fe in diluted seawater. The concentration of 10 mg of FAC in 10% seawater produces a tremendous increase in the cationic zone/tailing of [55]Fe, amounting to 80.7% of the total [55]Fe present (without FAC only 1.2% of [55]Fe is in the cationic tailing zone). However, after aging 27 days, this zone dropped to only 0.4% in favour of the anionic zone (48.4%). At FAC concentrations of 100 mg dm^{-3} the anionic [55]Fe zone amounted to 9%, and after 27 days it amounted to 79.6% of the total [55]Fe.

Differences in the distribution of [55]Fe zones in the 10% seawater-humic substances systems, when the same concentrations of different humic substance samples are used, are presented in Figure 6. There is no doubt that the FAC sample produces the highest amount of the anionic [55]Fe zone but all samples of the humic substances significantly influence the distribution of the [55]Fe electrophoretic zones. Besides that, the distribution of the [55]Fe zones in very aged systems (24 days) is always significantly different than in the freshly prepared or in the 1 and 7 day old systems.

[51]Chromium. [51]Cr introduced into diluted seawater as Cr(III), is distributed between three electrophoretic zones: one immobile zone and both anionic and cationic tailing zones. The distribution of electrophoretic zones is given in Table IX.

By the addition of humic substances to diluted seawater the distribution of ^{51}Cr species changed. In Figure 7 the distribution of ^{51}Cr species in 10% seawater - fulvic acid (Canet) is presented as a function of both the concentration of FAC and the aging of the system. The influence of FAC on the distribution of ^{51}Cr zones is especially

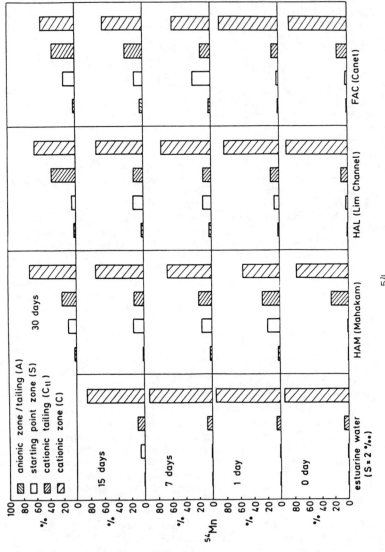

Figure 4. Distribution of ^{54}Mn zones in the estuarine water – humic substances system.

Table VI. Distribution of electrophoretic zones of ^{55}Fe in 10% sea-water – humic acid (HAM) systems (%)

HA conc. (mg dm^{-3})	t(days)	A	S	C	Total cpm
100	0	1.5	21.6	76.9	16,000
	1	2.1	27.2	70.7	27,000
	7	2.0	36.0	52.0	41,000
	24	36.3	63.4	0.3	22,000
50	0	0.9	13.0	86.1	14,000
	1	0.6	16.2	83.2	40,000
	7	3.0	31.0	66.0	11,000
	24	23.7	75.8	0.5	15,000
20	0	0.7	17.3	82.0	36,000
	1	0.5	14.5	85.0	19,000
	7	0.8	27.0	72.2	28,000
	24	36.7	63.0	0.3	9,000
10	0	0.7	21.2	78.1	9,000
	1	1.0	18.8	80.2	9,000

Table VII. Distribution of electrophoretic zones of ^{55}Fe in 10% sea-water – humic acid (HAN) systems (%)

HA conc. (mg dm^{-3})	t(days)	A	S	C	Total cpm
100	0	2.8	28.5	68.7	6,000
	1	4.4	28.5	67.1	12,000
	7	4.3	37.9	57.8	15,000
	24	8.8	90.8	0.4	30,000
50	0	1.6	27.3	71.1	7,000
	1	1.6	17.2	81.2	13,000
	7	1.0	18.1	80.9	32,000
	24	6.4	86.1	7.5	22,000
20	0	2.7	63.2	34.1	4,000
	1	0.1	31.4	68.5	6,000
	7	0.4	18.4	81.2	23,000
	24	7.8	69.7	22.5	6,000
10	0	1.3	34.5	64.2	5,000
	1	0.7	13.2	86.1	14,000
	7	1.7	20.5	77.8	5,000
	24	5.8	85.9	8.3	4,000

A = anionic zone and/or anionic tailing; C = cationic zone and/or cationic tailing; S = zone at the starting point of the electro-phoretic strip

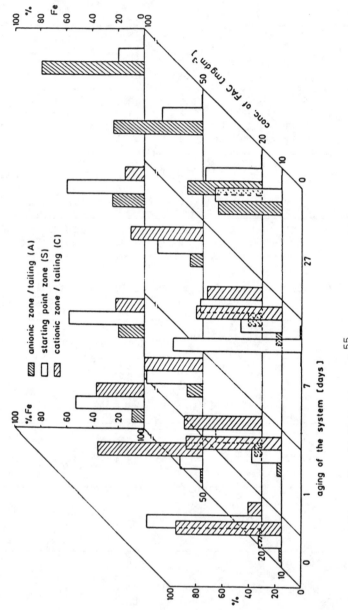

Figure 5.   Distribution of ^{55}Fe zones in the 10% seawater – fulvic acid (FAC) system.

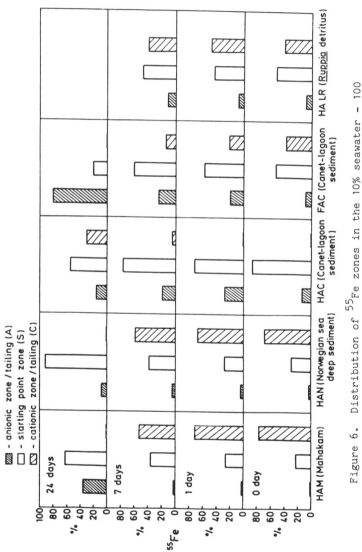

Figure 6. Distribution of ^{55}Fe zones in the 10% seawater – 100 mg humic substances systems.

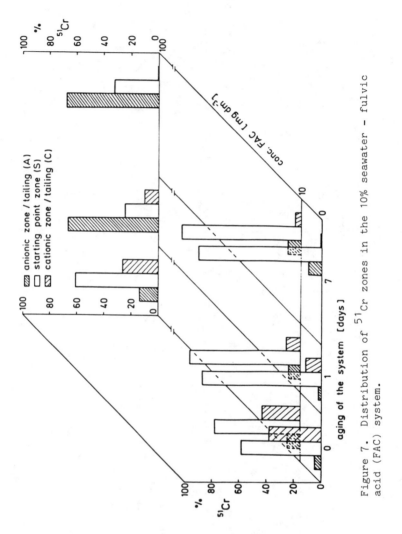

Figure 7. Distribution of ^{51}Cr zones in the 10% seawater – fulvic acid (FAC) system.

Table VIII. Distribution of electrophoretic zones of ^{55}Fe in 10%
seawater - humic acid (HAC) systems (%)

HA conc. (mg dm^{-3})	t(days)	A	S	C	Total cpm
200	0	13.9	77.8	8.3	82,000
	1	22.0	67.8	10.2	82,000
	7	29.2	60.1	10.7	74,000
	24	9.5	59.3	31.2	118,000
100	0	13.0	86.8	0.2	47,000
	1	28.0	71.8	0.2	63,000
	7	19.0	75.6	5.4	68,000
	24	15.6	54.5	29.9	101,000
50	0	9.3	88.8	1.8	4,000
	1	21.8	77.8	0.4	4,000
	7	15.9	78.7	5.4	5,000
	24	7.7	47.9	44.4	12,000
20	0	4.7	81.9	13.4	3,000
	1	14.2	81.2	4.6	2,000
	7	13.4	70.9	15.7	3,000
	24	6.1	51.6	42.3	5,000
10	1	15.6	81.2	3.2	2,000
	7	8.2	45.0	46.8	3,000

Table IX. Distribution of ^{51}Cr species in 10% seawater

Aging of the system (days)	Distribution of ^{51}Cr species (%)		
	C	S	A
0	37.4	57.9	4.7
1	11.4	87.1	1.6
7	0.3	90.3	9.4

A = anionic zone and/or anionic tailing; C = cationic zone and/or
cationic tailing; S = zone at the starting point of the electrophoretic strip

pronounced at higher fulvic acid concentrations and in older systems.
    The distribution of ^{51}Cr species in the 10% seawater - humic
substance systems (100 mg dm^{-3}), when different humic substances
samples were used, is presented in Figure 8. From all the humic substance samples used, FAC had the strongest influence on the redistri-
bution of ^{51}Cr species, producing the highest amount of anionic zone/
tailing.

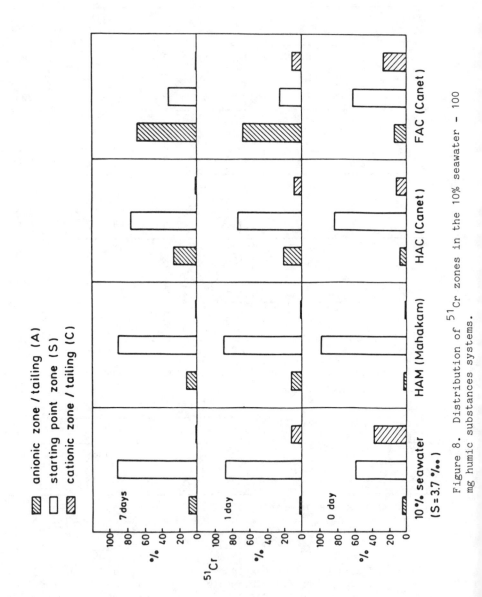

Figure 8. Distribution of ^{51}Cr zones in the 10% seawater - 100 mg humic substances systems.

Adsorption of ^{109}Cd on suspended humic acids. Recognizing the property of humic acids as adsorbents for metal ions, suspensions of humic acid (HAL, HAM and commercial "Fluka" samples) were prepared at 35 mg dm^{-3}. ^{109}Cd was added to that suspension and the adsorption of ^{109}Cd was measured. The results are expressed as K_d values. K_d value is the ratio of the cadmium concentration in solid and liquid phase given in cm^3 g^{-1}. As seen in Figure 9 the adsorption of ^{109}Cd is rather strong on the humic acid. The K_d values decrease as the solubility of the humic acid samples increase.

In Figure 10, some K_d values for the adsorption of ^{109}Cd on smectite (pure inorganic support), natural river sediment and a mixture of the same sediment and HAL in the ratio 1:1 are presented. The presence of humic acid in the sediment suspension significantly increases the adsorption of ^{109}Cd.

Discussion

Most of the divalent metals studied behave in diluted seawater as cations when introduced into seawater as metal (II) ions. According to our results in 10% seawater, 99.5% of ^{109}Cd, 98.2% of ^{65}Zn, 92.1% of ^{54}Mn and 61.8% of ^{210}Pb are in cationic forms giving defined electrophoretic zones. In estuarine water (S = 2%o) 93.5% of ^{54}Mn is in the cationic form of defined electrophoretic mobility. In 10% seawater the cationic zone of ^{109}Cd is a mixture of the monochloro complex, free Cd^{2+} and the dichloro complex (up to 7%) (8). The cationic electrophoretic mobility of ^{65}Zn corresponds to the electrophoretic mobilities of free divalent cations (35,36). The immobile zone could be related to a particulate form of Zn or to Zn adsorbed on particles (37). The cationic zone of ^{210}Pb might be a mixture of all cationic species of Pb present in seawater at pH values of 8.0 (Pb^{2+}, PbCl$^+$, PbOH$^+$ and PbHCO$_3^+$) as reported in the literature (38-44). The immobile zone of ^{210}Pb should correspond to all uncharged lead species present in seawater and to the precipitated Pb. The cationic zone of ^{54}Mn might be attributed to the free Mn^{2+} (45), the cationic tailing to positively charged hydrolytic products of Mn(III) and the immobile zone to hydrolyzed products, especially in older systems (46).

The addition of humic substances to seawater generally produces an increase of the immobile zone and the formation of anionic tailing of the investigated divalent radionuclides (8,9). In diluted seawater, if the immobile zone does not exist, it appears when humic acid is added.

The effect of the HAL sample in 10% seawater on the distribution of particular metal-radionuclide zones is demonstrated as follows: 1) disappearence of the ^{65}Zn cationic zone, increase of the immobile zone and formation of the anionic tailing up to 3%;   2) disappearance of the cationic zone of ^{210}Pb, increase of the immobile zone and formation of the anionic tailing up to 4%;   3) decrease of the cationic zone of ^{109}Cd and formation of the immobile zone and an almost negligible amount of anionic tailing (8).

Comparing the distribution of ^{54}Mn zones in 10% seawater and 10% seawater-humic acid (HAM sample) it is evident that there is a decrease of the amount in the ^{54}Mn cationic zone, the cationic tailing increases (up to 20% at the HAM concentration of 200 mg dm^{-3}) as well as the immobile zone (up to 20%) and the anionic tailing zone (up to 5%). However, at higher pHs it is possible to get an anionic zone of

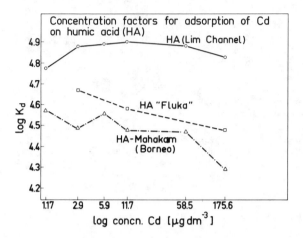

Figure 9. Concentration factors (log K_d) for the adsorption of
Cd on different suspended humic acids in the river water. Concentration of humic acids: 35 mg dm^{-3}. Particle size between 0.45
and 32 um.

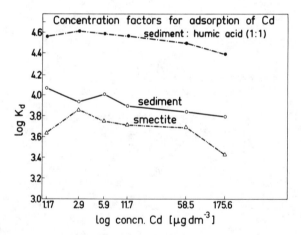

Figure 10. Concentration factors (log K_d) for the adsorption of
Cd on different supports in the river water. Concentration of
suspended particles 35 mg dm^{-3}. Particle size between 0.45 and
32 um.

the [54]Mn-humic acid complex. This was experimentally confirmed when the HAM sample was added to river water; at the HAM concentration of 200 mg dm^{-3} at pH 10.6, approximately 30% of [54]Mn was in the anionic form and only 0.1% was cationic. At the same humic acid concentration in river water but at a pH of 8.0, only 3.6% of [54]Mn was anionic and 55% remained in the cationic zone.

In estuarine water the interaction of [54]Mn and the three samples of humic substances was examined: the strongest influence on the re-distribution of [54]Mn species is by the HAM sample followed by FAC, while the HAL sample has the smallest influence. As said before, the HAL sample is the least soluble of all three humic substance samples. Therefore it seems that Mn is bound more to the dissolved part of humic substances (lower molecular weight fraction) and less to the undissolved and the immobile part. As seen in Figures 2, 3 and 4, the binding of [54]Mn to humic acid is stronger in the estuarine water than in seawater, owing to the lower content of other metals which can compete with manganese for binding sites on humic substances.

As demonstrated earlier ([45]) when [54]Mn was complexed with EDTA in seawater, the electrophoretic "picture" of the distribution of [54]Mn zones showed the spread of the radioactivity from the cationic to the anionic values without any electrophoretic zone defined. In other words, at the intermediate EDTA concentrations, which were not high enough to complex all manganese present, both Mn-EDTA complexes and free Mn^{2+} were present at each position of the electrophoretic strip moving in opposite directions and resulting in the spread of [54]Mn between the maximal cationic and anionic electrophoretic mobilities. That finding supports the idea that the cationic tailing of [54]Mn in estuarine or seawater-humic substance systems might be partly attributed to the formation of the Mn-humic substance complex (besides positively charged hydrolytic products of Mn(III) which are present to some extent in the estuarine or diluted seawater without the addition of humic substances).

Although we did not experiment with [54]Mn in the 10% seawater-HAL sample system, from the experiment performed in estuarine water, it can be predicted that the HAL sample would also bind [54]Mn less strongly than the HAM sample. Therefore the sequence of binding of divalent metals to the humic acid would be as follows: Pb > Zn > Cd > Mn. These results are in agreement with published stability constants data ([16],[17],[47]).

According to our results it appears that all investigated divalent radionuclides are bound more to higher molecular weight fractions of the humic substances used (except Mn, which forms positively charged species), forming either uncharged complexes or are adsorbed on undissolved fractions of humic substances, and less to lower molecular weight fractions of humic substances, which is demonstrated by the formation of negatively charged metal species.

In natural waters, iron concentrations are commonly several orders of magnitude greater than the equilibrium solubility of iron hydroxide ([48]). Two of the chemical species postulated to account for this phenomenon are: 1) fine colloidal particles of iron hydroxide (possibly associated with colloidal organic matter) and 2) dissolved complexes of iron with naturally occurring organic substances ([49]). During estuarine mixing processes, iron colloids or organic complexes are flocculated due to the increasing salinity. Sholkovitz and co-workers ([6],[26]) demonstrated the close association of Fe, Mn, Al and P

with both river-dissolved humic substances and seawater-flocculated
humates. On the other side, Fox (28) found that dissolved humic acid
and soluble iron appear to be chemically unassociated in estuaries
despite their coincidental removal. According to our results, ^{55}Fe in
diluted seawater is almost completely present in the immobile electrophoretic zone, but the addition of humic substances forms the
anionic species, which are specially abundant when fulvic acid is applied.

Another interesting influence of humic substances is recognized:
in the presence of humic substances iron becomes solubilized. Hydrolytic species of ^{55}Fe are adsorbed on the walls of experimental vessels but by the addition of humic substances to the system, the number of cpm (counts per minute) tremendously increased in the same
volume of solution applied for the electrophoretic experiment (see
Tables VI-VIII). Even at lower humic substance concentrations the
cationic zone of ^{55}Fe was found, which amounted depending on the
humic substance sample, up to 86% of the total ^{55}Fe applied to the
electrophoretic strip. The amount of these cationic ^{55}Fe zones decreases by increasing the humic substance concentration and by aging
of the system (except for HAC humic acid sample) in favour of the
anionic zone or the immobile zone. This zone might be attributed to
the positively charged iron colloids, which were "available" for
electrophoretic experiments when humic substances were added to seawater. Comparing these results with the results obtained for divalent
metals, it seems that the immobile zone of ^{55}Fe could be partly attributed to ^{55}Fe bound to humic substances by any possible mechanism
(complexation or adsorption). In the electrophoretic experiments we
can not distinguish between ^{55}Fe in the hydrolytic species and ^{55}Fe
bound to humic substances.

Hydrolytic species of iron are also able to adsorb humic substances on their surfaces and the negatively charged humic substances can adsorb additional Fe cations. From our results we can at
least conclude that iron forms negatively charged complexes with
humic substances and that humic substances solubilize iron when they
are present in diluted seawater. If we compare the quantity of ^{55}Fe
in the anionic zones depending on the humic substance used, we get the
following sequence of the humic substances FAC > HAC > HAM > HAN.
This sequence corresponds to the solubility of the humic substance
samples, and it seems that a fraction of humic substances having
lower molecular weight is responsible for the formation of Fe-humic
complexes.

Chromium(III) is present in seawater probably as $Cr(OH)_2^+$ (50).
Although the hexavalent state of chromium is thermodynamically the
most stable valent state of chromium in seawater, it can be reduced
in marine organisms or by some reducible materials. Cr(III) is easily
oxidized to Cr(VI) in the presence of manganese oxides, but organic
complexes of Cr(III) are not oxidized and the trivalent state is
stabilized (51). Thus chromium can be present in seawater as any form
of inorganic Cr(III), Cr(VI), and as organic species. According to
our results, ^{51}Cr (introduced to the diluted seawater as Cr(III)) is
affected by the presence of humic substances forming anionic species.
From the three humic substance sample studied, the FAC sample produces the most abundant anionic species followed by the HAC and the
HAM sample.

Comparing the influence of different humic substances on the behavior of ^{55}Fe and ^{51}Cr in 10% seawater it is evident, that at the same humic substance concentration and aging of the system, the ^{51}Cr anionic species are more pronounced for all humic substance samples studied, although the concentration of chromium was for almost three orders of magnitude higher than that of iron. That is specially evident in the 10% seawater-fulvic acid system.

The immobile zone of ^{51}Cr is more abundant for the HAM sample, followed by the FAC. But for the HAC sample, the immobile zones of both radionuclides are equal. Consequently, the cationic zone is more abundant for ^{55}Fe although in 10% seawater, without addition of humic substances, the cationic zone of ^{51}Cr amounted to 37% at t = 0 days, and for ^{55}Fe was negligible. Although from our results it seems that fulvic acid is more effective in the binding of ^{51}Cr than of ^{55}Fe there is still some doubt as to whether some chromate is also formed and since it is negatively charged, it could contribute to the anionic zone/tailing of ^{51}Cr. However, results of Nakayama et al. (51) support the idea that most of the anionic chromium might be attributed to a Cr(III) complex with humic/fulvic acid, because the organic complexes of Cr(III) are not oxidized.

The adsorption experiments show that undissolved humic acid (the HAL sample) has a very high ability to adsorb ^{109}Cd. Other humic acid samples, owing to their higher solubility, have lower abilities to adsorb Cd. For the adsorption experiments which were performed in river water and in seawater or estuarine water, the K_d values should be lower due to the presence of other macro- and microconstituents of seawater competing with Cd for adsorption sites on the negatively charged surface. These data are in agreement with the results of electrophoretic experiments showing that cadmium is adsorbed on undissolved or precipitated humic substances in seawater and that only small amounts of Cd is found in the anionic form as the Cd-humic substance complex.

By knowing the sequence in which metals are sorbed on humic acid (52) we could predict that on our humic substance samples, Fe, Pb and Cr will be sorbed more than Cd and Zn, and that Mn will be sorbed less than Cd.

Our results show that the adsorption on suspended matter is increased in the presence of suspended humic acid (Figure 10). This is supported by the finding of other authors (53,54) showing that organic flocculant coatings greatly affect the cation exchange capacities of sediment and suspended matter.

Conclusions
─────────

The interaction of divalent metal radionuclides with humic substances is particularly demonstrated by the formation of uncharged species and/or adsorption on undissolved or precipitated humic substances and less by the formation of negatively charged species. Nevertheless, the amount of cationic forms of all investigated divalent metals decrease in the presence of humic substances, while some other species are formed having different electrophoretic mobilities than particular metal ions in seawater or estuarine water. The concentrations of humic acid at which observable changes in speciation of trace metals occurred are rather high when comparing the humic substance concentrations in the open oceans. However, higher humic substance concentrations are possible in shallow waters, closed areas or estuaries with high biological activity.

For trivalent metals the influence of humic substances present in diluted seawater is shown by the formation of negatively charged species, which might be attributed to metal-humic substance complexes (especially pronounced for fulvic acid). The presence of humic substances, even at lower concentrations, solubilizes iron. This is partly evident by the formation of the cationic species, which are not found in estuarine or seawater unless humic substances are added. These effects might also be attributed to the stabilization of iron colloids in water rich in humic substances.

When present in undissolved states, humic substances act as adsorbants for trace metals. The presence of suspended humic substances increases the adsorption ability of suspended matter and sediment.

Acknowledgments

This work is devoted to the memory of Prof. H.W. Nürnberg, who unfortunately left us too early. The financial support by the Self--Managed Authority for Scientific Research, SR Croatia, Yugoslavia, is gratefully acknowledged. The authors are greatful to G. Cauwet for the supply of humic substances (HAC, FAC, HAN, HAM and HALR samples).

Literature Cited

1. Mantoura, R.F.C.; Riley, J.P. Anal. Chim. Acta 1975, 76, 97-106.
2. Harvey, G.R.; Boran, D.A.; Chesal, L.A.; Tokar, J.M. Mar. Chem. 1983, 12, 119-132.
3. Huljev, D. M.Sc. Thesis, University of Zagreb, Zagreb, 1970.
4. Rashid, M.A. Soil. Sci. 1971, 111, 298-305.
5. Schnitzer, M.; Khan, S.U. "Humic Substances in the Environment"; Marcel Dekker: New York, 1972.
6. Sholkovitz, E.R. Geochim. Cosmochim. Acta 1976, 40, 831-845.
7. Nriagu, J.O.; Coker, R.D. Environ. Sci. Technol. 1980, 14, 443-446.
8. Musani, Lj.; Valenta, P.; Nürnberg, H.W.; Konrad, Z.; Branica, M. Estuar. Coastal Mar. Sci. 1980, 11, 639-649.
9. Musani, Lj.; Nürnberg, H.W.; Valenta, P.; Konrad, Z.; Branica, M. Thalassia Jugosl. 1981, 17, 71-81.
10. Schnitzer, M.; Ghosh, K. Soil. Sci. 1982, 134, 354-363.
11. Raspor, B.; Nürnberg, H.W.; Valenta, P.; Branica, M. Mar. Chem. 1984, 15, 217-230.
12. Raspor, B.; Nürnberg, H.W.; Valenta, P.; Branica, M. Mar. Chem. 1984, 15, 231-249.
13. Schnitzer, M.; Skinner, S.I.M. Soil Sci. 1966, 102, 361-365.
14. Schnitzer, M.; Hansen, E.H. Soil. Sci. 1970, 109, 333-340.
15. Buffle, J.; Greter, F.L.; Haerdi, W. Anal. Chem. 1977, 49, 216-222.
16. Mantoura, R.F.C.; Dickson, A.; Riley, J.P. Estuar. Coastal Mar. Sci. 1978, 6, 387-408.
17. Whitfield, M.; Turner, D.R. In "Proc. Int. Expert Discussion, Lead-Occurence, Fate and Pollution in the Marine Environment"; Branica, M.; Konrad, Z.; Eds.; Pergamon Press: Oxford, 1980.
18. Torres, R.A.; Choppin, G.R. Radiochim. Acta 1984, 35, 143-148.
19. Hatcher, P.G.; Rowan, R.; Mattingly, M.A. Org. Geochem. 1980, 2, 77-85.

20. Wilson, M.A.; Barron, P.F.; Gillam, A.H. Geochim. Cosmochim. Acta 1981, 45, 1743-1750.
21. Duursma, E.K. In "Chemical Oceanography"; Riley, J.P.; Skirrow, S.; Eds.; Academic Press: London, 1965; Vol. I, pp. 433-473.
22. Ogura, N. Mar. Biol. 1972, 13, 89-93.
23. Frimmel, F.H. Vom Wasser 1977, 49, 1-10.
24. Stuermer, D.H.; Harvey, G.R. Deep-Sea Res. 1977, 24, 303-309.
25. Tipping, E. Chem. Geol. 1981, 33, 81-89.
26. Skolkovitz, E.R.; Boyle, E.A.; Price, N.B. Earth Planet. Sci. Lett. 1978, 40, 130-136.
27. Moore, R.M.; Burton, J.D.; Williams, P.J.; Young, M.L. Geochim. Cosmochim. Acta 1979, 43, 919-926.
28. Fox, L.E. Geochim. Cosmochim. Acta 1984, 48, 879-884.
29. Desai, M.V.; Ganguly, A.K. "Report Bhabha Atomic Research Center, B.A.R.C.-488"; Bombay, 1970.
30. Kononova, M.M.; Balachirova, N.P. Soviet Soil. Sci. 1960, 4, 1149-1155.
31. Pučar, Z. Anal. Chim. Acta 1957, 17, 476-484.
32. Faguet, D. Doctorat de Specialite Oceanologie, l'Universite de Perpignan, Perpignan, 1982.
33. Nissenbaum, A.; Kaplan, I.R. Limnol. Oceanogr. 1982, 17, 570-582.
34. Schnitzer, M.; Skinner, S.I.M. Soil. Sci. 1969, 108, 383-390.
35. Marazović, Lj.; Pučar, Z. Rapp. Comm. Int. mer Medit. 1972, 20, 701-703.
36. Kozjak, B.; Marinić, Z.; Konrad, Z.; Musani-Marazović, Lj.; Pučar, Z. J. Chromatogr. 1977, 132, 323-334.
37. Piro, A.; Bernhard, M.; Branica, M.; Verži, M. In "Radioactive Contamination of the Marine Environment"; IAEA: Vienna, 1973; STI/PUB 313, pp. 29-44.
38. Zirino, A.; Yamamoto, S. Limnol. Oceanogr. 1972, 17, 661-671.
39. Dyrssen, D.; Wedborg, M. In "The Sea"; Goldberg, E.D., Ed.; Wiley-Interscience: New York, 1975; Vol. 5, pp. 181-195.
40. Stumm, W.; Brauner, D.A. In "Chemical Oceanography"; Riley, J. P.; Chester, R., Eds.; 2nd edn., Academic Press: London, 1975; Vol. 1, pp. 173-234.
41. Florence, T.M.; Batley, G.E. Talanta 1976, 23, 179-186.
42. Lu, J.C.S.; Chen, K.Y. Environ. Sci. Techn. 1977, 11, 174-182.
43. Sipos, L.; Raspor, B.; Nürnberg, H.W.; Pytkowicz, R.M. Mar. Chem. 1980, 9, 37-47.
44. Sipos, L.; Valenta, P.; Nürnberg, H.W.; Branica, M. In "Proc. Int. Expert Discussion, Lead-Occurence, Fate and Pollution in the Marine Environment"; Branica, M.; Konrad, Z., Eds.; Pergamon Pres: Oxford, 1980.
45. Musani-Marazović, Lj.; Pučar, Z. Mar. Chem. 1977, 5, 229-242.
46. Beneš, P. J. Inorg. Nucl. Chem. 1967, 29, 2889-2898.
47. Pagenkopf, G.K. In "Organometals and Organometalloids. Occurence and Fate in the Environment"; Brinckman, F.E.; Bellama, J.M., Eds.; ACS SYMPOSIUM SERIES No. 82, American Chemical Society: Washington, D.C., 1978; pp. 372-387.
48. Stumm, W.; Morgan, J.J. In "Aquatic Chemistry"; Wiley: New York, 1970; pp. 238-299.
49. Perdue, E.M.; Beck, K.C.; Reuter, J.H. Nature 1976, 260, 418-420.
50. Elderfield, H. Earth Planet. Sci. Lett. 1970, 9, 10-16.

51.   Nakayama, E.; Kuwamoto, T.; Tokoro, H.; Fujinaga, T. Anal.
      Chim. Acta 1981, 131, 247-254.
52.   Kerndorf, H.; Schnitzer, M. Geochim. Cosmochim. Acta 1980, 44,
      1701-1708.
53.   Rashid, M.A. Maritime Sediments 1969, 5, 44-50.
54.   Pillai, T.N.V.; Desai, M.V.M.; Mathew, E.; Ganapathy, S.;
      Ganguly, A.K. Current Sci. 1971, 40, 75-81.

RECEIVED October 31, 1985

INDEXES

Author Index

Subject Index

A

Production and indexing by Karen McCeney
Jacket design by Pamela Lewis

Elements typeset by Hot Type Ltd., Washington, DC
Printed and bound by Maple Press Co., York, PA

RECENT ACS BOOKS

"Coulombic Interactions in Macromolecular Systems"
Edited by Adi Eisenberg and Fred E. Bailey
ACS SYMPOSIUM 302; 282 pp; ISBN 0-8412-0960-X

"Mineral Matter and Ash in Coal"
Edited by Karl S. Vorres
ACS SYMPOSIUM SERIES 301; 552 pp; ISBN 0-8412-0959-6

"Equations of State: Theories and Applications"
Edited by K. C. Chao and R. Robinson
ACS SYMPOSIUM SERIES 300; 608 pp; ISBN 0-8412-0958-8

"Xenobiotic Conjugation Chemistry"
Edited by Guy Paulson, John Caldwell, and Julius J. Menn
ACS SYMPOSIUM SERIES 299; 368 pp; ISBN 0-8412-0957-X

"Strong Metal-Support Interactions"
Edited by R. T. K. Baker, S. J. Tauster, and J. A. Dumesic
ACS SYMPOSIUM SERIES 298; 238 pp; ISBN 0-8412-0955-3

"Chromatography and Separation Chemistry:
Advances and Developments"
Edited by Satinder Ahuja
ACS SYMPOSIUM SERIES 297; 304 pp; ISBN 0-8412-0953-7

"Natural Resistance of Plants to Pests"
Edited by Maurice B. Green and Paul A. Hedin
ACS SYMPOSIUM SERIES 296: 244 pp; ISBN 0-8412-0950-2

"Microelectronics Processing: Inorganic Materials
Characterization"
Edited by Lawrence A. Casper
ACS SYMPOSIUM SERIES 295; 444 pp; ISBN 0-8412-0934-0

"Nutrition and Aerobic Exercise"
Edited by Donald K. Layman
ACS SYMPOSIUM SERIES 294; 150 pp; ISBN 0-8412-0949-9

"Historic Textile and Paper Materials: Conservation
and Characterization"
Edited by Howard L. Needles and S. Haig Zeronian
AVANCES IN CHEMISTRY SERIES 212; 464PP; ISBN 0-8412-0900-6

"Multicomponent Polymer Materials"
Edited by D. R. Paul and L. H. Sperling
ADVANCES IN CHEMISTRY SERIES 211; 354 pp; ISBN 0-8412-0899-9

For further information contact:
American Chemical Society, Sales Office
1155 16th Street NW, Washington, DC 20036
Telephone 800-424-6747